PRACTICAL HOLOGRAPHY

FRONTISPIECE HOLOGRAM 'Alice's White Knight', after Tenniel's original illustration. Model by Jim McIntyre of Holocraft Ltd. Silver-halide hologram by Applied Holographics Plc, on material supplied by Ilford Ltd.

NOTE This hologram has been left uncropped in order to allow the reader to see the whole model. If you mask the hologram down to the central portion (approximately 80 mm square) you will see the image as it was originally conceived.

COVER HOLOGRAM 'Cat's Whiskers' by David Pizzanelli, with assistance from Mike Burridge, Kevin Baumber and Rob Munday. Hologram embossed by Roger Dorey and Terry Neal, and sponsored by See 3 (Holograms) Ltd, Jayco Holographics Ltd and The Royal College of Art Holography Unit.

NOTE Although the two holograms will produce an image of sorts with almost any lighting, the best possible image will be produced by a compact-source light situated above the hologram and at an angle of incidence of 45°.

PRACTICAL HOLOGRAPHY

Graham Saxby

PRENTICE HALL

New York · London · Toronto · Sydney · Tokyo

First published 1988 by
Prentice Hall International (UK) Ltd
66 Wood Lane End, Hemel Hempstead,
Hertfordshire, HP2 4RG
A division of
Simon & Schuster International Group

© 1988 Prentice Hall International (UK) Ltd

All rights reserved. No part of this publication may be reproduced, stored in a retrieval system, or transmitted, in any form or by any means, electronic, mechanical, photocopying, recording or otherwise, without the prior permission, in writing, from the publisher.
For permission within the United States of America contact Prentice Hall Inc., Englewood Cliffs, NJ 07632.

Printed and bound in Great Britain at
The University Press, Cambridge

Library of Congress Cataloging-in-Publication Data

Saxby, Graham.
  Practical holography.

  Bibliography: p.
  Includes index.
  1. Holography. I. Title.
TA1540.S36 1987   621.36′75   87-12649
ISBN 0-13-693797-7

British Library Cataloguing in Publication Data

Saxby, Graham
  Practical holography.
  1. Holography
  I. Title
  744   QC449

ISBN 0-13-693797-7

1 2 3 4 5   91 90 89 88 87

ISBN 0-13-693797-7

Contents

Foreword xiii
Preface xv
Acknowledgments xviii

PART I   **PRINCIPLES OF HOLOGRAPHY**   1

Chapter 1   **What is a Hologram?**   3
Stereoscopy   3
Defining the problem   5
The problem solved   6
Interference   7
Diffraction   11
Amplitude and phase gratings   15

Chapter 2   **How Holography Began**   16
References   20

Chapter 3   **Light Sources for Holography**   21
Propagation on electromagnetic waves   22
Oscillators   24
Properties of light sources   24
Atoms and energy   26
Stimulated emission   27
The three-level laser   28
Q-switching   31
Ruby lasers for holography   32
The four-level laser   32
Mirrors and windows in CW lasers   33
Types of neutral-gas laser   34
Ion lasers   35
Other types of laser   37
References   39

v

Chapter 4 The Basic Types of Hologram 40
Laser transmission holograms 42
Reflection holograms 45
Image-plane holograms 48
Slit transfer (rainbow) holograms 50
Achromatic transfer holograms 53
Single-beam transfer holograms 55
Holographic stereograms 55
Alcove holograms 59
Other types of holograms 59
References 59

Chapter 5 Color Holography 61
Basic principles 61
Illumination 62
Color anomalies 63
Cross-talk 64
Diffraction efficiency of a color hologram 65
Color shift 66
Pseudocolor holography 67
References 67

Chapter 6 Materials, Exposure and Processing 68
Sensitometric requirements for holographic emulsions 70
Beam intensity ratio 70
Constituents of a developer 70
Bleaches 72
Colloidal-silver processes 74
Other processes 74
References 75

PART II PRACTICAL DISPLAY HOLOGRAPHY 77

Chapter 7 Making Your First Hologram 79
The laser 80
Laser safety 80
The need for stability 81
The beam expander 81
Film support 82
Film 83
Safelight 84
Choosing a suitable subject 85
Layout of components 85
Final preparations 87
Loading the film 87
Exposure 88
Processing 88
Assessing the test 89
Changing the image color 90
The pseudoscopic image 91
An 'orthoscopic' real image 92
Displaying your holograms 93

Chapter 8	**Single-beam Techniques** 95	
	Optical configurations 95	
	Construction of a frame system 96	
	Spatial filtering 99	
	Making an electrically-operated shutter 104	
	Safelights 105	
	Index-matching 106	
	Exposure and preparation for processing 107	
	Processing 107	
	Assisted single-beam configurations 109	
	Single-beam real-image techniques 109	
	Multi-exposure techniques 112	
	The transfer principle 112	
	Transmission master holograms 116	
	Further frame techniques 119	
	Overhead reference beam 119	
	Interferograms using frame set-ups 121	
	Copying holograms 122	
	Mounting and displaying 122	
	Trouble-shooting 123	
	References 127	
Chapter 9	**360° Holograms** 128	
	Cylindrical holograms 128	
	Conical holograms 130	
	360° holograms with plates 132	
	360° flat holograms 135	
Chapter 10	**Introducing Further Beams** 136	
	When do we need extra light? 136	
	Plain-glass beamsplitters 136	
	Metallized beamsplitters 139	
	A larger platform 140	
	Additional equipment 140	
	Setting up with an overhead reference beam 142	
	Checking the beam intensity ratio 143	
	Double-beam object illumination 143	
	Fiber-optics light guides 144	
	Polarization considerations: use of a half-wave plate 144	
	Reference 145	
Chapter 11	**Deep-image Reflection Holograms** 146	
	Types of optical table 146	
	Choosing a suitable site 147	
	Building a sand table 147	
	Supporting the optical components 149	
	Steadiness checks 150	
	Optical set-ups for reflection holograms 152	
	Basic two-beam layouts 153	
	Multi-beam layouts 154	
	Backlighting, shadowgrams and black holes 156	
	Matching beam paths 157	
	Large reflection transfer holograms 157	
	References 159	

Chapter 12 Transmission Master Holograms 160
Stability requirements 160
Optical equipment 161
Gravity bases 162
Plate and transfer holder 164
Collimators 166
Lens or mirror? 166
Full-aperture transmission master holograms 166
Processing a transmission master 170
Restricted-aperture master holograms 170
Slit transfer master holograms 174
Holographic collimators 177
Reference 179

Chapter 13 Transfer Holograms 180
Basic principles for transfer holograms 180
Aperture restriction 181
Manipulation of cut-off effects 183
Multiplexing 184
Full-aperture transmission transfer holograms 185
Slit transmission transfer holograms 187
Achromatic images 195
References 199

Chapter 14 Focused-image Holograms 200
Parallax in a focused image 200
Perspective in a focused image 201
Lenses for focused-image holograms 201
Full-aperture focused-image holograms 202
One-step rainbow holograms 204
Focused-image holograms using an optical mirror 206
Astigmatic one-step focused-image holograms 208
Fourier-transform holograms 209
References 212

Chapter 15 Home-made Optical Elements 214
Subject illumination 214
Liquid-filled lenses 220
Holographic optical elements (HOEs) 228
References 234

Chapter 16 Portraiture and Pulse-laser Holography 235
The need for a pulse laser 235
Safety considerations 236
Optical requirements 237
Energy requirements 238
Special problems with holographic portraiture 239
Lighting for portraiture 240
Exposure and processing 242
Making the transfer 243
Double-pulse holography 243
Other uses for pulse-laser holography 244
References 245

CONTENTS

Chapter 17 Holographic Stereograms 246
　　The multiplexing principle 246
　　Making a multiplexed hologram 247
　　Cross holograms 250
　　Computer-drawn images for holographic stereograms 253
　　Alcove holograms 257
　　References 259

Chapter 18 Holograms in Color 260
　　Natural or simulated color? 260
　　Difficulties inherent in natural-color holography 261
　　Practical set-ups for color reflection holography 262
　　Color rainbow holograms 263
　　Pseudocolor reflection holograms 264
　　Multicolor transmission holograms 268
　　A note on 'pre-visualization' 271
　　References 272

Chapter 19 Non-silver Processes 273
　　Dichromated gelatin: principles 274
　　Photopolymers 277
　　Photothermoplastics 278
　　Photoresists 278
　　References 280

Chapter 20 Embossed Holograms 281
　　Production of originals 284
　　Holographic recording of the initial image 284
　　Making the photoresist master hologram 285
　　The embossing process 287
　　References 289

Chapter 21 Display Techniques 290
　　The basic types of hologram and their display 290
　　General displays 292
　　Displays to accompany lectures and informal presentations 298
　　Exhibitions 299
　　References 303

Chapter 22 The Photography of Holograms 304
　　Reflection holograms 305
　　Transmission holograms 308
　　Viewpoint and parallax 310
　　Unusual holograms 311
　　Presenting slides 311
　　Copyright 311

PART III HOLOGRAPHY EARNS ITS KEEP 313

Chapter 23 Holography in Measurement and Displacement Analysis 315
Conventional measuring techniques 315
Holographic interferometry 316
Holographic contouring with two wavelengths 329
Summary of applications 331
References 332

Chapter 24 Data Storage, Processing and Retrieval 333
Data storage 333
Data processing 334
Computer-generated holograms 338
Data retrieval 343
References 345

Chapter 25 Other Applications of Holography 347
Holographic optical elements 347
Holography in biology and medicine 349
Acoustic and microwave holography 351
Conclusion 352
References 352

APPENDIXES

1 Lasers and Safety 354
Warning notices 355
Avoiding accidents 355
Ion lasers 356
Laser goggles 356
Pulse lasers 356
The laser itself 357
References 357

2 Mathematical Matters 358
Formation and reconstruction of a hologram 358
Zone plates 362
The lens laws 364
Axial magnification 365
F-number and parallax angle 367
Effects of shrinkage during processing 368
Modulation and contrast 370
Designing a liquid-filled lens 372
References 374

3 The Fourier Approach to Image Formation 375
Fourier series 378
Fourier transform 381
Reciprocal relationship of x-space and Fourier space 385
The Fourier convolution theorem 388
Two-dimensional objects 391

CONTENTS

4   **The Holodiagram**   393
    References   397

5   **Bragg Diffraction**   398

6   **The Reproduction of Color**   402
    The Young-Helmholtz model for color perception   402
    The CIE chromaticity diagram   403
    References   405

7   **Geometries for Creative Holography**   406
    Recording geometries for multicolor holograms   406
    Locating the hinge point and illumination axis   408
    Rainbow hologram: multi-strip set-up   408
    Rainbow hologram: multiple reference-beam set-up   411
    Reflection hologram: multiple reference-beam set-up   412
    A worksheet for a multicolor white-light transmission hologram   414
    References   416

8   **Monomode Optical Fibers in Holography**   418
    Multimode Fibers   418
    Single-mode propagation   419
    The use of monomode fibers in holography   419
    Launching the beam   419
    Making a fixed-ratio fiber coupler   420
    Variable couplers   421
    Practical use in holography   421
    References   422

9   **Fringe Stabilization**   423
    Error detector   424
    Expanding the fringes   425
    Comparator and amplifier   429
    Transducer   429
    References   431

10 **Processing Formulae**   433
    Transmission holograms   433
    Reflection holograms   435
    Stain removal   440
    Colloidal-silver developers   441
    Chemical blackening of reflection holograms   442
    References   442

11 **Books, Periodicals, Research Publications and Courses**   443
    Books   443
    Periodicals   446
    Research publications   446
    Courses in holography   446
    Computer link-ups   447

12 Suppliers of Equipment, Materials and Information 448
Chemicals 448
Collimating mirrors 448
Computer data information services 448
Courses on holography 449
Cross holograms 449
Embossing 449
Electronic components 449
Fringe stabilizers 450
General optical equipment for holography 450
Halide holograms 450
Holographic cameras 450
Ion laser tube repairs 450
Lasers (HeCd) 451
Lasers (HeNe) 451
Lasers (ion) 451
Lasers (Ruby pulse) 451
Materials 451
Optical fibers for HeNe light 452
PCB holders 452
Pinholes 452
Research-grade optical equipment 452
Safety goggles 452
Scaffolding clamps 453
Spatial filters and fiber-optics beam launchers 453
Spotlights 453
Stands and bases 453
Tables 453

Glossary 454
Author Index 481
General Index 483

Color Plates are located as follows:

Plates Nos.	*Facing pp.*
1–10	60–61
11–17	188–189
18–23	236–237
24–31	268–269

Foreword

Looking at holograms is exciting. Bobbing our heads around, probing the new visual spaces holography offers, can be a captivating experience. Indeed, most of us vividly recall the first hologram we ever saw, and many more we have seen since. And making holograms is even more exciting! Watching our own entirely new images come 'out of the soup' and light up in the beam brings a real thrill of exploration and discovery. It must be similar to what photography provided in the last century, and amateur radio earlier in this one. No matter how humble the set-up, every new hologram is an adventure. And as a holographer's understanding and insights develop, and his or her techniques advance, the holograms become ever more ambitious and intriguing.

For these are still the early days of holography. Despite its emergence more than twenty-five years ago in its modern laser form, most holographers are largely self-taught, and many fall away when their understanding fails to help them past the various challenges facing this complex new medium. There are books and papers to be found, certainly. Holography has stimulated some of the most profound theoretical work in optics. And there are even a few 'how to' books that show exactly where to put the laser, the mirrors, and so forth. But there are only a handful of educators who have brought their skill to a comprehensive approach to making, understanding, and improving holograms. I find this volume a singularly responsible work along these lines. Graham Saxby clearly describes what to bring together, and how to use it to make high quality holograms. And with an experienced educator's flair, he conveys the insights necessary for the student to take the next steps alone. Especially important is the ability to diagnose flaws in holograms, and to experiment with corrective actions. This can be especially challenging because holography draws together such diverse concepts from optics, photography, chemistry, and mechanical engineering. Further, the author's experience proves valuable in guiding the student toward set-ups and images that will work well enough the first time to provide an encouraging toe-hold in this mysterious territory.

The efforts of scientists, artists, inventors, and advanced amateurs are combining to change the 'look' of holography on a regular basis. Embossing technology has brought white-light transmission 'rainbow' holograms into the direct experience of hundreds of millions of people over the past few years. And with the requirement that they be visible in almost any light, a new genre of 'shallow depth' image design has developed. Pre-exposure swelling is still being refined to expand the palette of reflection holographers. And new materials are offering new options for all kinds of holograms. The scope of holographic imaging is being widened by refinements in pulsed laser holography, especially for portraiture, and in holographic stereography, which promises to make outdoor scenes, computer graphics, medical images, and other traditionally non-holographic content available in ultimate '3-D' form. From a fundamental point of view, these are all simply 'holograms'. But from a practical point of view, they involve different equipment and skills, and habits of thought, requiring a broad ranging concept-based education. Saxby has been tracking these with careful readings of the literature, backed up by interviews with the innovators involved, the essences of which are also offered here.

Where the trail of holography will lead next is anybody's guess. Armchair holographers had best stick to the current fare of science fiction, but for fellow 'hands-on' types, I warmly commend these pages. There are real adventures aplenty ahead!

Stephen A. Benton
Spatial Imaging Group
Massachusetts Institute of Technology
Media Laboratory, Cambridge, Mass.

Preface

For so youthful a technology, holography seems to have generated a record number of papers: more than six-and-a-half thousand, at a recent count. Even discounting near-repetitions, reviews and the more trivial papers, there are still more than a thousand which contain material of importance. Yet in the same time a mere dozen books have appeared on the subject, none of them covering it in anything like a comprehensive way, especially where practical techniques are concerned.

Even books plainly intended for the scientist and technologist are oddly reticent about practical details. Yet before such people can get down to work they need to know all kinds of practical things, some of them fairly mundane: how to set up a spatial filter; whether optical path matching is necessary (and how to check it); the highest laser power that can be used without damaging aluminized mirrors; what happens if retroreflection occurs in a pulse laser; how to adjust a Fabry-Pérot etalon; even where to find a 400 mm diameter collimating mirror for under £1000. Such practical books as do exist (almost all of them out of print at the time of writing) are mostly written at a do-it-yourself level that is unlikely to appeal to either the technologist or the professional; furthermore, writers on the artistic aspects of holography have almost always avoided any reference to the techniques they use, which is equally frustrating to the aspiring holographic artist.

This book attempts to fill this gap. It avoids mathematical theory, as this is already covered by existing books; for the same reason it does not go into detail over the analysis of interferometric data, and it mentions laser speckle interferometry only parenthetically. What it does cover is the *practical* aspect of holography, at all levels. The artist can find out the table geometry that can realize a particular concept; the aspiring professional is taken step by step through the simple foolproof set-ups to the most complex table geometries; and the technologist who wants to use holography as a research tool can find out how to set up a spatial filter, as well as why one is necessary, the kind of power that will damage an aluminized mirror (anything over 1 W), how to match optical paths (use a piece of string), what happens when the beam of a pulse laser is retroreflected (it blows the ends off the ruby rod), and so on.

The layout of the book is as logical as possible, given a subject with so many ramifications. To avoid having too many breaks in the flow, the important asides have been placed in boxes so that they can be read separately at some convenient point. Byways that are too lengthy to fit into boxes have been consigned to appendixes at the end of the book. Those readers with a pathological aversion to mathematics, ie about 90% of the likely readership (to which may be added the author), will be relieved to discover that the mathematics has also been quarantined in an appendix, and if they want to glue the pages together to stop the numbers and letters from getting out they may do so provided they have bought the book and not borrowed it.

The greater part of the book is concerned with practical holography for display purposes, and progresses from very simple single-beam holograms that a child of ten can make to techniques for producing multicolor art holograms of any desired degree of complexity. The final part of the book is concerned with applied holography. Here the treatment is confined to fairly broad outlines, with the accent on practical set-ups. Applied holography is well covered in the literature, and rather than attempt a long-winded rehash of what has been written much better by other people, I have kept to fairly general descriptions and have provided references to the more important original papers or books on the various aspects of the subject. Now that libraries are able to offer computer datalinks that can obtain access to abstracts of original material within seconds, and facsimiles of the complete papers within a day or two, this should not inconvenience anyone whose appetite has been whetted by a brief description of some application.

The early part of the book deals with the principles of holography, and in dealing with image formation I have used the Fourier approach rather than the more commonly-used Huyghens wave model. It is of course true that a rigorous Fourier treatment of image formation involves some neat mathematical footwork. Perhaps this is the reason why the Fourier model for image formation has until recently been eschewed by student textbooks on optics. But the difficult part – the computation of a two-dimensional Fourier transform – is no trouble at all for a modern computer; and writers of textbooks are at last beginning to get round to the Fourier approach, which provides an elegant and deeply-satisfying metaphor for the way in which both photographic and holographic images are formed. As the concepts involved can be grasped intuitively as well as logically, the whole model can be understood and applied entirely without the use of mathematics. Furthermore, once the fundamental concept of frequency space has been grasped, it is possible to look at an object and perceive it in a new and enriched way. Appendix 3 sets out the principles of Fourier optics; but it is necessarily somewhat brief, so a reading list is included for those who wish to follow up this fascinating approach.

There are a good many allusions to and comparisons with photography. In the early days of holography its practitioners were seldom trained photographers, and even now many professional display holographers come from a background of the fine arts. However, more and more professional photographers are beginning to realize that a knowledge of holography is becoming a necessary part of their stock-in-trade, and students working on visual communications courses are beginning to treat holography as just another medium for visual expression that touches and often overlaps the plastic arts, graphics and, most of all, photography. This is a healthy sign; the gee-whiz attitude to holographic images is disappearing as people become familiar with the medium, and images of taps and telephones are no longer exciting. Display holography has come of age, and is fit to take its place alongside photography and other media of visual communication.

Holography has bred its own jargon, some of it exciting and some of it deplorable. Much of the vocabulary of holography has come from the area of communications theory, so that, for example, there is some theoretical justification for calling a reference beam a 'carrier beam' and a Fresnel hologram a 'single-sideband hologram', as the terms are not inappropriate; but to describe the reconstruction of a holographic image as 'interrogating the hologram' is simply pretentious. The use of the word 'replay' to describe the same thing may be slangy, but it is graphic, alive and unambiguous. At the end of the book there is a glossary giving formal definitions and some suggestions as to usage. The recommendations are those of standard English, but are not compulsory: if you want to call your holographic table a camera there is nothing to stop you from doing so (except possibly the confusion of your audience, who already have a very good idea of what a camera is). When a technical term appears for the first time in the text and is printed in italics, this indicates that there is a formal definition in the glossary. American spelling has been adopted throughout, for the simple reason that it is easier for English readers to read this than for American readers to cope with English spelling conventions.

One of the recurrent nightmares of people who write books about rapidly-developing technologies is that some new discovery may render all or most of their text irrelevant before it appears in print. Indeed, this has happened to me on more than one occasion. However, in the past few years we have all learnt quite a lot about holography, and at the time of writing, holographic techniques seem to have reached something of a plateau. But new developments *are* appearing: for example, the use of monomode optical fibers and of fringe stabilization by servomechanical methods may before long make kitchen-table holography a reality; and the (by no means infeasible) appearance of a semiconductor laser the size of a penny, operating in the yellow-green region of the visible spectrum with a continuous power of 100 mW, with a pulse capability, a coherence length of 2 m, and a price-tag of only a few tens of pounds, would at a stroke render the whole of Chapter 3 obsolete — and we should all be dancing in the streets. Seriously, though, it *is* necessary to keep up with new techniques and discoveries, and for this reason Appendix 11 includes a list of publications where articles and papers on holography are likely to appear.

SI units are used throughout the book (see Glossary under *Multiples and submultiples* and *Scientific notation*). SI stands for *Système International d'Unités*, and is the only system of units now accepted world-wide. However, a number of 'rogue' units have persisted for various reasons. In this book the most important is the centimeter (cm), which is 0.01 m, and which appears where the use of 'correct' units would be unnecessarily pedantic. Some photographic sizes, though standardized in millimeters, are basically inch measurements, and have been left as such. In the text, measurements in mm are usually given parenthetically in inches for the benefit of those who (like the author) still measure by handspans and thumb-lengths.

Now let's get down to making some holograms.

G.S.

Acknowledgments

> While the Beaver confessed, with affectionate looks
> More eloquent even than tears,
> It had learnt in ten minutes far more than all books
> Could have taught it in seventy years.
> LEWIS CARROLL, *The Hunting of the Snark*

If I were to include in this list everyone who has given help and encouragement during the preparation of this book, it would include almost every holographer with whom I have had the pleasure of conversing or corresponding, as well as many whom I know only through their published papers. In many cases their information has made it possible to include practical advice that would have been impossible to give had I had to rely on my own limited experience. In particular I should like to thank Steve Benton, Nick Phillips, Ron Graham and David Pizzanelli for their valuable advice in both the early and late stages of the preparation of the book, Louise Sanders, who was responsible for the editing and Ron Decent, Production Director; Jeff Blyth, David Jackson, 'Hari' Hariharan, Rich Rallison, Fred Unterseher and Jody Burns for their help in their own specialist areas; Peter Miller, John Kaufman, Martin Richardson, Sam Moree and Dan Schweitzer for creative techniques; Steve McGrew and Suzanne St Cyr for help with complicated table geometries; and the specialist staff of the Central Electricity Generating Board's Engineering Laboratory, the National Physical Laboratory and Rolls-Royce plc for information on techniques in applied holography. I am much indebted to Jim McIntyre, Kevin Baumber, Mike Burridge, Rob Munday and the Royal College of Art Holography Unit, See-3 (Holograms) Ltd, Jayco Ltd, Applied Holographics plc and Ilford Ltd for their generous donations of time, materials and expertise in the preparation and production of the holograms in the book: without their generosity it would have been a good deal more expensive than it is. My thanks are also due to the authorities of The Polytechnic, Wolverhampton, for providing time and facilities for me to carry out essential research work. I must not forget Martin Gardner, whose 'Annotated Alice' and 'Annotated Snark' suggested the idea of heading each chapter with a quotation from the work of that mathematician, philosopher, photographer and storyteller extraordinary, Lewis Carroll.

I should also like to thank the staff of the New York Museum of Holography, in particular Scott Lloyd, for their help and generosity during my stay in New York, and all the holographers in that extraordinary city who went out of their way to make me welcome; also Steve Benton (again) and Bill Molteni, who did the same for me in Boston. I owe a special debt of gratitude to Tung Jeong, whose Second International Symposium on Display Holography at Lake Forest, Illinois, enabled me to meet, and pick the brains of, almost every practicing holographer in the world at that time. Any felicities you may find in the text are theirs. The mistakes are my own.

G.S.

PART I
PRINCIPLES OF HOLOGRAPHY

CHAPTER

1

What is a Hologram?

'I'm afraid I can't put it more clearly', Alice replied very politely, 'for I can't understand it myself, to begin with.'
LEWIS CARROLL, *Alice in Wonderland*

To the physicist, a hologram is a record of the interaction of two beams of coherent light which are mutually correlated, in the form of a microscopic pattern of interference fringes. To the informed layperson, it is a photographic film or plate which has been exposed to laser light and processed in such a way that when illuminated appropriately it produces a three-dimensional image. Less well-informed people often seem to think of a hologram simply as some sort of three-dimensional photograph. Certainly, both photography and holography make use of photographic film or plates, but that is about all they have in common. The most important difference is in the way the image is produced. The nature of the optical image produced by a camera lens can be described fairly accurately using a simple geometric or ray model for the behavior of light, whereas the holographic image cannot be described by the ray model. Its existence depends on diffraction and interference, which are wave phenomena. Photography has been around a long time, more than a century and a half, whereas holography, in the form in which we know it, has existed for little more than two decades; and it is barely forty years since the first, tiny, hologram was made.

Stereoscopy

When you take a photograph, the image you get is two-dimensional. If you look at it at an angle the only change you will see will be a foreshortening of the image. If the image is a face looking at the camera, the eyes appear to be fixed on you, and will remain so when you move to one side. This phenomenon seems so unnatural that it is often remarked on with surprise, although the trick was used by painters long before photography was thought of. In contrast, when you look at a sculptured head the eyes look at you only when you see the head from the front; the image is plainly endowed with depth. Each of your eyes sees a slightly different image by virtue of its slightly differing viewpoint; the impression of depth comes from the brain's interpretation of these differences. Soon after photography came on the scene, the idea of presenting two photographic images taken from appropriate viewpoints, one for each eye, led to the invention of the stereoscope by Herschel, and its commercial realization by Wheatstone. Since then the popularity of stereoscopic presentations has waned, but stereoscopy has more than earned its keep in metrology, photogrammetry, and, most notably, in photographic reconnaissance and aerial survey.

Aesthetically, there has always been something fundamentally unsatisfactory about two-image stereoscopy. For one thing, quite a large number of people, perhaps as many as one in five, have poor stereoscopic perception; indeed, one in twenty has none at all. Yet such people have no difficulty in telling whether they are looking at a real object or a flat photographic record of it. The reason for this lies in what is termed *parallax*, the change in appearance of an object with a change in viewpoint. Even with the head of the viewer fixed and one eye closed, the small movements of the eye made in scanning the image are sufficient to verify the solidity (or absence of solidity) of what is being examined. With stereoscopic pairs there is no live parallax: when you watch a stereoscopic film, everybody in the auditorium gets exactly the same view as you do. If you move your head sideways when you are looking at a stereoscopic illustration in a book, your viewpoint of the scene does not change. The eyes of a portrait remain fixed on you, for example and you may experience a loss of reality somewhat similar to that which you experience when you are listening to a binaural recording of an orchestra on headphones and, as you turn your head, the whole orchestra moves with it.

One attempt to overcome this lack of parallax, and at the same time to avoid the necessity for wearing special glasses, is the *parallax stereogram* or *panoramagram*. In the original version of this, a camera is moved round the subject (or the subject is rotated on a plinth) while photographs are taken at intervals of a few degrees. The resulting negatives are then printed on a single sheet of print material, using an optical printer which interlaces the images in narrow vertical strips. After processing, the print is mounted under a fine lenticular screen which allows only one image to be seen from any one viewpoint. The result can be viewed without glasses and provides genuine, if rather jerky, horizontal parallax. Such *autostereograms* still enjoy some measure of success in the picture-postcard industry, and from time to time cameras embodying similar principles appear on the amateur market. The images they produce are none too successful, as anyone who has seen them will testify. Horizontal parallax is certainly present, but figures and, even more, backgrounds, appear like cardboard cut-outs. In addition, the whole perspective appears shallow, as if the subject were standing in front of a painted background. The reason for this is that in the optical image formed by a camera lens the scale of the image in an axial direction is the square of its scale in a lateral direction (see pp. 365–7). If we reduce the image as compared with the object by a factor of 10 we reduce its depth by a factor of 100. Only at 1 : 1 scale is the depth revealed truthfully, and this is why the only convincing photographic stereograms are of table-top creations. However, fine stereograms *are* possible: the work of the Parisian photographer Bonnet in this medium, with full-size images and using as many as 36 separate views over a large angle of view, has to be seen to be believed.

Any single photograph is, of course, two-dimensional, yet the optical image that produced it was three dimensional (Fig 1.1). If you form a full-size image of

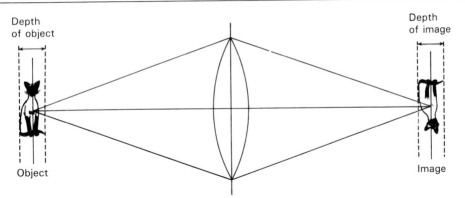

FIG 1.1 *The optical image produced by a lens is three-dimensional; a ground-glass screen placed at any plane within the depth of the image will produce a two-dimensional image with the part cut by the plane appearing sharp.*

an object such as a flower with a camera lens, and catch this optical image on a ground-glass screen, when you move the screen forward and backward you will see different parts of the image coming into focus successively. The image has the same depth as the object (if the scale is 1 : 1). If you remove the ground-glass screen you will be able to see the image hanging in the air; it is inverted, but has full vertical and horizontal parallax (though if the diameter of the lens aperture is less than the distance between your eyes you will not be able to see the image stereoscopically). Thus inside every camera there is a three-dimensional image struggling to get out. This, of course, is not news; Leonardo da Vinci knew it, and so did Galileo. But is there any way of recording this image so that we can see it as it really is, in three dimensions?

Defining the Problem

Stereoscopic photography, as we have seen, provides only a partial answer. Other systems involving, for example, vibrating screens, have been tried with varying degrees of success. But the only wholly successful technique has been to move away from photography altogether and to look at the problem in a different light — or rather, with a different model for the behavior of light.

It is the *ray model* that is generally used in photographic optics. It enables us to describe the formation of an optical image, to calculate the *depth of field* and angle of view, and even to design lenses. But the ray model is limited. For example, it does not correctly describe the fine detail of the optical image of a point object formed by a lens, even a theoretically-ideal lens, nor does it predict that the fine detail in a photographic image progressively disappears as the aperture is closed down. Nor can the ray model tell us anything about *polarization*, nor describe the special qualities possessed by *laser* light. For an explanation of these phenomena we need to think of light as *electromagnetic radiation* propagated through space in the same manner as radio waves. Indeed, light does possess all the characteristics of radio waves; the only difference is in its much higher frequency.

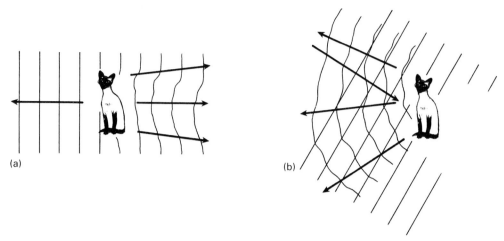

FIG 1.2 *A plane wave diffracted by a transparent object* (a) *or an opaque one* (b) *becomes a highly complicated wave which contains information about the object.*

The problem can be defined when we analyze what happens when we 'see' an object in terms of a wave model. When a beam of light falls on an object, the transmitted or reflected light is modified by it, so that the *wavefronts* are quite complicated (Fig 1.2).

Now, the only information the eye receives about the object is contained in the part of the wavefront intercepted by the pupil of the eye. As long as the eye remains stationary the shape of the intercepted wavefront remains unchanged, and the appearance of the object also remains unchanged. But if the viewpoint is changed, the eye intercepts a different portion of the wavefront: the information carried by the new portion is different, and a different view is seen. This is the clue to the working of stereoscopic vision: the two eyes intercept different portions of the object wavefront, and thus see two different views. Stereoscopic photographs contain part of the information contained in the two portions of the wavefront; though only a part, it is sufficient to provide the illusion of depth; however, the only way to provide *all* of the information is to provide a reconstruction of the entire wavefront. If this can be done, the experience of the viewer will be precisely the same as if the object itself were present. The question is, how *can* this be done?

The Problem Solved

Holography provides the answer. A *hologram* is a complete record of the information, and when correctly illuminated it will generate a replica of the object wavefront, enabling the viewer to see an *image* which in every respect replicates the object, with full parallax.

There is one proviso. The wavefront from the object gives us information not only about the object but also about the illuminating source. In order to record the

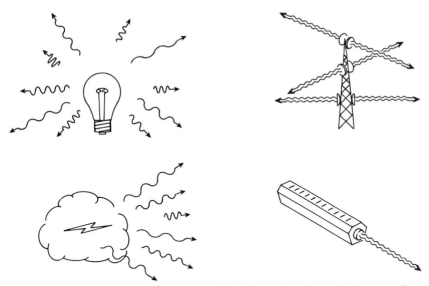

FIG 1.3 (a) *Randomly emitted light contains all wavelengths emitted more or less at random, like the random radio waves emitted by an electrical storm.* (b) *The well-disciplined beam of a laser contains only a very narrow band of wavelengths, which remain in phase for a considerable distance, like the beam of radio waves emitted by a broadcasting station.*

object information uncontaminated by information about the light source (which would otherwise swamp it) the illuminating light must contain no information at all, that is, it must consist of *plane wavefronts*. Such wavefronts would be produced only by a *monochromatic* (ie single-wavelength) point source at *infinity*. A filament lamp is plainly out of the question: it is an extended source with a whole spectrum of *wavelengths*. Laser light, on the other hand, conforms closely to the requirements. Light from a filament lamp and light from a laser may be compared respectively to the longer-wave radiation from an electrical storm and from a radio beacon. In each case the former emit radiation (light and radio waves respectively) that is random, containing all frequencies and traveling in all directions, whereas the latter emit well-disciplined beams which contain only a narrow band of frequencies and can be made highly directional (Fig 1.3). They also possess an important property known as *coherence*, which is discussed more fully on pp. 24–6.

Interference

In order to understand the way in which a hologram encodes the object wavefront it is necessary to introduce the concept of *interference*. When two or more sets of waves travel through the same space they interact, and if the waves are of the same wavelength the interference pattern, as it is called, is regular and predictable. You may have seen demonstrations of interference patterns using water waves in a ripple tank. Interference also occurs with electromagnetic waves. If you have a VHF radio in your car, you may have noticed that sometimes when you are passing through a town the sound from your speakers begins to fade up and down in pulses. You may also have had the frustrating experience of being stopped at traffic signals where you are receiving no sound at all. What is happening is this: at the same time as you receive a signal directly from the transmitter, you are also receiving the same signal after it has been reflected from a tall building, and the two signals, at this location, are in *antiphase*, that is, crests (positive voltage pulses) from the transmitter are coinciding with troughs (negative voltage pulses) from the echo off the building. As the lights go green and you move off, you enter a region where the crests (and the troughs) from the two sources are synchronized (in phase); you are now receiving a doubly-reinforced signal and the sound is loud and clear. Another meter or two farther on, and it fades again; and so on (Fig 1.4).

An experiment with interference fringes The phenomena we have been discussing are called respectively *destructive* and *constructive interference*. (Do not confuse this with radio 'interference' from other stations or from thunderstorms.) The first demonstration of the interference of light waves was by Thomas Young early in the nineteenth century, using a pair of narrow slits, now known as Young's slits, illuminated by sodium light, which contains only a narrow band of wavelengths. Because the slits had to be very narrow and the light source itself had to be masked by a further slit, the interference pattern known as *Young's fringes*, was very weak, and Fresnel suggested using an optical device now known as a *Fresnel biprism*, which is a pair of very shallow prisms made back to back from a single piece of glass. This device causes the two halves of a beam of light to overlap. If you use laser light for this purpose you will get strong interference effects. When you make your first multi-beam holograms you will need a *3° prism*; this acts like one half of a Fresnel biprism, and can also be used for the experiment described below. You might like to try this experiment yourself; it will give you considerable insight into the way a hologram codes the information contained in the object wavefront (Fig 1.5).

You will need a concave lens of focal length approximately -6 mm; this will be a useful item when you start making holograms. Fix the lens to the output port of the laser, and place the prism in the beam with its thicker edge bisecting the disk

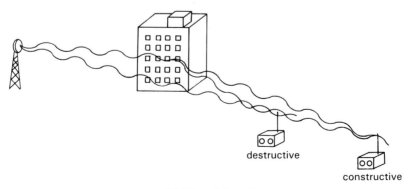

FIG 1.4 *Multi-path interference.*

WHAT IS A HOLOGRAM?

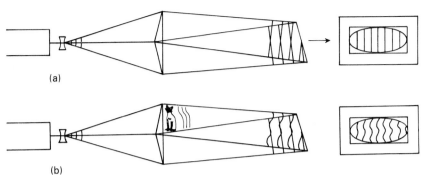

FIG 1.5 *Fresnel biprism experiment. (a) When both beams are undisturbed the inference pattern is a row of parallel straight fringes. (b) When an object is inserted, the pattern becomes distorted. The distortion contain all the information about the object wavefront.*

of light. Place a large piece of white card about 1.5–2 m away, at a distance where the two D-shaped beams overlap. Now turn the card about a vertical axis until the overlapping patch is well drawn out (Fig 1.5). You should be able to see a series of vertical dark and light bands. The dark bands correspond to regions where the two wavefronts are interfering destructively; the light bands correspond to constructive interference. The spacing of the bands will be of the order of a millimeter or so. They are called *interference fringes*, and you will notice that they are straight and parallel.

Now comes the crucial demonstration. Take a fine sewing needle, embed one end in a cork so that it sticks up obliquely, and place it in one of the two halves of the beam; then examine the fringe pattern again. You will see that the fringes are still vertical overall, but are now kinked in a regular sort of way. This effect is a direct result of the disturbance of the wavefront by the object, and the amount of displacement of a fringe at any point is directly proportional to the disturbance, or change in *phase*, of the *object beam* wavefront relative to the phase of the undisturbed or *reference beam* wavefront. If you now remove the needle and replace it by a more complicated object such as a glass animal, you will see that the pattern has become much more fragmented. Nevertheless, it has still recorded the disturbance in the object beam faithfully. What is more, all you need to do in order to record this information is to position a photographic plate or film in place of the screen and, when it has received sufficient exposure, remove and process it. You then have a record which contains *all* the information about the object wavefront. This photographic record is a hologram*. The term was coined in 1948 by Denis Gabor,

---

* By analogy with the word 'photography', G L Rogers coined the term 'holography' to describe the technique; unfortunately the word 'holograph' had already been pre-empted by the literary fraternity, to mean a document in the author's handwriting. However, the term 'holograph' as an alternative to 'hologram' is accepted by the Oxford English Dictionary, and it may well eventually become the accepted term for a representative image, with the term 'hologram' being used, as with 'photogram', for completely abstract images. Other formations by analogy with photography include '*hologrammetry*', '*holomicrography*', etc.

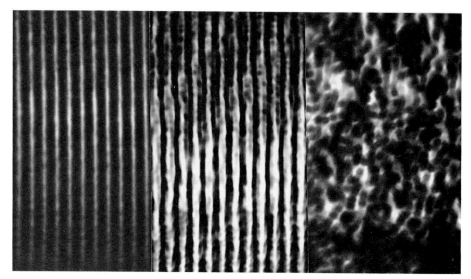

FIG 1.6 *Fringes formed using the configuration of* FIG 1.5. (a) *Both beams undisturbed;* (b) *the change in pattern when a sewing needle is placed in one of the beams;* (c) *the pattern when a glass animal is placed in one of the beams.*

the prefix 'holo' coming from a Greek word meaning 'whole' and the syllable 'gram' from another Greek word, conventionally signifying a drawing or other visual representation (Fig 1.6).

If you do not have a 3° prism you can carry out the same experiment by putting a large sheet of glass into the beam at a very shallow angle to deflect one half of the beam across the other half. The system is known as *Lloyd's mirror*, and is an exact replica (except for scale) of the VHF radio two-path interference phenomenon with which we began this section. You were, in fact, driving right though a huge fringe pattern (Fig 1.7).

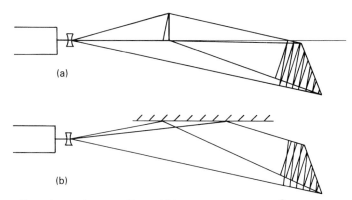

FIG 1.7 *Two alternatives to a Fresnel biprism.* (a) *A single 3° prism;* (b) *Lloyd's mirror. The geometry is the same as that of multi-path distortion of a radio signal.*

Reconstruction We have now recorded the information carried by the object wavefront, frozen into the hologram fringe pattern; but can we retrieve it? The answer is yes, and the method is surprisingly simple. If the developed hologram is placed back in its original position and illuminated by just one of the original beams, *both* beams will emerge (there is also a third emergent beam which we will leave till later). So if we illuminate the hologram with the original reference beam, which we may now call the *reconstruction beam*, we obtain a replica of the object beam, an *image beam*, which emerges from the hologram as a continuation of the object beam. This procedure is usually called *replaying*, by analogy with sound recording. As this image beam is identical with the object beam, an observer looking along it will see what appears to be the object itself, in full realism, and, what is more, in full parallax, as a shift in viewpoint causes the eye to intercept a different part of the image wavefront. The upper part of the hologram records the view of the upper part of the object, the right side of the hologram the right-hand view of the object, and so on. It is just like looking through a window at the object; and if you cut the hologram in half you do not destroy the image, but merely restrict the view. However, since the aperture of the hologram is reduced you also reduce the image resolution, much in the same way as reducing the diameter of a camera lens reduces the resolution of the optical image.

Now, although the fringe pattern we obtained in our $3°$ prism experiment is a genuine hologram it is not a very interesting one, as there is very little parallax, and the direct beam is very close to the image. If we increase the angle between the reference and object beams, the direct beam will be well out of the way of the image beam. We can also use more interesting objects. They need not even be transparent; we can reflect the object beam off opaque subject matter, and by placing the holographic plate or film close to the object we can get a hologram which offers a wide parallax angle.

But before we go any further, we should look a little more closely at what goes on when a hologram is replayed. How does this set of irregular fringes turn one beam into three when it passes through them? What is really going on? The optical phenomenon concerned is called *diffraction*, and it plays a key role in the reconstruction of the image.

Diffraction

You will certainly be familiar already with some of the manifestations of diffraction. The iridescent colors of butterflies and dragonflies, the flashing hues reflected from a CD or videodisk recording, the tail of a peacock, the rainbow colors of the little cards and bumper stickers in souvenir shops, all of these produce their colors by diffraction. When light waves (or, for that matter, any waves) pass through a narrow grating, they spread out in well-defined directions. The simplest possible form of grating is called a *cosine grating*, and if you pass a beam of laser light through it, three beams emerge. One is a continuation of the original beam; the other two, symmetrically on either side, emerge at an angle which depends on the wavelength of the light and the number of light/dark cycles per millimeter, usually called the *spatial frequency*. The spacing of the grating cycles is the reciprocal of the spatial frequency, and is called the *spatial period* (Fig 1.8).

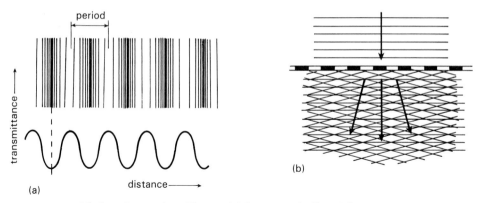

FIG 1.8 (a) *A cosine grating. The spatial frequency is 1/spatial period.* (b) *When a laser beam passes through a cosine grating, three beams emerge.*

A photographic record of fringes produced by the 3° prism in the absence of any disturbing object is in fact a cosine grating, and if it is illuminated by a laser beam it behaves in precisely the manner described. If the light is reflected from the surface of the grating instead of being transmitted, the effect is the same. All the objects and creatures that show iridescent colors do so because they have regular patterns of scales or of fine grooves in their surfaces. The reason for the colors is that the angle of diffraction depends on the wavelength: blue light has a shorter wavelength than green light and is diffracted less than green light, which in turn is diffracted less than red light. Thus white light, which contains all visible wavelengths, undergoes *dispersion* into the hues of the spectrum by the pattern of scales. However, as lasers have only one wavelength, diffraction of a laser beam by a cosine grating produces only three small spots of light (Plate 1).

The Fourier model for diffraction

If two cosine gratings of differing spatial frequencies are superimposed, each will produce its own independent pair of spots, the spacing of the spots being directly proportional to the spatial frequency of the grating. Most diffraction gratings that are manufactured for use in physics teaching or research purposes are made initially from glass sheets that have been ruled with opaque lines; such gratings are known as *square gratings*, from their transmittance profile. A square grating illuminated by a laser beam produces a whole row of spots spaced at regular intervals (Fig 1.9)

The technique of *Fourier analysis*, a mathematical procedure developed by Joseph Fourier at the beginning of the nineteenth century, shows (among other things) that a square grating is identical with the sum of (ie the superposition of) a series of cosine gratings having equal increments of spatial frequency (eg 10, 30, 50, 70, etc cycles/mm. In fact, any regular grating, in one, two or even three dimensions, can be constructed from a *Fourier series* of cosine gratings of differing spatial frequency and orientation. Moreover, the 'grating' does not even have to be a regular figure, but can be literally any figure; it can still be described as a spectrum

WHAT IS A HOLOGRAM?

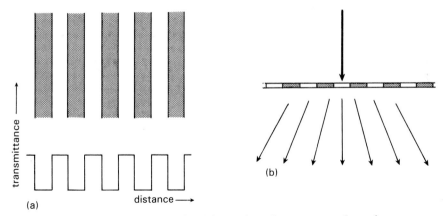

FIG 1.9 (a) *A square grating.* (b) *When a laser beam passes through a square grating, a whole series of beams emerges.*

of spatial frequencies. Fourier developed a mathematical technique, now known as a *Fourier transform*, to perform this operation. Although he evolved his model in connection with thermodynamics, its main use in the first half of the twentieth century was in communications technology. However, in the 1950s it became clear that the Fourier model could be applied to the quantitative evaluation of optical images, and the Fourier-based concept of the *optical transfer function* led to a revolution in the understanding of the performance of lenses, and eventually to new insights into lens design. The advent of the laser in the early 1960s made it possible to demonstrate that a lens actually produces in its rear focal plane an optical Fourier transform of an object in its front focal plane, thus confirming the validity of *Fourier optics*. You will find a fuller treatment of the Fourier model for imaging in Appendix 3.

In the experiment with the 3° prism we disturbed one of the beams with a sewing needle. What happened to the interference pattern? Well, the disturbing object was a single opaque bar, an object mathematicians call a *top hat function* from its transmittance profile. This function has a readily-calculable spatial frequency content (see Box above), and produces a well-described diffraction pattern (Fig 1.10).

It can be shown mathematically that if the disturbed and undisturbed wavefronts are combined and recorded on film, the interference pattern on the processed film will diffract a laser beam so as to produce a continuation of the same disturbed wavefront, propagated away from the recorded pattern; and so it proves in practice. For more complicated diffraction fields such as the interference pattern caused by a glass animal, and even for the light reflected from an opaque object, the same holds true: the beam diffracted away from the hologram replicates the original object beam. For the more mathematically-minded reader there is a proof of this in Appendix 2.

So far, we have been considering the behavior of light when the distance between the object and the hologram is large enough for the various diffracted

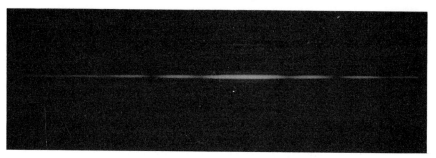

FIG 1.10 *The diffraction pattern produced when a laser beam passes through a single slit (a top hat function).*

components of the object beam to have sorted themselves out. This type of far-field diffraction is known as *Fraunhöfer diffraction*, and it is not particularly difficult to analyze by Fourier methods. In general, however, when we make a hologram we have the recording material quite close to the subject matter, and the diffracted wavefronts are all mixed together. This is known as *Fresnel diffraction* (it was Fresnel who first analyzed this situation). In a *Fresnel hologram* (sometimes called a *sideband hologram*) the information is not localized, but is distributed over the whole emulsion surface.

The information in a hologram

The Fourier model tells us that all the information about the subject matter of a hologram is coded in its diffraction field. By looking at its *far-field diffraction pattern* you can get the following information about an object:

1. The spatial frequencies of the cosine-grating component that are present (from the distances of the pairs of spots from the *optical center* of the pattern).
2. The relative *amplitude transmittances* or reflectances of the components (from the brightness of their spots).
3. Their orientation (from the orientation of the pairs of spots).

But it is impossible to tell *where* the dark and light bars of the gratings are (ie their spatial phase) because this information depends on the relative phase of the components of the diffracted wavefront. Phase information disappears when a diffraction pattern is recorded, for one simple but frustrating reason: every light detector, photochemical, photoelectronic or biological, records only the *time-averaged intensity* of the field. This is proportional to the square of the amplitude; when you take the square root to get back to the amplitude you have no way of knowing whether the answer is positive or negative (Fig 1.11). The presence of a reference beam, however, preserves this information, as it gives us something we can measure the phase of the object beam against. *The phase of the object beam relative to that of the reference beam is coded in the displacement of the fringes.* So a hologram is, indeed, a total record of the wavefront diffracted by the subject matter; and, as we

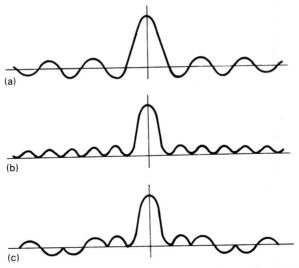

FIG 1.11 *If the function (a) is squared, the function (b) results. It is, of course, entirely positive; you cannot now be sure that the original function was (a) or (c) or, indeed, one of many other functions.*

have seen, its illumination by a replica of the reference beam reconstructs the wavefront itself.

Amplitude and Phase Gratings

There is just one more item. In the mathematical analysis of the behavior of a hologram in Appendix 2 it is assumed that it is the *transmittance* of the hologram that codes the object information. Now it is an unfortunate fact that if you operate in the *linear region* of the conventionally-processed emulsion (see p. 370), less than 3% of the replay beam actually goes to make up the image beam. We say that the *diffraction efficiency* is less than 3%. However, if the grating that forms the hologram consists of variations in *refractive index* rather than transmittance, it is possible to make all of each fringe contribute to the diffracted beams; by converting all the developed silver back into silver bromide (which happens to have a very high refractive index) we can raise the diffraction efficiency to better than 30%, and in some types of hologram to nearly 100%. There are a number of ways in which we can do this, and these are described in later chapters.

CHAPTER

2

How Holography Began

'Ahem!' said the Mouse with an important air. 'Are you all ready? This is the driest thing I know. Silence all round, if you please! "William the Conqueror, whose cause was favoured by the Pope, was soon submitted to by the English, who wanted leaders, and had been of late much accustomed to usurpation and conquest. Edwin and Morcar, the earls of Mercia and Northumbria – "' 'Ugh!' said the Lory, with a shiver.
LEWIS CARROLL, *Alice in Wonderland*

There seems to have been no particular reason why holography should have been such a late developer. Although holograms today are invariably made using a laser as the light source, they can – and have been – made (with certain restrictions) with other light sources, even, most improbably, with white light. The theoretical principles underlying holography could well have been worked out as early as 1816, when Auguste Fresnel clothed Thomas Young's 1802 theory of diffraction and interference (in terms of Christiaan Huyghens's 1678 wave model) with the respectable garment of mathematical rigor; at about the same time, the first experiments that resulted in photography were being carried out (Thomas Wedgwood, 1802; Nicéphore Nièpce, 1816). In 1856 Scott Archer discovered how to produce a light-sensitive material coated on glass. The monochromatic property of the golden-yellow sodium flame was well known, and it would at that time have been just possible to make a Denisyuk-type reflection hologram of, say, one of the new-fangled postage stamps (Fig 2.1).

But the history of technology makes it plain that inventions appear only when contemporary culture is ready for them. The principles of holography in fact came together only in the late 1940s, and for a purpose nobody, with hindsight, would ever have anticipated: the improvement of the quality of electron-microscope imagery. The idea came to Denis Gabor quite suddenly: it seems, while waiting for a game of tennis on Easter Day 1947. At that time he was an electrical engineer working for British Thompson-Houston in Rugby. He spent the rest of the year working with his assistant Ivor Williams on his 'new microscopic principle'. As it was not possible at the time to generate beams of electrons that would be sufficiently well-behaved for his requirements, Gabor carried out his experiments using visible light from a filtered mercury arc. Because of the limited coherence of his source his holographic images were restricted to transparencies little larger than a pinhead. The

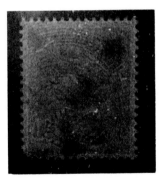

FIG 2.1 *A Denisyuk (single-beam reflection) hologram made using a sodium-lamp source.*

first successful hologram bore an image containing the names of Huyghens, Young and Fresnel. Gabor's two papers (1, 2) for which he was subsequently to be awarded a Nobel prize, were published respectively in 1948 and 1949; but it was to be fifteen years before there were any further useful developments published.

One of the main troubles was that Gabor's optical configuration (see p. 40), in which the light source, object and photographic plate were in line, had a serious flaw. It certainly re-created the object wavefront, as Gabor had predicted; but the genuine (virtual) image was obscured by a spurious real image in line with it (Fig 2.2).

The inability to produce a coherent beam of electrons put the originally-proposed application in electron microscopy out of court, and before long new insights into transfer function theory had made possible the improvement of electron microscope images to such an extent that the holographic principle became irrelevant to them. Only a few workers continued to work on the subject, notably Hussein El-Sum (3) at Stanford University. Yet, unknown even to Gabor, important things were happening. Behind closed doors at the University of Michigan, Emmett Leith and Juris Upatnieks were working on radar imagery. Familiar with Gabor's papers, they had noted that *synthetic-aperture radar* records had many features in common with holograms, and that when recorded in the correct format they could be reconstructed optically. Radar signals had the same order of coherence, wavelength for wavelength, as the mercury arc. Furthermore, radar engineers had already noted, and solved, the spurious-image problem, correctly recognizing it as an *aliasing artifact*.

This was a troubled period for holography. But as Leith said (4) in 1981, 'As holography actually shrank in one area, it grew in the other ... In a manner of speaking, it went underground ... Optical holography ... gave rise to new data-processing systems. What was pertinent to holography in the field of radar was not available to the optical holographer, because radar research was a classified area.' So Leith and Upatnieks worked on optical holography themselves, and quickly solved the problem of the spurious real image by displacing the reference beam (see p. 41–2), moving the unwanted image out of line.

Meanwhile, in the Soviet Union, Yuri Denisyuk was experimenting with an

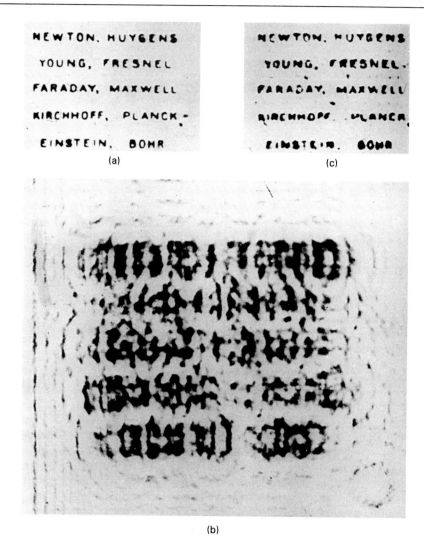

FIG 2.2 *One of Gabor's first holograms:* (a) *shows the original object, (a tiny transparency),* (b) *shows the hologram, much magnified;* (c) *shows the reconstructed image. Photographs courtesy of National Physical Laboratory.*

optical configuration that was radically different from Gabor's. In this configuration the reference and object beams were incident on the photographic plate from opposite sides. This was achieved by placing the plate between the light source and the subject matter, so that the portion of the reference beam not absorbed by the emulsion passed through and was reflected back from the subject, forming the object beam. This configuration produced interference planes that were much more nearly parallel with the surface of the plate than those of transmission holograms: the latter were like a venetian blind, the former more like the pages of a book. Because a photographic process for producing natural colors which had been developed by Gabriel Lippmann in the 1980s (*Lippmann photography*) showed superficial similarities to Denisyuk's principle, he modestly called his holograms *Lippmann holograms* and they are sometimes still called by this name, though more often known as *Denisyuk* or *single-beam holograms*. By 1962 Denisyuk had succeeded in producing holograms in which the image could be reconstructed using a point source of white light (5). This was a considerable advance on other configurations which required a monochromatic reconstruction beam.

The appearance of a workable laser in 1962 gave holography the impetus it needed. Its importance centered round the large increase in coherence length – centimeters or even meters rather than fractions of a millimeter. It now became possible to make holograms of solid objects. Leith and Upatnieks produced the first *laser transmission hologram* of a solid object (a model railway engine) in 1963 (6) (Plate 2), and Denisyuk began to produce holograms of art objects in the same year (7) (Plate 3).

After this holography began to develop rapidly. A good deal of the progress consisted of small improvements in optical components, holographic emulsions and processing methods, allied to a growing mastery of the techniques by practitioners. But there were also three advances that were so significant as to introduce a whole new dimension (in a figurative sense) to the applications of holography.

The first was the discovery that if an object was subjected to stress between two holographic exposures, any distortion that resulted would be contoured by *secondary fringes* (see pp. 316 ff); the first published paper on holographic interferometry, by Robert Powell and Karl Stetson (8), appeared in 1965. The importance of this discovery to measurement science and to stress and vibration analysis has been incalculable.

The second advance had its main impact in the field of creative holography. As long as holography was confined to producing *virtual images*, ie images lying behind the hologram, it could be argued by jaundiced critics that a hologram provided no more than a vicarious experience of the original subject matter. But in the mid-1960s it began to be appreciated that by using an appropriate reconstruction configuration the virtual image could be suppressed and the spurious real image reconstructed instead. This image could be used as the object for making a second, or *transfer hologram*. By restricting the vertical parallax, Stephen Benton produced in 1968 a *transmission hologram* which could be replayed using white light (9). The principle of transfer images was quickly extended to reflection holograms; thus holograms could now be produced with an intermediate stage. Just as in creative photography, it now became possible to introduce artifacts into the final hologram. Creative holography was born.

The third advance was in the commercial field. In 1974 Michael Foster (10) introduced a method for duplicating holograms mechanically by using them in the same way as audiodisks. It thus became possible to mass-produce holograms at very low cost, holograms which, turned into reflection holograms by an aluminium backing, could be used in textbooks, art publications and publicity handouts, and on credit cards as a security device.

The past two decades have seen many more advances in holographic technologies, such as live portraiture, natural color and *holographic stereograms* made from movie frames and computer graphics. These will all be dealt with in later chapters.

References

1. Gabor D (1948) A new microscopic principle. *Nature*, **161**, 777–8
2. Gabor D (1949) Microscopy by reconstructed wavefronts. *Proceedings of the Physical Society of America*, **197**, 454–87.
3. El-Sum H M A and Kirkpatrick P (1952) Microscopy by reconstructed wavefronts. *Physical Review*, **85**, 763.
4. Leith E N (1981) The legacy of Denis Gabor. *Holosphere*, **10**, 5–7.
5. Denisyuk Yu N (1962) Photographic reconstruction of the optical properties of an object in its own scattered radiation field. *Journal of the Optical Society of America*, **53**, 1377–81
6. Leith E N and Upatnieks J (1964) Wavefront reconstruction with diffused illumination and three-dimensional objects. *Journal of the Optical Society of America*, **54**, 1295–301.
7. Denisyuk Yu N (1963) On the reproduction of the optical properties of an object by the wave field of its scattered radiation. *Optics & Spectroscopy* **18**, 152–7.
8. Powell R L and Stetson K A (1965) Interferometric vibration analysis by wavefront reconstruction. *Journal of the Optical Society of America*, **55**, 1593–8.
9. Benton S A (1969) A method of reducing the information content of holograms. *Journal of the Optical Society of America*, **59**, 1545.
10. Foster M (1974) Quoted in private communication with Benton S A (Feb 1976). Benton refers to it in *Proceedings of the Society of Photo-optical and Instrumentation Engineers** **532**, 8–15.

---

* This Society is invariably called by its initials (SPIE) and will be so indicated in references throughout this book.

CHAPTER

3

Light Sources for Holography

> 'I engage with the Snark – every night after dark –
> In a dreamy delirious fight:
> I serve it with greens in those shadowy scenes,
> And I use it for striking a light.'
> LEWIS CARROLL, *The Hunting of the Snark*

Light as an Electromagnetic Phenomenon

Over the past two millenia there have been many theories purporting to explain the nature of light; most have failed as theories because they could not explain all of its behavior. Modern philosophers of science argue that the concept of a theory of light (or, indeed, of any physical phenomenon) is logically unsound, and that it makes more sense to consider the 'theories' as models or metaphors which represent certain aspects of the behavior of light under certain circumstances. These models can then be used within their limits to make useful predictions. It is of course good sense to choose the simplest model for a given investigation. For most of this book a comparatively simple model, the *Huyghens wave model*, is used: this model represents light as traveling in the form of *transverse waves*. If these waves are considered to be the propagation of an electromagnetic disturbance (Maxwell's *electromagnetic model*) then the behavior of light can be described using the same equations as are used to predict the behavior of radio waves, and *optoelectronic phenomena* such as the propagation of light within *monomode optical fibers* (see Appendix 8) can be predicted. The transverse wave models predict all the phenomena of holography and the optical properties of holograms, but are unwieldy when attempting to describe what happens at the *principal focal plane* and the image plane of a convex lens. For this we use the powerful and elegant *Fourier model* (see pp. 12–5 and Appendix 3). However, neither of these models can satisfactorily describe the way in which light is generated, nor what happens when it is absorbed, say, by a photographic emulsion (these are known as *photoelectronic phenomena*). For this we need to use the *quantum model*, which represents light energy as being generated not continuously, but as large numbers of pulses of electromagnetic energy called *photons*.

These models must not be thought of as mutually incompatible. The Fourier approach is a slightly different way of looking at the transverse wave model which makes the discussion of the nature of image formation more straightforward; similarly, the quantum model does not contradict the electromagnetic model but subsumes it, so that it can describe not only the propagation of light but also the action of a laser and the various photoelectronic effects.

Propagation of Electromagnetic Waves

You are probably aware that a magnetic field is always generated around an electrical conductor carrying a current by the passage of a current. You can easily check this by putting a small compass needle close to the conductor with its axis of rotation parallel to the conductor. When you switch on the current the needle is deflected in such a way that it lies on an imaginary circle drawn round the conductor. If you place a second compass needle a short distance away and similarly aligned, it will also be deflected, apparently simultaneously. However, the deflection is not quite simultaneous: the second needle begins to move a fraction of a second later than the first. This delay occurs because the magnetic field does not appear instantaneously throughout space; it is propagated outwards from the source at approximately 300 000 000 m/s* (186 000 miles/s)(Fig 3.1). As a magnetic field created in this way always has an electric field associated with it, we call this propagation electromagnetic radiation. The electric and magnetic fields are perpendicular to one another, and are propagated simultaneously, or *in phase*.

Let us now suppose that our conductor is carrying an alternating current (AC). This changes direction 100 times a second, so that a full cycle occurs 50 times a second (60 in the United States). We say that the frequency (strictly, *temporal frequency*) of the AC is 50 cycles per second, or 50 hertz (abbreviated Hz). These cycles are being propagated into space at a fixed speed, so that at any instant a single cycle will be spread out over a distance which can be calculated. As the frequency

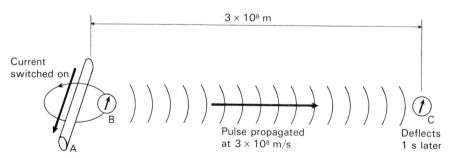

FIG 3.1 *The propagation of an electromagnetic disturbance. The conductor (A) is switched on. The compass needle at B deflects at once. The needle at C, 3×10^8 m away, deflects 1 s later.*

---

* For brevity, this figure is usually written 3×10^8, where the number 8 represents the number of zeros that come after the 3. This method of writing numbers is known as *scientific notation*, and is used throughout this book.

becomes higher, the wavelength becomes shorter. You can find the wavelength by dividing 3×10^8 by the frequency; for 50 Hz AC it works out as 6000 km.

How to describe a wave

The simplest kind of wave has a curved profile that is described as being *sinusoidal*, because if you want to derive the shape of it by plotting its graph, the equation of the curve is of the form $y = A \sin Bx$. (A and B are constants). All the electromagnetic waves we shall be considering are sinusoidal. We have seen in the main text that the frequency, which is the number of complete cycles passing a given point in a second, is related to the wavelength, which is the distance between successive crests, and the relationship is:

$$\text{frequency} \times \text{wavelength} = \text{velocity}$$

Frequency is given the symbol f for radio waves and ν (nu) for light waves; wavelength is given the symbol λ (lambda) for both radio and light waves. The velocity of propagation in empty space is given the symbol c, and is equal to 3×10^8 m/s. We also need to know the *amplitude* (strictly, peak amplitude, if it is necessary to distinguish it from *instantaneous amplitude*) of the wave, which is the maximum deviation of the wave from zero (it is actually measured in volts per meter (V/m) but this is not important for our purposes). The wave is always described in terms of its electric component (or *electric vector*); the magnetic component (or *magnetic vector*) is in phase with this, and the direction of its field is at right angles to the electric field, clockwise to it when looking down the direction of propagation (Fig 3.2). There is a fuller treatment of the subject in Appendix 2.

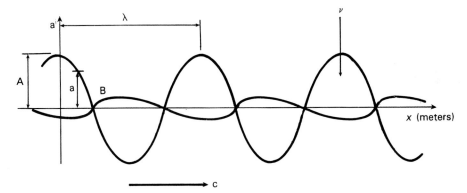

FIG 3.2 *An electromagnetic wave. A is the (peak) amplitude of the wave; a is the instantaneous amplitude; c is the velocity of propagation, 3×10^8 m/s in empty space; λ is the wavelength, or distance between successive crests; and ν is the frequency (number of wavecrests passing a point in 1 s). B represents the associated magnetic wave, out of the plane of the paper; it is clockwise to the electric wave by $90°$.*

Oscillators

By using electronic devices called *oscillators* we can produce electromagnetic waves with frequencies much higher than 50 or 60 Hz, and correspondingly shorter wavelengths. These are called radio waves, and frequencies between approximately 100 kHz and 100 MHz are used for sound broadcasting. The sound signal is superimposed on the 'carrier' wave in the form of modulations in amplitude (AM) or frequency (FM). The conductor that radiates the signal is called a transmitting aerial, and its dimensions need to be matched to the wavelength being transmitted. The receiver also has an aerial, which picks up the coded signal (the electromagnetic radiation induces a flow of current in it); the receiver decodes the signal, amplifies it, and turns it back into sound energy via loudspeakers.

Still higher frequencies, of the order of 1 GHz (10^9 Hz), are used to carry television broadcasts, the much higher information content of pictures being more readily accommodated by high-frequency waves. As the frequency rises, so the radiation becomes more directional: the parabolic aerials used to relay TV programs to local transmitters begin to resemble optical devices.

The wavelengths corresponding to the frequencies we have been discussing go from 3 km for 100 kHz to a few centimeters for UHF relay stations. These very short waves are called *microwaves*, and are so highly directional that they can be beamed to satellites hundreds of kilometers away. Microwaves represent the highest frequencies we can produce using conventional electronic circuitry. If we could raise the frequency to about 5×10^{14} Hz the wavelength would be 600 nm, which is in the visible spectrum! Of course, we cannot produce a conventional oscillator that can do this, but by stimulating atoms to emit waves at these frequencies we can produce visible radiation as a continuous wave. The device that achieves this is called a *laser*, and its name is an acronym of its method of operation: Light Amplification by Stimulated Emission of Radiation. This is the light source used for making holograms.

Properties of Light Sources

Holograms can be made only by using light sources that have rather special properties. Certain sources based on *gas-discharge* tubes possess these, but to so small a degree that they can be used only for holograms of flat originals or for making copies of other holograms under certain restricted circumstances. However, the light produced by a laser possesses the required properties in large measure. They are known collectively as 'coherence', and separately as *spatial* and *temporal coherence*:

Spatial coherence Of the two qualities this is the easier to define. It is the degree to which a beam of light appears to have originated from a single point in space. Spatial coherence is inversely proportional to the apparent diameter of the source. Sunlight has fairly high spatial coherence, as the sun subtends only about one-third of a degree at the earth's surface; it is possible to make spotlights with even higher spatial coherence. The usual method is to use a compact-filament lamp with the filament focused by a lens to form a small concentrated image, the size of which can be further restricted by an iris diaphragm or pinhole. (Fig 3.3).

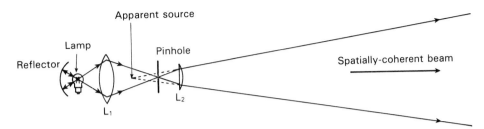

FIG 3.3 *Optical arrangement for producing a spatially-coherent beam of white light. The filament is at the center of a hemispherical reflector, and a small image of the filament is produced at the pinhole by the lens L_1. The lens L_2 controls the divergence of the beam.*

Most lasers produce beams which have not only very high spatial coherence, but also a high degree of *collimation*, ie the divergence of the beam is very small (typically better than 1 part in 1000).

Temporal coherence In order to define this term we need to use the more sophisticated model for the behavior of light that was mentioned earlier. This model first appeared at the turn of the century, in order to describe anomalies in photoelectric and photochemical phenomena. A large discrepancy had been found between the predicted and actual energy of a light beam, and it was suggested by Albert Einstein that luminous sources could be thought of as emitting light in the form of discrete pulses of radiation (for which he coined the name of 'photons') rather than as continuous waves. Max Planck subsequently showed that many other phenomena could also be modelled more exactly if all types of energy were to be treated in the same way (the quantum model); and that if the energy of a quantum was taken as being proportional to its frequency, many questions about the structure of matter could be clarified. We shall be making use of this assumption when we come to consider how a laser works.

Temporal coherence is the extent to which all the photons in a light beam are of the same frequency (or wavelength). The spread of frequencies in a beam of electromagnetic radiation is called the *bandwidth*; in communications terminology it is the spread of frequencies taken up by a transmission channel, say 98 MHz ± 0.02 MHz for a typical FM sound radio transmission. With light we usually speak of wavelength rather than frequency, so we tend to describe the bandwidth of a typical helium-neon laser as, say, 632.8 nm ± 0.002 nm. From such figures it is possible to calculate the distance over which the photons remain substantially in phase, and this distance is called the *coherence length*. For white light it is only a few hundred nanometers; for a filtered mercury arc it is a few millimeters; but for a laser it may be from a centimeter or two up to 20 meters or more, depending on the type of laser and the optics used (Fig 3.4).

It is the coherence length of the light source that determines the available depth of the subject space in making a hologram. A (hypothetical) light source that emits light of a single wavelength or frequency is said to be 'monochromatic', and this

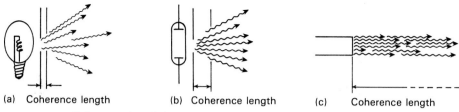

FIG 3.4 *Coherence length for* (a) *a filament lamp,* (b) *a gas discharge tube,* (c) *a laser.*

term, while not strictly descriptive of laser light, will be used throughout. The more rigorous term *quasi-monochromatic* will be used to describe light sources such as sodium and filtered mercury arcs with a bandwidth of 0.1–15 nm.

Polarization A further property a light source must possess for the formation of good interference fringes is polarization. A source is said to possess *linear* or *plane polarization* when the electric component of its oscillation lies in a single plane. The magnetic component lies in a plane perpendicular to this. In general the light from a laser is linearly polarized, though some types of laser have what is called *random polarization*.

You will see in the next chapter why we need these special properties in making holograms. Let us now look at how a laser generates such a very well-behaved beam of light.

Atoms and Energy

Filament lamps, gas discharge tubes and lasers all emit photons, and though the light from them differs greatly in its qualities, the photons all originate in a similar manner. They are emitted by the individual atoms which make up the light source in question. An atom emits a photon only when it rearranges its structure, in general changing from a less stable state to a more stable state. There are many ways in which an atom can do this, but it is constrained by certain rules. To see how these rules operate we will look at a simplified version of the *Bohr model* for atomic structure. Niels Bohr visualized an atom as being made up of a central nucleus of protons and neutrons surrounded by three-dimensional orbitals in which electrons are to be found. Each orbital represents a fixed energy state, and as no intermediate states are permitted by the model, they are called *quantum energy levels*. Each orbital or 'shell' can contain only a certain number of electrons, the permitted numbers being fixed by the laws of quantum physics. Thus the innermost shell can contain only two electrons, the next one eight, the third one eight again. When the atom is in its normal state, the *ground state*, the electrons will all occupy the shells which represent the lowest energy possible. Thus in the neon atom, which has ten electrons, the two innermost shells are completely filled, and in the magnesium atom (13 electrons) the two innermost shells are filled and there are three electrons in the third shell.

In a filament lamp the electrical energy absorbed by the filament causes the atoms in it to vibrate violently. Some of this energy of vibration is converted into electromagnetic energy, which is radiated away. The radiated waves form a *continuous spectrum* of frequencies, the energy distribution of which is related to the *thermodynamic temperature* (also called the 'absolute temperature') of the filament. Thus at a temperature of 3400 *kelvins* (abbreviated 3400 K) a filament radiates mainly infrared and red wavelengths, even though it looks white (eg tungsten-halogen lamps in a photographic studio). At 6000 K the peak radiation is in the yellow-green region (daylight). The close match between thermodynamic temperature and perceived hue leads to the notion of *color temperature*, used to describe the spectral energy distribution of *incandescent light sources* and some other sources (such as the xenon arc) which closely match them.

So-called cold light also originates from atoms, but in a somewhat different way. Consider a glass tube filled with hydrogen gas. Hydrogen is the simplest of elements: its atoms consist of a single proton and a single electron. If we put energy into this gas, say by passing an electric discharge through it, we shall excite some of the electrons present into higher-energy quantum states, from which they will quickly drop back to the ground state, emitting a photon. The highest possible *energy state* for hydrogen is where the electron escapes completely; this is known as *ionization* (an atom that has lost one or more electrons is called an *ion*) and an electron returning from this state to the ground state emits a photon in the ultraviolet region. There are also levels below the ionization state which emit an ultraviolet photon. But the electrons need not return directly to the ground state; they may return, say, to quantum level 2, with the lost energy corresponding to a photon in the visible spectrum, or to level 3, with the lost energy corresponding to the emission of a photon in the infrared region. The possible transitions, which are not many for such a simple atom, give a number of discrete wavelengths. If the light emitted by a hydrogen-filled discharge tube is dispersed by a prism it appears not as a continuous spectrum but as series of lines (known as a *line spectrum*), the wavelengths of which can be predicted from the Bohr model. Sources which emit this type of spectrum are called *line sources*. Unfortunately, the lines are not infinitely narrow, as we should like them to be for holography. While they are emitting the radiation the atoms themselves are moving quite rapidly, so that the bands of radiation are broadened by the *Doppler effect*; the bands become wider still if the temperature or the density of the gas is increased. This *Doppler broadening* also occurs when the atoms in solids are excited so that fluorescence occurs. However, not all the lines in solids are broadened in this manner. With some elements, notably chromium and some *lanthanides*, a few energy levels are not spread at all. This forms the first clue to laser operation.

Stimulated Emission

The second clue comes from the work of Einstein. In 1917 he showed that if an electron went from a higher to a lower energy state, emitting a photon in the process, then if that photon passed close to another electron in a similarly-excited state, the second electron would also fall to its lower-energy state in the same way, emitting

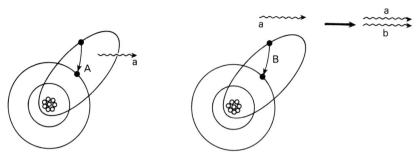

FIG 3.5 *Stimulated emission of a photon. If a photon a spontaneously emitted by an excited electron A encounters an appropriately-excited electron B it stimulates it to fall back to the ground state, emitting a photon b of the same frequency and phase as the stimulating photon.*

an identical photon. What was more, that second photon would travel in the same direction as the first, and would have the same polarization and the same phase. He called this phenomenon *stimulated emission* (Fig 3.5).

This is precisely the kind of event we are looking for. If we have a large number of atoms, all in the same excited state, and we introduce a photon of just the right wavelength traveling in just the right direction, it will trigger a chain of emissions of photons from the excited atoms, and build up a beam of photons all of the same wavelength, phase and polarization, and all traveling in the same direction: a beam of *coherent light*. This is indeed achieved provided the number of excited atoms exceeds the number of ground-state atoms; this condition is called *population inversion*. Under these conditions, a photon of the right wavelength entering the space will cause a chain reaction. One photon enters the chamber; millions leave it in a collimated coherent beam. Forty-three years after Einstein's prediction, the first laser light was seen, in California.

The Three-level Laser

The first *lasing* medium to be tried out (and still one of the best) was chromium. It was chosen because it has a partially-stable or *metastable* energy level above the ground state, the energy difference between it and the ground state corresponding to a photon in the visible region (694 nm, in the deep red). In the metastable state the electron remains in the excited state until disturbed by quantum fluctuations or some external disturbance: for something of the order of a microsecond, a comparatively long time in atomic terms. If we take a sort of cross-section of the energy levels involved, we can get a picture of the energy spectrum of chromium – at least, the part of it that is relevant (Fig 3.6).

In order to raise the energy level of an electron from the ground state to the precise level *A* in Fig 3.6 we would have to supply it with an exact amount of energy, otherwise nothing would happen. Fortunately, a little above level *A* is a group of energy levels *B* which are so close together that they form what is effectively a continuous band (an *energy band*), and to raise an electron into this band requires a

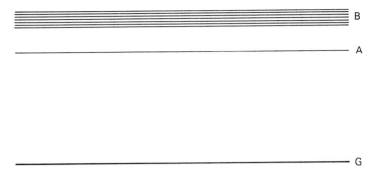

FIG 3.6 *The relevant energy levels of a chromium atom. G represents the ground state, or minimum possible energy state. A is a metastable energy level somewhat higher than the ground state and B is a group of energy levels close enough together to form an energy band.*

much less determinate amount of energy, an amount such as might be given by a powerful electronic flash directed into the material. Now band *B* is not stable, so the electron quickly loses energy (through vibrations which appear as heat) and falls to the metastable level *A*. Here it is a target for stimulated emission.

Of course, it is not quite as simple as that, because the flashtube contains an immense range of levels of energy; very few of the chromium electrons would actually receive the right amount of energy to find themselves in energy band *B*. So the chromium atoms are embedded in a matrix of aluminum oxide crystal. This absorbs all the energy from the flash and hands it on in the right quantities to the chromium atoms, whose electrons are thus raised to the right energy band. The aluminum oxide is referred to as the *pumping* medium.

So, with the aid of the pumping medium, we have a population inversion. Now, if a stray photon-producing decay occurs, and the photon happens to be traveling in the right direction, stimulated emission will begin. Or will it? If each photon were to gather up another photon every hundredth of a millimeter, we might well imagine a single photon gathering up millions of companions in a transit of a meter or so. But, again, it is not as simple as that. To begin with, one photon is not nearly enough to begin the chain reaction, and only a tiny fraction of the spontaneously-released photons are traveling in the right direction; and our chromium-doped rod can be only a few centimeters long. What should we do?

The answer again turns out to be straightforward: we put a mirror at each end. Now the photons will traverse the rod many hundreds of times, each time gathering up more photons. One of the mirrors can be made slightly leaky, and the laser radiation will emerge through it in a pulse lasting for roughly the duration of the flash, ie about 1 ms. This type of laser is called a *three-level laser*, as three energy levels are involved (Fig 3.7). Most three-level lasers are *solid-state lasers*.

It takes quite a lot of energy to power a chromium laser. The most common energy source is a xenon flashtube, much like the flashtube used by photographers, but straight. An elliptical cylinder mirror surrounds the rod and flashtube, which are situated respectively at the two foci of the ellipse (Fig 3.8).

As lasing action occurs only when there are more excited atoms than ground-

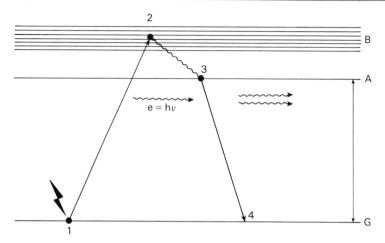

FIG 3.7 *Action of a three-level laser. An electrical or light energy stimulus (pumping) raises the energy of an electron in the ground state G (1) into energy band B (2). From there it loses energy spontaneously and falls to energy level A (3) until stimulated by a photon of precisely the right energy e. It then falls to G (4), emitting an identical photon, in phase with the first (lasing action).*

state atoms the energy input to a three-level laser has to be considerable. Although in theory a three-level laser could be operated continuously, so much of the input energy would reappear as heat that the crystal would rapidly overheat and be destroyed. Because of the short duration of the laser action, such a device is called a *pulse laser*.

Chromium laser elements have to be made as single crystals. They are usually referred to as 'ruby rods', as their chemical composition (aluminum oxide doped with chromium) is the same as that of ruby. However, the chromium content is only about 8 parts per million, which makes the 'ruby' decidedly anemic.

Now, a *ruby laser* of this type *can* be made to produce holograms without any modification, and was so used in the early days of holography. Unfortunately, the

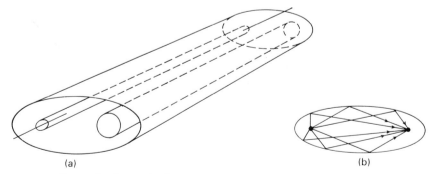

FIG 3.8 *In a ruby laser, the laser rod and the flashtube are placed at the conjugate foci of an elliptical cylinder (a). All rays emanating from one focus are reflected through the other focus (b).*

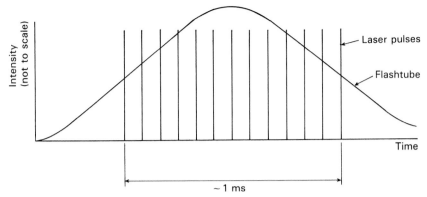

FIG 3.9 *A simple ruby laser produces a series of pulses over roughly 1 ms.*

range of its applications is severely limited, largely because its light is squandered in the form of a succession of mini-pulses spread out over a millisecond or so. The reason for this is that in a three-level laser stimulated emission can occur only when more than half of the atoms in the lasing medium are in the appropriate excited state; as soon as the number of excited atoms falls below the number of ground-state atoms more energy is absorbed than emitted, and laser action is quenched. This happens repeatedly throughout the flash duration, resulting in a number of spikes of laser action (Fig 3.9).

Q-switching

The output of a simple ruby laser is low, typically about 0.04 *joules* (J). By a process known as *Q-switching* it is possible not only to confine the laser action to a very short period (25–50 ns) but also to dramatically increase the power of the pulse. This can be achieved by the use of one of two systems, called respectively *passive* and *active switching*. For both systems the 'leaky' mirror is separated from the ruby rod and the switching device is situated in the intermediate space.

In the passive system a cell of a bleachable dye such as rhodamine is used. This dye is opaque to light of the laser wavelength, so that losses in the *optical cavity* are so great that no laser pulse can be formed until the population of excited atoms is high enough for the amplification to become greater than the losses (in telecommunications parlance the 'Q' or gain factor becomes greater than unity) and laser action begins. The dye bleaches virtually instantaneously and the pulse is emitted. As soon as the pulse is over the dye regains its opacity.

The active type of Q-switch employs a *Pockels cell*. This is basically a crystal of a substance which exhibits *birefringence* when an electric field is applied to it. When linearly-polarized light is passed through a Pockels cell, the application of a suitable voltage causes a phase difference of one-quarter of a wavelength between the two transmitted components. The result is *circular polarization* of the beam. To use the Pockels cell as a shutter, it is placed between a linear polarizer and the leaky mirror. When energized, the cell produces a quarter-wave phase shift in one of the transmitted components; when this passes through the cell a second time after reflection its undergoes a further quarter-wave phase shift. This amounts to a half-wave delay, which results in a rotation of the plane of polarization of the beam of 90° to the original direction. With this polarization the beam will not pass through the polarizer on its return. If the Pockels cell is switched on as the flashtube is fired, and switched off towards the end of the flash, a giant pulse results.

Both passive and active switching have their advantages. The former is cheap, and improves the coherence length of the beam; the latter, although comparatively expensive, permits double and triple pulses. A *Fabry-Pérot etalon* (see pp. 36–7) can be used to increase coherence length.

Ruby Lasers for Holography

As the power output of a single-mode ruby laser is low, modern pulse-laser designs pass the beam through two further ruby rods of large dimensions, known as *amplifiers*; these are also excited to the point of laser action. These amplify the beam by a factor of about 100: the final pulse has a typical energy of 0.1–10 J. The instantaneous power is thus 4–400 MW. Plainly this sort of power is not to be treated casually; the precautions needed in pulse-laser holography are set out in Chapter 16, and recommendations for safety are given in Appendix 1.

Ruby lasers operate at a wavelength of 694 nm in the red region of the spectrum, close to the long-wavelength limit of visual perception. The extremely short duration of the pulse makes it possible to produce holograms of subjects moving at up to 4 m/s; the possibility of double- and triple-operation adds further scope for the use of this laser in holographic interferometry (Chapter 22). When used for more general holography, pulse lasers avoid the need for heavy and expensive isolation tables, and conditions can resemble those of a normal photographic studio. However, ruby lasers are expensive, and require a bulky energy storage capacitor bank and special precautions not only for the operator but also for the laser itself.

The Four-level Laser

Although pulse lasers are nowadays routinely used for making holograms, they need special optical equipment and a good deal of expertise in their use, and this, along with their cost, puts them out of the reach of most private holographers. What is needed is a laser that produces a continuous beam, with a power output that is sufficiently low to require neither special optics components nor tedious and restricting safety precautions. The problem with a three-level laser is that it takes a great deal of energy to raise more than half of the atoms in the lasing medium into the energy band at the same time in order to achieve a population inversion, and this energy reappears as heat. Fortunately, a solution is not hard to find. All we need is an element whose atoms have a low-energy state that is unstable and is normally completely empty. For such an element, the raising of even a single electron to a higher energy level constitutes a population inversion. Provided the lower level is unstable, any electron falling to it from a higher level will quit it immediately,

LIGHT SOURCES FOR HOLOGRAPHY

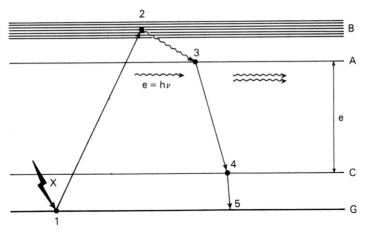

FIG 3.10 *Action of a four-level laser. The pumping stimulus X raises the energy of the electron from the ground state G (1) to the energy band B (2). It loses energy and falls to the metastable energy level A (3). A photon of energy e stimulates the electron energy to fall to level C (4), emitting a photon of the same energy and phase. Level C is unstable, and the electron immediately returns to the ground state (5).*

leaving the level empty again. The energy required to drive a *four-level laser* system is much less than that required to drive a three-level system, and such a laser can be operated continuously without overheating. Most low-powered *continuous-wave (CW) lasers* are *gas lasers*, ie the pumping and lasing media are both gaseous. As the lasing medium is not ionized this type of laser is called a *neutral-gas laser*, in contrast to an *ion laser* (see below). Neutral-gas lasers produce only one wavelength at a time (*single-line emission*).

Mirrors and Windows in CW Lasers

The mirrors in a CW laser are not both flat. One (or each) is made slightly concave, so that the photon transits are kept aligned within the tube. If both mirrors are curved, they should form parts of an imaginary ellipsoid surrounding the tube. If one curved and one plane mirror are used, the curved mirror should be the end surface of an ellipsoid of major axis twice its distance from the plane mirror. As the mirrors need to be very efficient they are not metallized, but are optically coated with a number of layers of alternately low and high refractive index; the thickness of these coatings is such that the reflected light waves interfere constructively. Such *dielectric mirrors*, apart from being very efficient, are selectively reflective; ie they reflect only a very narrow band of wavelengths over a fairly narrow range of angles. Light of other wavelengths, or coming from a non-axial direction, simply passes through. In low-power lasers the mirrors are sealed to the ends of the tube, but in medium-power (15–50 mW) lasers they are separate, and capable of fine adjustment. In high-power lasers there is room to insert further optical components into the optical cavity.

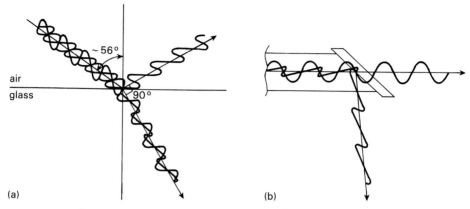

FIG 3.11 *Brewster-angle reflection and transmission. When unpolarized or randomly-polarized light is incident on a surface at an angle such that the reflected and transmitted rays are orthogonal* (a) *the reflected beam is totally s-polarized and the transmitted beam partially p-polarized.* b *A Brewster-angle window has virtually 100% transmittance to the p-polarized component of the beam.*

Laser light is linearly polarized, a property which is important in the making of a successful hologram. Unfortunately, if the laser windows are perpendicular to the beam the direction of polarization wanders randomly, and this is undesirable in holography. The polarization can be stabilized by the use of *Brewster-angle windows*. The way these function is ingenious. When light falls on a glass surface obliquely, some of it is transmitted and some is reflected. Both parts are partially polarized, the reflected beam being *s-polarized* and the transmitted beam *p-polarized*. At a certain angle of incidence, about 56° for optical-quality glass, the reflected and transmitted beams are orthogonal; at this angle the degree of polarization is total for the reflected beam and a maximum for the transmitted beam. This angle is called the *Brewster angle* after Sir David Brewster, who first described the phenomenon. In a laser, as the light passes back and forth through the windows it becomes more completely polarized at each transit, so that by the time it finally emerges it is nearly 100% polarized (Fig 3.11).

Types of Neutral-gas Laser

1 Helium–neon (HeNe) The *helium-neon laser*, usually abbreviated to HeNe, is the most common type of gas laser. The tube contains helium gas at a pressure of about 1 torr and neon at a pressure of about 0.1 torr (a torr is a unit of pressure equivalent to 1 mm of mercury or 1/760 of an atmosphere). The main purpose of the helium is to act as a continuous reservoir of energy (supplied by an electric discharge) to the neon. As this laser is the one that is best suited to general-purpose holography it is worth discussing in some detail.

HeNe lasers as used for holography operate at a wavelength of 632.8 nm, with a power ranging from 0.5 mW to 100 mW. They can also be operated at several other wavelengths, including infrared and green (545 nm). The coherence length is greater at lower powers, varying from about 1 m at 0.5 mW to about 100 mm at 100 mW. The randomly-polarized type are unsuitable for serious holography, as the direction of polarization is an important factor in obtaining optimum image quality. A laser with Brewster-angle windows has a somewhat lower output than its randomly-polarized equivalent, but it has a completely stable plane of polarization and can be used with plain glass *beamsplitters*.

Although in a HeNe laser both the helium and neon appear to be at such a low pressure they are nevertheless at a much higher pressure than the very small amounts of free helium and neon in the atmosphere. Neon atoms are comparatively large, but helium atoms are so small that they slowly diffuse out through the walls of the tube. Their loss gradually deprives the neon of its energy source. For this reason HeNe lasers are initially overfilled with helium, and their output power rises during their first few thousand hours of use. A HeNe laser rated at 15 mW could well be producing as much as 24 mW in its prime. The actual life of a laser tube depends to some extent on the amount of helium leakage as well as on other factors such as cathode poisoning, and it is less harmful to keep a HeNe laser switched on for long periods of time than it is to keep other types of laser switched on.

HeNe lasers are the cheapest gas lasers, and among the most reliable: the tube has a minimum life of 20 000 h, which is the equivalent of keeping it switched on continuously for more than two years – though that would perhaps be overdoing things. The diameter of the beam on emergence is between 1 and 2 mm. The main disadvantage of the laser is its low power; also, anything over 20 mW it has too short a coherence length to make successful holograms of subjects that are more than a few centimeters deep, unless the table geometry is very carefully worked out.

2 Helium–cadmium (HeCd) The *helium-cadmium laser* produces a beam with a wavelength of 442 nm in the blue-violet region of the visible spectrum. HeCd lasers have a range of output powers similar to that of HeNe lasers, and are suitable for making holograms on a variety of non-silver materials, which in general are sensitive only to short-wave radiation. The coherence length is less than that of a HeNe laser, typically 20–50 mm, and the tube life is also somewhat less, of the order of 5000 h. This is because of the progressive deposition of cadmium metal on the laser windows.

Ion Lasers

1 Argon-ion (Ar^+) Neutral-gas atoms mostly have energy levels that are too close together to produce laser radiation in the visible spectrum (HeNe and HeCd are notable exceptions); however, once a gas is *ionized*, ie has one or more electrons completely removed from its constituent atoms, a whole new range of energy levels becomes available – a simple example being the energy difference between the removal of one electron and the removal of two. The *argon-ion laser* (usually abbreviated to Ar^+) employs such transitions. It can produce coherent light at a number of wavelengths, the most important being 488 nm (green) and 514.5 nm (blue). Much higher powers are obtainable from Ar^+ than from neutral-gas lasers: typically 0.5–1 W, making this laser the workhorse of professional transmission holography. There are two disadvantages to their use in holography, apart from their cost. The first is that even at low powers they require forced cooling, with ensuing problems of vibration and air currents, though in modern lasers with large plasma tubes and efficient water cooling these problems have largely been overcome. The second is that as they operate at a comparatively high temperature, there is serious Doppler broadening, the result being that the coherence length is very low, about 10 mm. In addition, the optical cavity formed by a well-separated pair of laser mirrors allows many wavelengths to satisfy the integral-multiple condition, so that a number of wavelengths are emitted simultaneously. There are several methods for dealing with this, the most effective being the use of a Fabry-Pérot etalon, which acts as a single-frequency selector (see Box).

The Fabry-Pérot etalon

As used in laser optical cavities, this essentially simple device consists of two extremely flat glass plates separated by rods of invar metal which produce a precisely-parallel gap. A quasi-monochromatic beam of light incident on the etalon will undergo multiple reflections within the cavity. Only light that has a wavelength that is an integral submultiple of twice the cavity width will emerge; light of other wavelengths will interfere destructively. You can experience an acoustic parallel when you walk down a narrow alley. You will hear the echo of your footsteps at

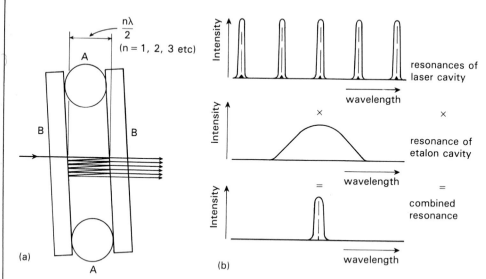

FIG 3.12 *Fabry-Pérot etalon. a A common form of the etalon is two optical flats BB separated by two accurately-machined invar steel rods AA. It resonates for wavelengths that are an integral submultiple of twice the separation of the plates, and can be tuned by turning through a small angle. b The large optical cavity of the laser produces a series of narrow-band frequencies; the etalon isolates one of these.*

a definite musical pitch, the wavelength you hear being twice the distance between the walls (Fig 3.12)

If you place a short-distance optical cavity such as a Fabry-Pérot etalon in a much longer optical cavity such as a pair of laser mirrors, you will get an exceedingly good frequency selector. To tune it to a particular wavelength you mount the etalon in such a way that you can move it through a range of angles about a vertical axis. This makes it possible to vary smoothly the effective gap between the faces of the etalon. If you rotate the etalon through a few degrees you can capture any of the wavelengths emitted by an ion laser, and suppress the remainder. In the process you increase the coherence length by several meters, the only penalty being a drop in the output power of the laser. This type of emission is known as *single-frequency emission*. Because of their sensitivity, etalons are usually thermostatically controlled in an 'oven' at about 50°C; they can be fine-tuned by varying the temperature of the oven. With care it is possible to get an etalon to resonate at several frequencies simultaneously, and (at least in theory) to get from a Kr^+ laser as many as four collinear beams – red, yellow, green and blue – from a single source. This has important implications for natural-color holography (see p. 63), though it has so far not been easy to achieve.

2 Krypton-ion (Kr^+) Another ion laser with applications in holography is the *krypton-ion laser*, usually abbreviated to Kr^+. Krypton is a heavy gas with many possible states of ionization. It can produce coherent beams of wavelengths ranging throughout the whole visible spectrum. The most useful line is in the red region at 647 nm, at a typical power of 250–750 mW. As this is reasonably close to the wavelength of the ruby laser (694 nm), it is useful for producing transfer holograms from pulse-laser master holograms. The higher light output of the Kr^+ laser makes it a better choice than the HeNe laser in professional reflection holography. Kr^+ lasers also have a useful output at 521 nm (green) and 476 nm (blue); they can thus be used for making color holograms (see Chapter 18) without the necessity for a second laser (see also Box above).

3 Other ion lasers Many metallic ions have transitions that are suitable for lasers. In the visible spectrum two important examples are copper and gold. Copper vapor lasers, in particular, can produce up to 40 W at 511 and 578 nm with a coherence length of about 20 mm; gold vapor lasers produce about 10 W at 628 nm. The output is usually in the form of 20–50 ns pulses with a repetition rate of 6–8 kHz. At the time of writing (1987) they do not seem to have been employed for holography, but they hold promise for stroboscopic work.

Other Types of Laser

Many types of CW laser can also be operated in a pulsed mode, though so far none of them seems to be suitable for holography. Monocrystalline aluminum oxide doped with lanthanide elements such as yttrium (yttrium aluminum garnet, or *YAG crystal*) can be used to change the wavelength of a laser. *Excimer lasers* use the high energy levels achieved when *noble gases* such as xenon and *halogens* such as chlorine form interatomic bonds when ionized at elevated temperatures, producing laser radiation in the ultraviolet. When such a beam is passed into cells of certain dyes, such as rhodamine (*dye lasers*), a large range of visible wavelengths can be produced, any of which can be selected out by a dispersing device called a *Littrow prism* (see Box). Excimer lasers can produce high power and a rapid repetition rate, and like copper vapor offer possibilities for stroboscopic holography. A *semiconductor laser* is a special kind of light-emitting diode. It produces a beam of light in the near

---

Littrow prism

A Littrow prism is a dispersion device which causes very little light loss, and is suitable for singling out spectral lines where the laser gain is low. The refracting angle of the prism is such that a beam of p-polarized light entering the prism at the Brewster angle suffers no surface reflection and strikes the second, glass–air surface at normal incidence. The second surface is coated to form a dieletric mirror, and acts as the plane mirror for a half-elliptical cavity (see p. 33). Its operation is shown in Fig 3.13.

By rotating the prism any desired wavelength can be chosen for *retroreflection*, so the device can be used to separate out one of a number of wavelengths, to eliminate unwanted strong lines such as the powerful HeNe line at 1152 nm in the infrared, or to fine-tune a dye laser. A Littrow prism is one-half of a *Brewster prism*, another dispersing prism designed for zero surface reflectance with p-polarized beams.

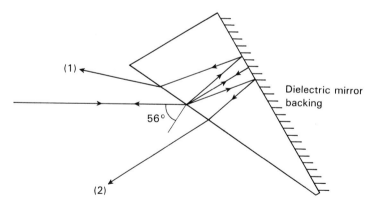

FIG 3.13 *Frequency-selective action of a Littrow prism. When a ray containing several frequencies enters the prism, higher frequencies (1) are refracted more, and lower frequencies (2) less, than the selected frequency, which alone is retroreflected. The prism also acts as a Brewster window.*

---

infrared with a divergence of about 15°, but the cone of emitted light is elliptical rather than circular, so that the beam appears to have originated from a line rather than a point. If the *astigmatism* of this beam is corrected by means of aspherical optics, a spatially-coherent beam can be obtained, and this has been used experimentally for making holograms (1, 2, 3), though so far not to any extent outside the Soviet Union. The main attraction of semiconductor lasers is that they are cheap and very small. They also operate at comparatively low voltages and at a similar power range to that of HeNe lasers. Semiconductor lasers operating at visible wavelengths exist, and it would be good to be able to forecast their employment as a standard light source for a new generation of amateur holographers.

References

1 Bykovskii Yu A, Evtikhiev N N, Elkhov V A and Larkin A I (1974). Recording of holograms with the aid of single-mode pulse semiconductor lasers. *Soviet Journal of Quantum Electronics*, **5**, 587–9.
2 Kalashnikov S P, Klimov I I, Nikitin V V and Semenov G I (1976) Recording of Fourier holograms using radiation emitted from pulse semiconductor lasers. *Soviet Journal of Quantum Electronics*, **7**, 946–9.
3 Vorob'ev A V, Elkhov V A, Klimov I I, Pak G T, Popov Yu M, Shidlovskii R P and Yashumov I V (1980) Recording of holograms using semiconductor laser radiation and a holographic selector. *Soviet Journal of Quantum Electronics*, **10**, 1557–8.

CHAPTER

4

The Basic Types of Hologram

> Of course, the first thing to do was to make a grand survey of the country she was going to travel through. 'It's something very like learning geography', thought Alice...
> LEWIS CARROLL, *Through The Looking Glass*

Holograms are of two main types: those which must be replayed using laser light or at least light exhibiting *partial coherence*; and those which can be replayed using white light. The latter have applications mainly in the world of display holography, whereas the former are usually confined to scientific and technological applications.

Gabor's original in-line layout

Denis Gabor (1) made his first holograms using a filtered mercury-arc lamp as the light source; it had a coherence length of a millimeter or so. Fig 4.1 shows how he managed to get round this difficulty. His object was a tiny circular transparency of opaque lettering on a clear background. The light diffracted by the lettering formed the object beam, and the light which passed undeviated through the clear part formed the reference beam. The photographic plate was 600 mm from the object, so the interference fringes that formed were sufficiently widely separated to be

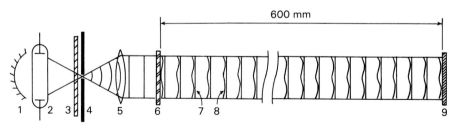

FIG 4.1 *Gabor's holographic system. 1, Reflector; 2, small-source mercury arc; 3, green narrow-band filter; 4, pinhole; 5, collimating lens; 6, transparency; 7, plane reference wavefront; 8, object wavefront; 9, photographic plate recording phase differences between 7 and 8 as an interference pattern. Exactly the same system, minus the transparency, was used for reconstruction.*

capable of recording on a normal photographic plate. Such a hologram is termed a *far-field* or *Fraunhöfer hologram*.

When the original light source was used to illuminate the hologram, a virtual image appeared in the position of the original object. Unfortunately, the view of the image was marred by the presence of a spurious real image in line with it. Nevertheless, *Gabor* or *in-line holograms* still find uses in industrial research, for example in particle counting (see pp. 343–5).

Early off-axis holography

The first attempt to get rid of the obtrusive real image was made by Emmett Leith and Juris Upatnieks (2) who simply moved it out of the way. Their arrangement moved the reference beam alongside the transparency, deflecting it by means of a thin prism to overlap the object beam. This off-axis beam, when used for reconstruction, resulted in displacement of the unwanted image by an angle equal to twice the angle between the reference and object beams, moving it far enough out of line for it to cause no further trouble (Fig 4.2) This arrangement is identical with the one used in Chapter 1 (p. 9) to illustrate the principle of holographic fringes.

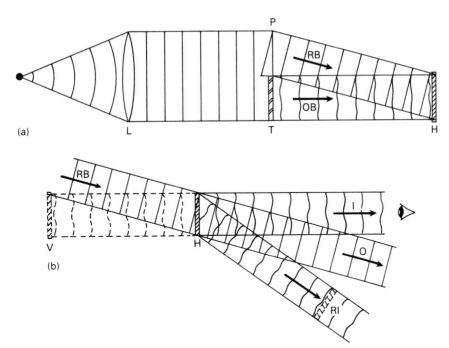

FIG 4.2 *The off-axis principle. When the incident beam is deflected by the prism P to form an off-axis reference beam RB, this interferes at an angle with the object beam OB from the transparency T at the holographic plate H (a). When the holographic image beam I is reconstructed by the reference/replay beam RB, the virtual image V is in its original position; the zero-order beam O and the spurious real image RI are moved out of line. Notice that the geometry is identical with that in* FIG 1.7(a).

> Because of the poor coherence of pre-1962 light sources, these early holograms were all of two-dimensional transparencies. It was only with the advent of the first practical laser that the holography of solid objects in three dimensions became possible.

Laser Transmission Holograms

We saw in Chapter 1 how the interaction between a reference beam and a beam disturbed by having passed through a transparent object generated an interference pattern which could be recorded in the form of a hologram. Now, there is no essential difference between a wavefront refracted by a transparent object and a wavefront reflected by an opaque one. Both are examples of diffraction by the object and both contain the entire information about the object, and it is encoded in the same way. There are numerous ways of providing the two beams, one of the simplest geometries being shown in Fig 4.3.

In this arrangement the mean angle between the object beam and the reference beam is roughly $45°$; at this angle the spacing of the interference fringes is about 1.4 times the wavelength or slightly under 1 μm (10^{-6} m). To record so fine a pattern demands an ultrafine-grain recording material. There is also a need for absolutely stable conditions during the exposure: a lateral displacement of the fringes of as little as one-tenth of the fringe spacing will noticeably affect the contrast of the fringes, and this will result in a low diffraction efficiency and a weak image reconstruction.

Reconstruction of the image   To produce the holographic image, all you have to do is to process the hologram and then place it back in the original reference beam. We have seen that if a hologram is illuminated by a replica of either of the two original beams alone, both of the original beams will emerge (this is proved in Appendix 2); thus, if you place the hologram back in the original reference beam both this beam and a replica of the original object beam will emerge (Fig 4.3).

If you look along the reconstructed object beam you will see a replica of the object, and as you shift your viewpoint you will see different perspectives of the object. Thus it appears to be three-dimensional; it is autostereoscopic, with full parallax. Because the light does not actually pass through the image, but only generates a wavefront that makes it appear as though the light had originated from the object, this image is called a virtual image. A virtual image is like the image you see in a mirror: the light does not actually pass through the mirror, but it does look as though it has come from behind it.

The real image   In contrast to the virtual image, an image that light has actually passed through is called a real image. The difference is that a real image can be caught on a screen placed in its plane. The image formed by a camera lens is a real image. When a hologram is replayed, there is a third beam diffracted in a different direction from the image beam. It produces a real image, on the viewer's side of the hologram, and it possesses some peculiar properties. For a start, its perspective is reversed; parts of the image which should be at the rear appear at the front, and

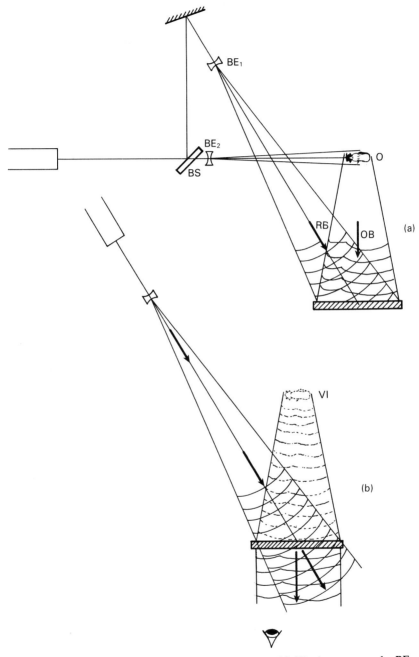

FIG 4.3 *A simple transmission hologram geometry. (a) The beam expander* BE_1 *provides the reference beam RB via the beamsplitter BS; the beam expander* BE_2 *that illuminates the object O forms the object beam OB. (b) Illuminating the processed hologram with a replica of the reference beam provides a virtual image VI occupying precisely the space of the original object.*

vice versa. Thus the holographic real image of a ball looks like a hollow cup, and the image of a cup looks like a ball. Moreover, parts which in the image appear to be in the rear cast shadows on parts which appear to be in front; and as you move your head to the right, you see more, not of the right side, but of the left side. Your brain has to cope with this anomalous information, and usually does so by perceiving the image as swinging round as the viewpoint moves sideways. The effect is particularly striking with images of human faces, and has been used in experimental psychology for studies of the nature of visual perception. The image is called a *pseudoscopic (real) image* (an image with normal perspective is said to be *orthoscopic*).

Because the emulsion has a finite thickness, the fringes that form the hologram run through its thickness somewhat like the slats of a venetian blind, ie roughly at right angles to the emulsion surface. This affects the diffraction efficiency, as the brightness of the image is partly determined by the *Bragg condition*. This states that when a beam of light passes through a thick grating (any grating that is several wavelengths in thickness is a thick grating), light of a given wavelength will emerge at high intensity if and only if the wavefronts emerging from each cycle of the grating have optical path differences that are an integral number of wavelengths, ie all the waves are in phase (the concept is discussed in more detail in Appendix 5)*. In a hologram of finite thickness, the Bragg condition is satisfied for the virtual image but not for the real image, which is consequently very dim. By turning the hologram through 180° out of its plane about a horizontal axis (a procedure we shall refer to hereafter as *flipping*), the Bragg condition is satisfied for the real image instead of the virtual image, and the real image can be clearly seen hovering in front of the hologram (Fig 4.4). Any type of hologram can be made to produce both an orthoscopic and a pseudoscopic image.

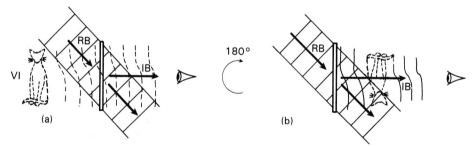

FIG 4.4 *Virtual and real images.* (a) *When illuminated by the reconstruction beam RB, the image beam IB is a replica of the original object beam, so that the viewer sees a virtual image VI in the position of the original object.* (b) *When the hologram is flipped so that the reconstruction beam becomes its conjugate, a real image is formed. It is pseudoscopic (ie has reversed parallax) and inverted.*

---

* Bragg diffraction is of overwhelming importance in all types of holographic imagery, both transmission and reflection, and you are urged to study Appendix 5 at an early opportunity. Over and over again, as you will discover as you read this text, the Bragg condition is brought into discussions of image formation, and an appreciation of its implications is essential in making top-quality holograms.

Reflection Holograms

The maximum angle of the fringes to the emulsion in a transmission hologram is 45° if the subject matter is directly in front of the photographic emulsion; at this angle the Bragg condition exerts sufficient effect to permit viewing by partially-filtered white light. If the reference beam is taken still farther round so that it falls on the emulsion from the side opposite to the object beam, white-light viewing is possible. Such a hologram is known as a *white-light reflection hologram*. The simplest form of reflection hologram is the single-beam or Denisyuk hologram. In this type of hologram the beam falls on the emulsion from one side, acting as the reference beam, and passes through the emulsion to be reflected back by the subject matter on the other side to form the object beam (Fig 4.5).

The fringes are formed by the *standing waves* generated when two beams of coherent light traveling in opposite directions interact (see Appendix 2). The fringes formed are sheets more or less parallel to the surface of the emulsion, like the pages of a book, these sheets are roughly one half-wavelength apart. Under these circumstances, Bragg diffraction is the controlling phenomenon in image formation. The spurious real image is totally suppressed, and the zero-order beam very nearly so. The diffraction efficiency can be very high indeed; in certain types of reflection hologram it can approach 100%. In addition, we can replay the hologram using white light. As with a dielectric mirror, a reflection hologram reflects light only within a narrow band of wavelengths, so if we illuminate it with a highly-directed beam of white light such as is given by a spotlight or a slide projector, the hologram will select the appropriate band of wavelengths to reconstruct the image, the remainder of the light passing straight through. Reflection holograms are usually sprayed black on the rear surface to absorb this and any other stray light (Fig 4.6).

The hue of a reflection hologram is not necessarily that of the laser beam used to make it. Indeed, it can be quite difficult to make a reflection hologram that reconstructs in its original hue. As a general rule, processing a holographic emulsion

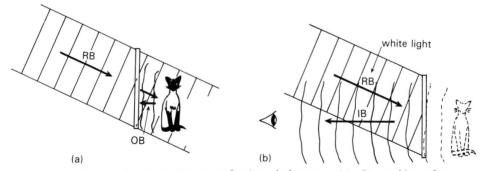

FIG 4.5 *Denisyuk (single-beam) reflection hologram.* (a) *In making the hologram the reference beam* RB *passes through the emulsion and illuminates the object to form the object beam* OB, *which is incident on the emulsion on the side opposite to the reference beam.* (b) *The hologram is replayed by a reconstruction beam* RB *on the same side as the viewer; the image beam* IB *is reflected from the hologram.*

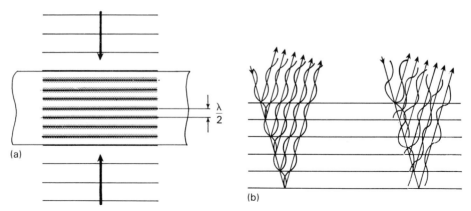

FIG 4.6 *Reflection hologram.* (a) *Two coherent wavefronts passing through an emulsion in opposite directions create a standing wave; the result is interference fringes one half-wavelength apart.* (b) *Light of wavelength equal to twice the fringe spacing is reflected from the fringes in phase, resulting in constructive interference and a strong reflection; light of other wavelengths suffers destructive interference.*

removes some of the material originally present in it, so that the image replays at a shorter wavelength than the one at which it was made. This is not necessarily a disadvantage, as for a given intensity a green image appears brighter than a red one. The manipulation of the hue of the image is an important factor in creative holography.

Phase holograms The earliest reflection holograms were very dim. When a hologram is developed and fixed using conventional photographic processing methods it is called an *amplitude hologram*, as the fringes modulate the amplitude of the transmitted beam. If the *photographic image* is examined under an electron microscope, the developed silver grains have the appearance of opaque black mopheads. They absorb almost all of the light that falls on them, and scatter the remainder randomly. The result is that what should be strongly-reflective layers in practice absorb most of the light that falls on them, so that the diffraction efficiency of a reflection hologram processed by ordinary develop-fix methods is usually less than 3%. However, if the silver layers can be turned into transparent layers, and if these have a high refractive index compared with that of plain gelatin, the emulsion will effectively have *two* sets of fringes, which are interleaved (Fig 4.7) and *all* the light will contribute to forming the holographic image.

By good fortune, the refractive index of silver bromide is very high, higher even than that of diamond, and this delays the waves traveling in it so that their phase almost matches that in the plain-gelatin fringes. Thus we need only to change the silver back into silver bromide after development, and we hall have the kind of interference mirror that is so efficient in a laser. In practice this process is quite simple; it is called *rehalogenation*, or, more simply, *bleaching*. The result is a *phase hologram*; in a phase hologram every part of the emulsion earns its keep.

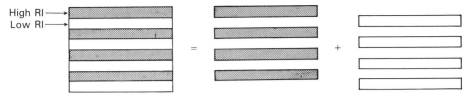

FIG 4.7 *The effect of a phase reflection hologram is as if two stacks of Bragg planes were acting independently. The stack of high refractive index* (RI) *planes delays the phase of the reflected waves so that it is more nearly in phase with the waves reflected from the low refractive index stack. The effect also occurs with transmission holograms.*

The Denisyuk configuration provides a simple and reliable system for this, but it will give satisfactory images only if the object is light-colored, shallow and mounted close to the holographic plate. This is because the light transmitted through the emulsion is attenuated by the sensitizing dyes in it. The ideal *beam intensity ratio* (see Box) for a reflection hologram is quite low, somewhere between 1.5 and 3. This means that the intensity of the object beam needs to be as high as possible, and with subject matter that is deep or not very bright it is difficult to achieve the optimum ratio. So to obtain bright reflection holograms of deep or dark objects it is necessary to split off part of the laser beam to illuminate the object independently using a *variable beamsplitter*, so that the beam ratio can be controlled (Fig 4.8).

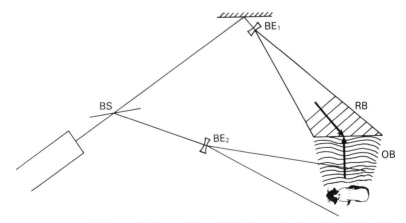

FIG 4.8 *The introducion of a beamsplitter* (BS) *allows the intensity of the object beam to be increased with respect to that of the reference beam* (RB). *If a variable beamsplitter is used the beam ratio can be optimized.*

Beam intensity ratio

It might at first seem that for the best possible interference pattern the intensities of the object beam and the reference beam should be equal. This is not so. If the were equal, the dark fringes would be represented in the hologram by 100%

transmittance, which is outside the linear part of the transmittance-vs-exposure curve of the emulsion. This would result in a lowering of the diffraction efficiency. In addition, the intensity of the object beam is not uniform, particularly if the object is close to the plane of the hologram. Furthermore, the average intensity of the object beam is much less than its maximum intensity; this is because it is made up of bright and dark speckles (*laser speckle*). Unless the reference beam has at least as high an intensity as the bright speckles, the fringe contrast will be low at these points, and the diffraction efficiency will suffer. The intensity of the reference beam is therefore set at a value where the minimum transmittance of the fringe pattern lies on the *linear region* of the transmittance-vs-exposure curve over the whole area of the hologram. Beam intensity ratios are usually measured with the meter pointed straight up each beam in turn, taking an average for the object beam over the whole area of the hologram. Optimum ratios are usually between 3 : 1 and 8 : 1 for a transmission hologram (though ratios as high as 30 : 1 have been used successfully), and between 1.5 : 1 and 3 : 1 for a reflection hologram. These are not the effective or true beam ratios, which can be measured by pointing the light-meter along the bisector of the angle between the two beams and using a polarizing filter with its polarizing axis parallel to the direction of the polarization of the reference beam; however, this somewhat tedious method is not in common use. With highly oblique beams much of the incident light is wasted by specular reflection from the surface of the holographic plate; in these cases the readings of the beam intensities must be taken behind a glass plate in the plateholder (see graph on p. 138, Fig 10.2).

Image-plane Holograms

A hologram that has been made with the object wholly in the plane of the emulsion (called an *image-plane hologram*) can be replayed using white light that is not spatially coherent. This applies even to transmission holograms. The reason is that the distance over which the light waves have to remain in phase is effectively zero, and the image does not shift no matter where the replay light comes from. Furthermore, the light used for making the hologram itself does not have to be particularly coherent. You can make a Denisyuk hologram of a flat object using the light from a low-pressure sodium lamp – even a yellow sodium street-lamp – which has a coherence length of less than 0.3 mm and is by no means a point source (Fig 2.1, p. 17).

Such holograms are two-dimensional and not very interesting. By using a fairly flat object in contact with the emulsion it is possible to make a Denisyuk hologram that will replay using almost any light. This is how the early hologram pendants were made: the images are small and shallow, and not very sharp under diffuse illumination. If we could find a way of placing the optical image of a shallow object right across the plane of the emulsion, we should be able to produce a passable three-dimensional image in a transmission hologram with a white, or at least only partially-filtered, replay beam. We can in fact do this by producing a real image either by means of a lens or another hologram. The image becomes the object for the hologram. When the image is produced by a lens the result is called a *focused-*

image hologram; when produced by another hologram it is called a *full-aperture transfer hologram*. Both types of hologram belong to the class of *image holograms*.

Focused-image holograms   A hologram is a record of the electromagnetic field in its plane; Nils Abramson (3) described it as 'a window with a memory'. When you replay a hologram, you see what the holographic plate saw, no matter how bizarre this may have been. So there is no reason why an optical image produced by a lens should not be used as the subject for a hologram. If you focus a camera on a near object at full aperture, then, with the shutter held open, if you open the camera back, you will be able to see the optical image hovering in the region of the film plane. It is inverted, certainly, but it has full and orthoscopic parallax. Of course, a photographic exposure of the usual kind will show sharply only the region of the image that is cut by the plane of the film, the remainder being blurred. But if you were to use coherent illumination of the subject, and to add a reference beam, you would be able to record all the missing information. The optical arrangement for a *focused-image hologram* is shown in Fig 4.9.

If the perspective of the final image is to be acceptable, the scale of the optical image needs to be close to 1 : 1 (see p. 201 and Appendix 2). The condition for this is that the distance from the object to the focusing lens should be approximately twice the focal length of the lens; the image will be on the other side of the lens at an equal distance.

You can also make a focused-image hologram using a concave mirror instead of a lens, but the image will be pseudoscopic as well as inverted, and if you try to get round this by viewing the hologram from the other side, you will find that it is now *pseudoptic* (reversed right to left) (see pp. 206–7).

Full-aperture transfer holograms   The pseudoscopic real image produced by a

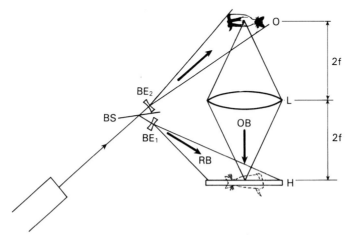

FIG 4.9 *Geometry of a full-aperture focused-image hologram. The object O is imaged (I) by the lens L in the plane of the hologram H. The reference beam RB is reflected from the beamsplitter BS.*

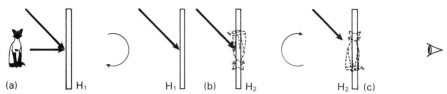

FIG 4.10 *Full-aperture transfer hologram. First, a laser transmission hologram H_1 is made (a). This is flipped, and the real pseudoscopic image is reconstructed to form the object for a second hologram H_2 in its plane (b). Finally, this hologram is again flipped for display (c); the image now becomes orthoscopic again.*

flipped hologram is not only full size, it is also pseudoptic. But if we make a second hologram using the real image as the object, then after processing the hologram we can flip it to get a hologram which is both orthoscopic (because it is doubly pseudoscopic) and *orthoptic* (because it is doubly reversed). To make a hologram of the real image we merely have to flip the master hologram (usually designated H_1 in diagrams) and place a second holographic emulsion in the plane of the real image, with a suitable reference beam. This produces a second transfer hologram with the image formed across its plane. The image is pseudoscopic, and needs to be flipped again to make it orthoscopic. Provided the image is shallow it is possible to use white light for the replay beam. Fig 4.10 shows the principle.

Slit Transfer (Rainbow) Holograms

In a full-aperture transfer hologram the depth of the subject matter is limited to a centimeter or two. Anything deeper than this will show color smear when looked at from slightly to one side. In 1968 Stephen Benton (4) fortuitously* found a way of getting over this difficulty. As binocular vision operates in a horizontal plane, the vertical parallax of a hologram is not very important, and if eliminated from the holographic image, its absence has only a minimal effect on the realism of the image. Benton therefore eliminated vertical parallax by making a transfer hologram in which the master was masked down to a narrow horizontal slit. In effect, this replaced the horizontal parallax by a holographically-generated diffraction grating. The result was that, when illuminated with white light, any point in the holographic image would be seen by light of only one wavelength. Thus in a *Benton* or *rainbow hologram*, a change in viewpoint in a vertical direction results only in a change in image hue and not in image perspective. This geometry is usually (though not necessarily) such that a progressive vertical change in viewpoint from above to below changes in the hue of the image through the visible spectrum from red to violet.

The way in which this happens is ingenious but simple. The layout is similar to that for a full-aperture transfer hologram, but the aperture of the master hologram is restricted to a narrow horizontal slit (conventionally designated S_1 in

---

* He was looking for a way of reducing the information content of a hologram as a step towards making it possible to transmit holograms by television.

diagrams). Thus, whereas the whole width of the hologram is used, coding all the horizontal viewpoints in the transfer (or final) hologram, only a single viewpoint is coded in the vertical sense. When the processed final hologram is flipped to give an orthoscopic image, the image of the slit is flipped too, and in a reconstruction by laser light the image is visible only when the viewer's eyes are precisely lined up with the slit, very much like looking through a letter-box; any small movement up or down on the part of the viewer, and the image disappears.*

However, if we illuminate the final hologram with white light, there will be an individual letter-box for each wavelength, and these letter-boxes will not coincide. Since red light is diffracted more than green light and green light more than blue, the red letter-box image will be higher than the green one, and the green one will be higher than the blue, with the other spectral hues in between (Fig 4.11).

The purity of the colors and the sharpness of the image in the vertical direction depend on how narrow the slit is. By varying the distance of the transfer plate from the master hologram the final image can be made real, virtual or partly real and partly virtual (ie across the hologram plane). The distance of the object from the master hologram determines the position in space of the real image of the slit. Rainbow holograms belong to the class of holograms known as *white-light transmission holograms*.

The process is very versatile. By varying the angle of the reference beam for the transfer hologram, the image hue as seen at normal viewing height can be controlled; in addition, several images can be stored in a single hologram and recreated in different hues either successively, at different viewing heights, or simultaneously, at a single viewing height. A small number of workers have brought the technique to a precise art, and their methods and geometries are described in Chapters 12 and 13 and in Appendix 7.

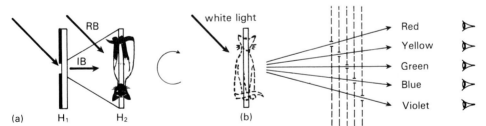

FIG 4.11 *Making a rainbow hologram. The master H_1 (a) is masked off to a narrow horizontal slit. This produces an image hologram H_2 in which the vertical information is replaced by a diffraction grating (a 'trivial' hologram produced by interference between IB and RB). When H_2 is flipped (b), the image of the slit is projected close to the eye of the viewer, and when H_2 is illuminated with white light, the image of the slit varies in its position according to the wavelength. The viewer sees the image in a spectral hue which depends on the height of the viewpoint.*

---

* Such *slit transfer holograms*, while used mainly for white-light transmission images, also have their uses in reflection work, chiefly for *multiplexing*. This is described on pp. 184–5.

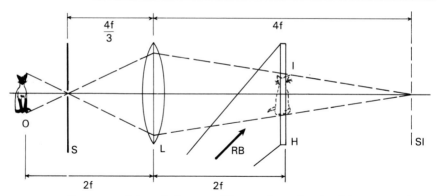

FIG 4.12 *One-step rainbow hologram. The configuration is the same as that of* FIG 4.9 *except for the horizontal slit S, which performs the same function as the slit in a rainbow hologram. The real image of the slit* (SI) *is approximately in the plane of the viewer when replayed (without flipping).*

The focused-image technique can also be used to make a one-step slit hologram. In a method described by Hsuan Chen and F T S Yu (5), the optical image of an object produced by a lens is limited vertically by a slit placed in the object space at a distance such that a real image of the slit is produced in the image space. The result is a one-step focused-image rainbow hologram (Fig. 4.12). This method appears to have been used first by Stephen Benton in 1968 and was subsequently patented. After the publication of Chen and Yu's paper the method was elaborated on by Benton *et al* (6, 7), who modified the geometry by placing the slit on the lens instead, with a strongly-diverging reference beam. When the hologram is replayed using a nearly-collimated reconstruction beam the virtual image of the slit becomes a real image on the viewing side of the hologram (Fig 4.13).

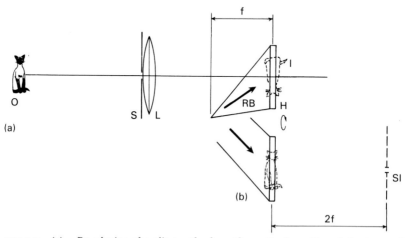

FIG 4.13 (a) *By placing the slit on the lens the optical quality of the image is improved. The reference source must be placed close to the hologram.* (b) *On replay using a much less divergent beam, the image of the slit* SI *is projected into the viewing space.*

Achromatic Transfer Holograms

Although the colors of a rainbow hologram can be used creatively, the idea of an uncolored or *achromatic* hologram has always been attractive. Full-aperture transfer holograms are almost achromatic, but have very limited depth. For deeper images, the difficulty is that each wavelength reconstructs the image (including the real image of the slit) at differing magnification and distance, 'red' wavelengths producing the smallest and closest, 'blue' the largest and farthest image. The *tip angle* of the line of images of the slit is about 35° to the normal to the hologram (ie about 65° to the hologram plane). In order to produce an achromatic image it is necessary to produce a number of spectra aligned in such a way that at each point there are red, green and blue images of the slit which precisely coincide (Fig 4.14).

Benton (8) succeeded in achieving this by making a *holographic optical element* which he called a 'diffractor plate'; when illuminated by collimated laser light this produced the effect of a number of point sources of light along the tip angle of the spectrum. He used this diffractor plate to produce multiple overlapping images in a first transfer hologram. When this hologram was used as a master for recording the final transfer hologram, this final transfer contained images of all the slits so that when it was replayed with white light the spectra overlapped to produce an achromatic image (Fig 4.15). The method is described in full on pp. 196–9. Plate 4 shows the well-known 'Head of Aphrodite' made using this method.

A much simpler method suggested in the 1960s by DeBitteto (9) and developed by Bazargan (10) uses a different approach known as dispersion compensation. The method is broadly similar to that adopted for correcting *chromatic aberration* in a lens system, in that an equal and opposite aberration is introduced into the system so that the dispersion is cancelled. In this system a holographic diffraction grating is made using the same beam angles as were used for making the original hologram. When this element is *rotated* (ie turned through 180° in its own plane) it produces a spectrum that is the reverse of that produced by the hologram, and the dispersion is cancelled. This method preserves vertical parallax. It was designed to be used with specially-made transfer holograms, but in fact works perfectly well with any laser

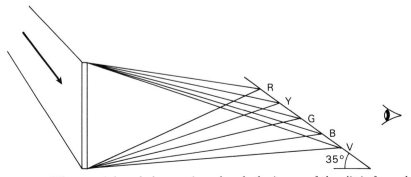

FIG 4.14 *When a rainbow hologram is replayed, the image of the slit is formed at different distances for different wavelengths. The angle the spectrum makes with the normal to the plate is call the 'tip angle' and is typically about 35°, depending on the original geometry.*

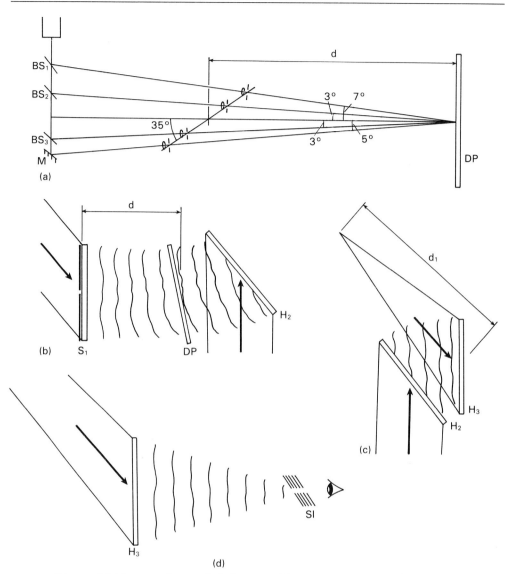

FIG 4.15 *Making an achromatic hologram with a holographic diffractor plate.*
(a) *Making the diffractor plate. The beams derived from the main beam by beamsplitters* BS$_1$, BS$_2$ *and* BS$_3$ *and the mirror* M *are directed towards the optical center of the plate angles to the normal as shown. The beams are expanded from sources (pinholes) aligned along the tip angle (35° to the normal).*
(b) *The diffractor plate* DP *(flipped) is interposed between the slit master* S$_1$ *and the second hologram* H$_2$, *which are angled to suppress reflections.*
(c) H$_2$ *(flipped) is used to project the real image of the diffractor plate into the plane of the final hologram* H$_3$ *along with the real image of the object.*
(d) *When* H$_3$ *is replayed (unflipped), about 12 slit images are projected in the same plane with overlapping spectra, resulting in an achromatic reconstruction.*
NOTE: *Distances and angles are not to scale in this figure.*

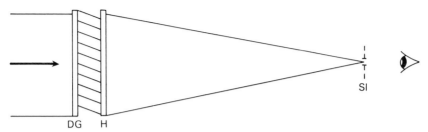

FIG 4.16 *Using a holographically-generated diffraction grating to compensate for dispersion. The grating DG has been made with the same geometry as the hologram H, but has been rotated through 180° so that its dispersion is in the opposite direction, giving a net dispersion of zero and an achromatic slit image SI. DG and H may be separated by a layer of poprietary cellular material which suppresses any higher-order spectra.*

transmission hologram made with the correct geometry (Fig 4.16). It is described in more detail on pp. 195 and 197.

Single-beam Transfer Holograms

It is possible to make image-plane transfer holograms using a single-beam arrangement. The image-forming light acts as the object beam and the directly transmitted light forms the reference beam. For full-aperture transfers the light passes through H_1 and H_2 in turn; for a slit transfer hologram a separate part of the beam illuminates S_1 and H_2. For a reflection master hologram the order of H_1 and H_2 is reversed (Fig 4.17).

Holographic Stereograms

The principle of the parallax stereogram was briefly discussed on p. 4. It depends on a set of photographic images that are displayed in such a way that from any view-

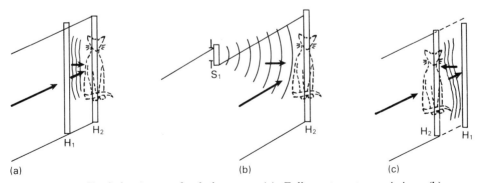

FIG 4.17 *Single-beam transfer holograms. (a) Full-aperture transmission; (b) rainbow transmission; (c) reflection. In each case the master (H_1) is flipped to produce the pseudoscopic real image. After processing, the final hologram H_2 is again flipped for white-light viewing.*

ing position the eyes of the viewer see only one image each, and that these two images are in all cases the correct two views for the two viewpoints. In photographic stereograms of this type it is necessary to print the various images in the form of interlaced strips behind a lenticular screen. This places severe difficulties on resolution.

If the photographic images can be replaced by holographic images of the photographs, many more views can be fitted into a given area and the lenticular screen can be dispensed with. There are two basic ways of achieving this. In both methods the images begin life as a more or less conventional set of motion picture frames. In one system these are taken with the camera moving past the subject along a horizontal bar; in the other the camera remains stationary while the subject is rotated on a plinth in front of it. In the first system, derived from an original idea by deBitteto (11) and developed by Stephen Benton's team (12, 13), the holograms are *multiplexed* onto a holographic emulsion, that is, the individual exposures are all distributed over the whole area of the emulsion. In the second system, due to Lloyd Cross (14), the holographic fringes for each ciné frame are localized in adjacent narrow vertical strips. The principles underlying the two techniques are shown in Fig 4.18, and the practical systems are dealt with fully in Chapter 17. The two types of stereogram have become known respectively as a *Benton stereogram* and a *Cross hologram.* The latter, having had some commercial exposure, has also acquired the names of *integral hologram, integram* and *Multiplex hologram*, the last rather unfortunately, as its images are *not* multiplexed. 'Multiplex' happens to be the name of Cross's corporation.

The main difference in viewing the two types of stereogram is that in the Cross hologram the viewer looks through a narrow strip hologram at a virtual image of the ciné frame at the center of the hologram (which is in the form of a cylinder or part of a cylinder), whereas in the Benton stereogram the viewer looks through a real image of a vertical slit that is well in front of the hologram at a holographic image of the ciné frame that is in or close to the plane of the hologram (which is usually flat).

The image in a Cross hologram exhibits several kinds of distortion (these are discussed in Chapter 17), and several workers (15, 16) have designed ingenious methods for minimizing these. The main trouble with a Cross hologram is an unavoidable mismatch between taking and viewing optics, which leads to images that are stretched out vertically. As the actual shooting time for a 120° image is 48 s, it is usual for the subject to make some limited movement during the exposure. Fig 4.19 shows the well-known 'Kiss II' in which Cross's assistant Pam Brazier is seen to wink and blow a kiss as the viewpoint moves from left to right. However, unless carried out with exaggerated slowness (or with a speeded-up camera) any gestures on the part of the subject may lead to further distortions (see pp. 250–1). Fig 4.19 shows the distortions inherent in the format very clearly. For comparison, Plate 5 shows a Benton stereogram of the TV personality Michael Rodd.

Being based on a motion-picture film, the holographic stereogram has the considerable artistic bonus of being able to take advantage of all the tricks of the movie business such as dissolves, montages and animation; and it is comparatively easy to make stereograms in color. Recently, a great deal of interest has been shown in very large stereograms which have originated in computer-drawn material (Molteni, 17).

THE BASIC TYPES OF HOLOGRAM

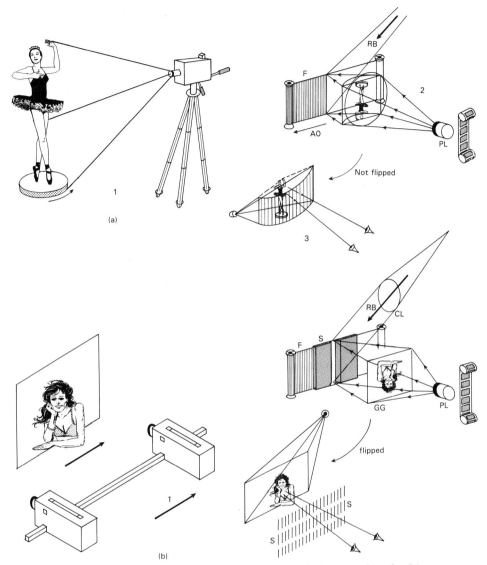

FIG 4.18 (a) *Optics of the Cross hologram. 1 Ciné photography of subject on slowly-rotating plinth. 2 Images of ciné frames are projected by projector lens PL onto anamorphic optics AO, which focus the images as a series of thin line holograms on file F. 3 Reconstruction: a virtual image of the subject is seen behind the line hologram, a different image being seen by each eye.*
(b) *Optics of the Benton stereogram. 1 Ciné camera slides on a rail parallel to subject plane. 2 Images of ciné frames are projected onto ground glass screen GG and holograms are made separately behind slit, using reference beam collimated by lens CL. 3 Hologram is displayed flipped: eyes see their appropriate image through real images of slits SS.*
NOTE: *Both these diagrams are simplified. The full set-ups are described in Chapter 17.*

FIG 4.19 *Two views of 'Kiss II' by Lloyd Cross and Pam Brazier. Notice the vertical distortion, a consequence of the astigmatic optics and the mismatch of taking and viewing optics. Photographs courtesy of the Museum of Holography, New York.*

FIG 4.20 *Artist's impression of a full-scale alcove hologram.*

Alcove Holograms

The most recent development in holographic stereograms is the *alcove hologram*, also from Benton's team. This stereogram is in the form of a large concave half-cylinder; a real image of a computer-generated three-dimensional image is projected to its center, and the background goes through the hologram, to infinity if required. The images are prepared from computer drawings which have been pre-distorted according to a program which pre-compensates for the geometric distortions introduced by the format (Benton, 18) (Fig 4.20).

Other Types of Hologram

There are several other types of hologram: these are mainly of interest to scientists and technologists, and are discussed more fully in the chapters devoted to applied holography.

1 In-line hologram (pp. 343–5)   This type of hologram uses an optical arrangement that is much the same as Gabor's original layout; it is used mainly for particle and droplet counting.

2 Fourier-transform hologram (pp. 209–12)   This type of hologram uses one or other of a number of possible optical arrangements which record the information present in an optical Fourier transform (see Appendix 3). While not particularly interesting aesthetically, Fourier-transform holograms have considerable applications in information processing and pattern recognition; they are also the only type of hologram that readily lend themselves to being drawn directly by a computer.

3 Holographic interferograms (pp. 316–29)   Holographic interferograms are of immense importance in non-destructive testing, quality control and flow analysis; they record the change in shape of an object between two testing conditions in the form of contour fringes. They are discussed in Chapter 22.

References

1 Gabor D (1951) Microscopy by reconstructed wavefronts II. *Proceedings of the Physical Society (London) B*, **64**, 449–69.
2 Leith E N and Upatnieks J (1963) Wavefront reconstruction with continuous-tone objects. *Journal of the Optical Society of America*, **53**, 1377–81.
3 Abramson N (1981) *The making and evaluation of holograms*. Academic Press, p 48.
4 Benton S A (1969) On a method of reducing the information content of holograms. *Journal of the Optical Society of America*, **59**, 1545.
5 Chen H and Yu F T S (1978) One-step rainbow hologram. *Optics Letters*, **2**, 1479–80.
6 Benton S A, Mingace H S Jr and Walter W R (1984) One-step white-light transmission holography. *Holosphere*, **12**, 7–9.
7 Benton S A, Mingace H S Jr and Walter W R (1979) One-step white-light transmission holography. *Proceedings of the SPIE*, **212**, 2–7.

8 Benton S A (1978) Achromatic images from white-light transmission holograms. *Journal of the Optical Society of America*, **68**, 1441.
9 deBitteto D J (1966) White-light viewing of surface holograms by simple dispersion compensation. *Applied Physics Letters*, **9**, 417–8.
10 Bazargan K and Waller-Bridge M (1985) A practical portable system for white-light display of transmission holograms using dispersion compensation. *Proceedings of the SPIE*, **523**, 24–5.
11 deBitteto D J (1969) Holographic panoramic stereograms synthesized from white-light recordings. *Applied Optics*, **8**, 1740–1.
12 Molteni W J Jr (1982) Black and white holographic stereograms. *Proceedings of the International Symposium on Display Holography* **1**, 15–26.
13 Benton S A (1983) Photographic holography. *Proceedings of the SPIE*, **391**, 2–9.
14 Cross L (1977) Multiplex holography. Paper presented at SPIE seminar on three-dimensional imaging, Aug 1977, but not offered for publication. The process was described in Benton S A (1976) Three-dimensional holographic displays. *Proceedings of the Electro-Optical Systems Design Conference*, 481–5.
15 Huff L and Fusek R L (1981) Optical techniques for increasing image width in cylindrical holographic stereograms. *Optical Engineering* **20**, 214–5.
16 Huff L and Loomis J S (1983) Three-dimensional imaging with holographic stereograms. *Proceedings of the SPIE* **402**, 38–50.
17 Molteni W J Jr (1985) Computer-aided drawing of holographic stereograms. *Proceedings of the International Symposium on Display Holography* **2**, 223–30.
18 Benton S A (1987) 'Alcove' holograms for computer-aided design. *Proceedings of the SPIE* **761**,

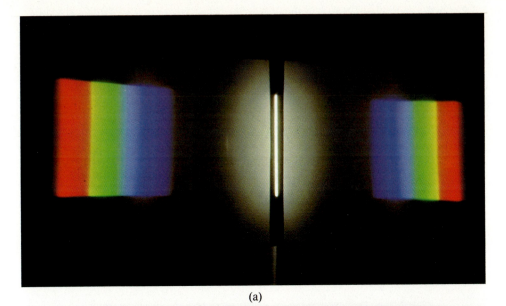

(a)

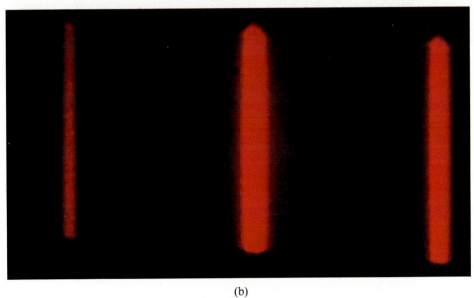

(b)

PLATE 1 *When a cosine grating is illuminated by a beam of white light, the diffracted light is dispersed into a spectrum (a). As laser light has only a single wavelength, diffraction by the grating produces only two lines, apart from the zero-order line (b). (See p. 12.) In both cases the excessively bright zero-order line has been attenuated by a strip of black card.*

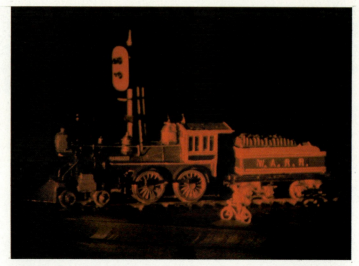

2

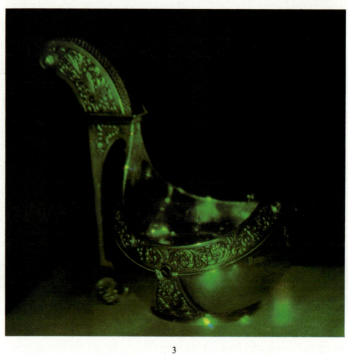

3

PLATES 2 AND 3 *Two of the earliest holograms of solid objects. Above, made around 1963 by Leith and Upatnieks, is an off-axis laser transmission hologram. Below is an early example of a Denisyuk hologram from Russia, processed by the colloidal-silver method (see p. 19).*

PLATE 4 *Head of Aphrodite, an achromatic white-light transmission hologram by Stephen Benton, made using a diffractor plate. The method is described on pp. 196–9.*

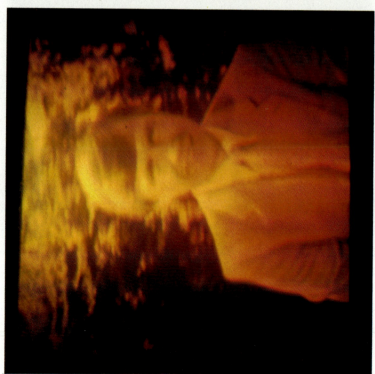

PLATE 5 *Two views of a Benton achromatic stereogram (see pp. 56–7) of the television presenter Michael Rodd. Notice the absence of off-colors and distortion. These two photographs can be treated as a stereoscopic pair.*

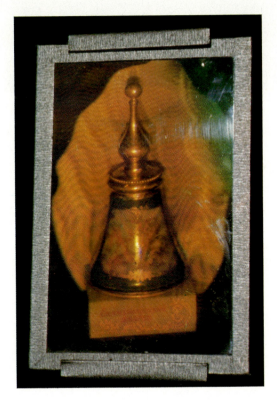

PLATE 6 *Russian holograms of antique art objects, made in natural color using up to five laser wavelengths (see p. 64). As they are reflection holograms they can be replayed using white light.*

PLATE 7 *Single-beam reflection hologram by Martin Richardson (see p. 111). In this type of semi-abstract hologram several images can be combined to give more than one perspective, so that changing the viewing position makes them appear to rotate relative to one another.*

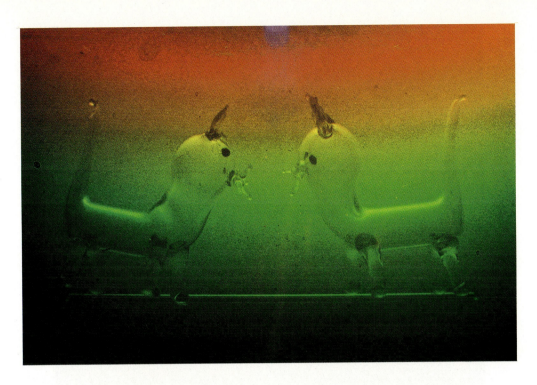

PLATE 8 *A small rainbow hologram made by the single-beam transfer technique described on pp. 116–8. This technique works best with translucent objects because of the back lighting.*

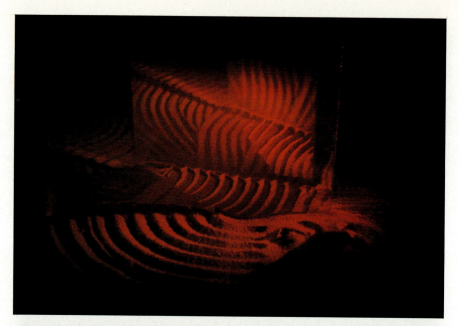

PLATES 9 AND 10 *Two early examples of creative holography. 'Thoughts', by Ken Dunkley, is an image of a sand table; the background contains further holographic images of the raked sand. Margaret Benyon's 'Picasso' is a three-dimensional comment on the well-known painting 'Les Demoiselles d'Avignon'. Photograph of 'Picasso' by Margaret Benyon (see p. 116.)*

CHAPTER

5

Color Holography

'Still, you're the right colour, and that goes a long way.'
'I don't care about the colour', the Tiger-lily remarked. 'If only her petals curled up a little more, she'd be all right.'
LEWIS CARROLL, *Through the Looking-Glass*

Basic Principles

In the early days of photography it was more than once declared that painting was now dead: nobody seemed to notice that the photographic images were blurred and lacking in color. Today, photographs are not only sharp but almost always in color; the black-and-white print is fast becoming obsolete, except for occasional pictures where the starkness of monochrome is exploited for its dramatic possibilities. Will a similar fate overcome the monochrome hologram? The answer so far is not entirely clear. The problems facing true-color holograms are to some extent parallel by those which faced color photography in its early days; but holography in color has a number of problems of its own. With a single exception, Lippmann photography (see p. 19), all color photography has followed the *Young–Helmholtz model* for color perception (see Appendix 6) by recording the red, green and blue portions of the spectrum separately, and presenting the respective images to the eye together in register.

In holography it is not as simple as that. There are no fewer than six specific difficulties associated with holography in natural colors. Some can be dealt with fairly easily; others may prove hard to overcome.

1 You have to use three beams of monochromatic light to make a color hologram; you thus need to align three laser beams along precisely the same path. There are no theoretical difficulties involved, but in practice such alignment is by no means easy.
2 Three laser wavelengths cannot span the gamut of colors associated with all possible subject matter. There is a fuller discussion of this problem in Appendix 6.
3 The monochromatic illumination can result in anomalies in color reproduction. The subject matter may not reflect the wavelengths concerned in a regular way. If you look at a deep red rose under a sodium lamp it looks black, because it absorbs the wavelength emitted by the lamp. Certain yellow objects can appear brown when illuminated by a mixture of red and green laser light which exactly matches their color as seen under white light.

4 The presence of fringes in the hologram corresponding to the three wavelengths means that the set of fringes formed by the red laser beam reconstructs not only the true red image, but also spurious red images from the sets of fringes formed by the green and blue laser beams. The same thing happens for the green beam, which reconstructs spurious red and blue images in addition to the true green image; and for the blue beam, which produces spurious green and red images. Of the nine images produced, three are genuine and form an accurately-superposed image in full color, and six are spurious; they occur in different positions and are of different sizes. Apart from the waste of light, these images overlap each other and often the genuine image too.

5 The three sets of fringes that make up the color hologram occupy the same volume; this is true for both transmission and reflection holograms. When more than one hologram is recorded in a single emulsion layer, its dynamic range (i.e. the available range of densities for an amplitude hologram or of refractive indices for a phase hologram) has to be shared between the sets of fringes. The result is that the diffraction efficiency of each hologram is reduced by a factor that is roughly proportional to the square of the number of holograms recorded.

6 There are difficulties associated with the display of true-color holographic images. In particular, a transmission hologram requires a laser set-up of the same type as was used to make the hologram. A reflection hologram can be displayed using white light, but any change in emulsion thickness as a result of processing will cause an unacceptable shift in hue: the 'red' image may even replay as a green one, the 'green' as blue, and the 'blue' as invisible ultraviolet.

All in all, this seems a formidable list; fortunately, the difficulties are not entirely insurmountable. Let us look at each in turn.

Illumination

The first problem is to get three laser beams along the same path to illuminate the subject, otherwise there will be odd-looking colored shadows and highlights. It is in fact not necessary, nor even desirable, for the beams to be of the same intensity, as holographic emulsions are not sensitive uniformly to the visible spectrum; in any case the three exposures are usually made separately. Fig 5.1 shows a typical arrangement for producing a combined beam.

The beamsplitter ratio (ie the intensity ratio of the two beams emerging from the beamsplitter) has to be 1 : 1, so that to get the beam ratio correct there will need to be a neutral-density filter in the reference beam, and perhaps further filters in the separate beams to make the three exposures more or less equal (this is because of possible reciprocity-failure effects). If an Ar^+ laser is used for the green and blue beams, it is possible to get it to emit at all possible wavelengths simultaneously. Only two of these (488 nm and 514 nm) are of high power, and filtering can balance these up and reduce the effect of the unwanted wavelengths. However, the coherence length is very low when the laser is used in this manner, typically about 10 mm; so it is more usual to give separate exposures for the green and blue using a Fabry-Pérot etalon and re-tuning the laser between the exposures. Ar^+ lasers suitable for color holography are a good deal more powerful than HeNe lasers, and if it is desired to

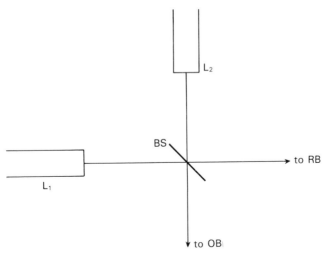

FIG 5.1 *Combining laser beams for color holography. The 1:1 beamsplitter BS combines the beams of the two lasers L_1 and L_2 to produce beams for both reference and object illumination.*

make the color hologram in a single exposure, the Ar^+ beam will need about 90% attenuation. In order to match the power of the Ar^+ beam a Kr^+ laser can be used instead.

As can be seen from Appendix 6, the second problem turns out to be a very small problem. The three beams of 488, 514 and 633 nm wavelength are capable of synthesizing all colors with the exception of highly saturated purples. A change to the somewhat weaker 477 nm Ar^+ line gives a significant improvement. The HeCd line at 442 nm is better still, but adds to the set-up the complexity and expense of a third laser. It would seem that the chances of synthesizing the more saturated colors are slim, yet in fact the worst case of 488, 514 and 633 nm give a greater range of color than can be achieved by either color photography or color TV, and a far greater range of intensity than can be achieved in the best *halftone* color printing.

As far as the multiple-laser situation is concerned, however, better news has come recently. As discussed on p. 37, given high-quality optical equipment and a good deal of time and patience it is possible, at least in theory, to adjust the etalon of a Kr^+ laser so that its cavity length is an integral multiple for three, and possibly even four wavelengths simultaneously: red, green and blue, and perhaps yellow too. This covers almost the entire *CIE chromaticity diagram*, giving much better color reproduction and making possible the fabrication of both master and transfer holograms with the minimum of fuss. But you would need a deep pocket to go with the time and patience: and there may be stability problems due to inter-line competition effects, and the special wide-band mirrors necessary have a limited life.

Color Anomalies

All three-color reproductions suffer to some extent from anomalies in hue: in photography from imperfect spectral characteristics of the dyes and in TV images

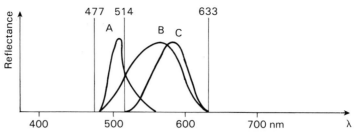

FIG 5.2 *Color anomalies using laser light. When illuminated by white light, an object with reflectance range A will appear blue-green; one with range B will appear greenish-yellow and one with range C will appear golden-yellow. If a hologram is made using the laser wavelengths shown, the reconstruction will show A and B as the same apple-green hue; C will appear black!*

from the lack of color saturation of the phosphors. In holography there is, as we have seen, little difficulty in synthesizing any color, no matter how saturated, with three suitably-chosen laser wavelengths. The reasons for the color anomalies are more subtle. They can perhaps be best understood by studying some analyses of the spectral reflectances of objects.

Let us consider some hypothetical objects (Fig 5.2). By white light, an object with reflectance range *A* will appear blue-green. An object with reflectance range *B* will appear yellow-green. An object with reflectance range *C* will appear golden-yellow. If a color hologram is made with these three objects included in the subject matter, using the three laser wavelengths 477, 514 and 633 nm, in the reconstructed image the hue of the images with the reflectance ranges *A* and *B* will appear the same shade of apple green, and that of the image whose object has the reflectance range *C* will appear black. Only if the range of wavelengths reflected by the subject matter considerably exceeds the spacing between two of the laser wavelengths will anything like a true color representation be achieved. This means that if a color hologram is to reproduce the colors realistically, the subject matter must provide a wide spectral reflectance range. Fortunately, most dyes and pigments do exhibit such a range, and natural-color holography seems to be set for success in the reproduction of artifacts such as antique vases and archaeological finds (Plate 6).

Cross-talk

The fourth problem is usually known as *cross-talk*, by analogy with communications technology. It occurs when two or more images are reconstructed by a common replay beam. In a three-color hologram the fringes which record the 'blue' image are closer together than those which record the 'green' image, which are themselves closer than those which record the 'red' image. Now, a 'green' replay beam will produce a green image in the correct position, and it will be the correct size. But it will also produce a green image from the 'blue' fringes, and as these fringes are closer together the diffraction will be greater, so that this spurious image is smaller and closer to the hologram. Furthermore, it will produce a green image from the

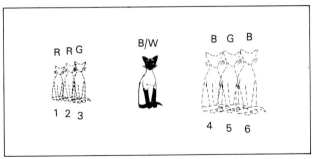

FIG 5.3 *The spurious virtual images in a color transmission hologram. They are from left to right: 1, 'blue' fringes replayed by red; 2, 'green' fringes replayed by red; 3, 'blue' fringes replayed by green; 4, 'green' fringes replayed by blue; 5, 'red' fringes replayed by green; 6, 'red' fringes replayed by blue. 1 is the nearest to the viewer, 6 the farthest away.*

more widely-spaced 'red' fringes, and this spurious image will be larger and farther away from the hologram. The red beam will produce a genuine image in the correct position and two spurious images from the 'green' and 'blue' fringes, both smaller and nearer to the hologram; and the blue beam will produce a genuine image and two spurious images from the 'red' fringes, both larger and farther away from the hologram. In addition to these six spurious images there are nine spurious real images. Fig 5.3 gives an idea of the positions of the virtual images, which are the ones most likely to be troublesome.

One way of bypassing the problem is to use widely-differing reference beam angles, so that the spurious images are generated (if at all) well out of the way of the genuine image. This, however, presents well-nigh insuperable difficulties in display, as the slightest error in alignment results in misregistration of the images; in addition, white-light reconstruction is out of the question. A better way is to make use of the thickness of the emulsion by using a very large angle of incidence for the reference beam. The fringes will then be formed close to 45° to the surface of the emulsion, and will be many wavelengths long, so that the Bragg condition will suppress most of the cross-talk. In a reflection hologram the spurious images will be totally suppressed, and white light can be used for reconstruction in full color.

Diffraction Efficiency of a Color Hologram

A hologram contains so much information that it seems scarcely possible to fit it into so small an area (or volume, to be more exact). It seems unlikely that there would be room to get in twice as much again. With present-day emulsions this is indeed so, though information capacity is steadily increasing. The main problem with color holography is that it is the same image that is being recorded at three different wavelengths, and therefore the fringes are parallel and overlap considerably; this tends to overload the information-carrying capacity of the emulsion much more than in multiple-image holograms. One solution is to use widely-separated reference

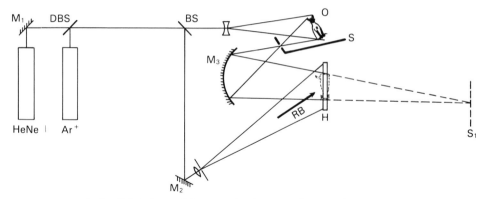

FIG 5.4 *A simplified focused-image color layout, after Hariharan et al (2). The two laser beams are combined via the mirror M_1 and dichroic beam splitter DBS. The object beam is focused by the optical mirror M_3 to form a full-size optical image in the plane of the hologram H. The slit S which restricts the mirror aperture is imaged at S_1, roughly the plane of the viewer's eye. The purpose of the slit is to brighten the image and reduce the optical aberrations of the mirror.*

beams, but this, as we have seen, raises serious difficulties in reconstruction. A better solution is to restrict the information content of the three holograms.

The obvious way to do this is to make separate holograms, one for each color, and to mount them in register. There is no great difficulty in doing this, and the results are gratifying. For most purposes two plates are sufficient, one recording the green and blue images and the other the red image. Hariharan (1) has gone one stage further, by eliminating about three-quarters of the vertical parallax (which plays little part in three-dimensional perception) with a slit, in a focused-image system using a wide-aperture concave mirror (Fig 5.4).

Color Shift

The last problem is the color shift that results from shrinkage of the emulsion during the processing of a reflection hologram. When a reflection hologram is viewed from either higher or lower than the optimum position the hue moves towards shorter wavelengths, and this can aggravate the problem; cutting down on the vertical parallax, as described above, limits the vertical viewing angle. The angle of the replay beam is even more important; if it is too high the image will be bluish, and if it is too low the image will be reddish. Fortunately, modern methods of processing (see Appendix 10) have been able to produce reflection holograms of high diffraction efficiency and low noise that reconstruct at the original laser wavelength and at the correct reference-beam angle. Transmission holograms do not present the same problems, as they are laser-illuminated; but color transfer holograms made from them show the usual characteristics of rainbow holograms and need to be displayed very carefully so that they are viewed from the correct angle (2).

As we have seen, the problems of color holography are by no means insoluble. It is fairly certain that art objects now being recorded in monochrome for museums and art galleries will before long be in full color. In the Soviet Union a xenon-ion laser has been used for one-shot color Denisyuk holograms at five simultaneous wavelengths, though without etalon, and with a rather weak yellow line. This seems likely to simplify straightforward record holography of colored objects to the point where color holograms of the Denisyuk type might well become routine. Certainly it is already possible to produce, for the first time, recipes — albeit tentative ones — for practical holography in natural color (Chapter 18).

Pseudocolor Holography

The techniques for producing holograms in any desired hue are now well established, and are discussed in Chapter 18. For reflection holograms the procedure consists of pre- or post-exposure swelling of all or part of the emulsion according to a pre-determined program. For transmission holograms the colors are produced by changing the reference beam angle for successive transfer exposures. These techniques require a good deal of planning in terms of table geometry. It is possible to produce realistic color images by these processes, but it must be emphasized that the colors are produced by control of the processes, and are in no way natural colors, any more than chemically-toned photographs are natural-color photographs. The geometries are discussed in Appendix 7.

References

1. Hariharan P (1978) Hologram recording geometry: its influence on image luminance. *Optica Acta*, **26**, 211–15.
2. Hariharan P, Hegedus Z S and Steel W H (1979) One-step multicolor rainbow holograms with wide angle of view. *Optica Acta*, **26**, 289–91.

CHAPTER

6

Materials, Exposure and Processing

> ('That's exactly the method,' the Bellman bold
> In a hasty parenthesis cried,
> 'That's exactly the way I have always been told
> That the capture of Snarks should be tried!')
> LEWIS CARROLL, *The Hunting of the Snark*

Several categories of light-sensitive materials can be used for making holograms, but the most widely-used of these is the *silver-halide emulsion*. When this is exposed to light a *latent image* is formed; it is made visible as a photographic image by the process of *development*. This image consists of black metallic silver, and in an ordinary black-and-white* photograph this is the negative. Fixation removes any undeveloped *silver halide*, and a final wash removes the remaining chemicals, leaving a stable photographic image. (The italicized terms above are defined formally, as usual, in the Glossary, but for those readers unfamiliar with the photographic process there is a fuller description in the Box.)

---

* The term 'black-and-white' is used to refer to a continuous-tone image in shades of grey, and *not* a binary (all-or-nothing) image. This is an accordance with common parlance; there is no point in being unnecessarily pedantic.

The silver-halide process

Silver chloride, bromide and iodide, collectively known as silver halides, are used in various proportions in photographic emulsions (there are two other halides: fluoride, which is soluble in water, and astatide, which is highly radioactive). In the manufacture of a photographic emulsion the silver halide is formed as a dispersion of exceedingly small particles (microcrystals) in gelatin. When coated on a glass or film base and allowed to dry, this becomes a thin, tough, transparent layer which is very sensitive to light of short wavelength (less than 500 nm). However, its sensitivity can be extended to cover the whole of the visible spectrum and even the near infrared by the addition of appropriate sensitizing dyes.

When light is allowed to fall on the emulsion its energy is absorbed by the silver halide particles, causing local disruptions of some of the bonds that hold the crystalline structure together and releasing free silver atoms within the body of the crystal. Above a certain critical energy enough silver atoms are released to form a stable speck of metallic silver, or latent image. The developer, which is used to turn the latent image into a visible photographic image in metallic silver, a process called *reduction*, is a solution containing a *reducing agent* which is capable of reducing silver halide to silver, but only in the case of those crystals which bear a latent image. Those which do not are unaffected by the developer. Developers have two main constituents: the developing agent, and an alkali; the latter is necessary because most developing agents work efficiently only at high pH values. The actual proportion of crystals that are reduced increases steadily throughout development, but the reaction is slow enough for it to be halted when the photographic image is judged dark enough. Fixation is carried out by immersion in a solution of a salt such as ammonium thiosulfate, with which the unchanged silver halide readily forms soluble complexes. These can be removed by washing in water.

In photography, the negative needs to have a wide range of densities, and is in general not concerned with spatial frequencies greater than 100 cycles/mm. The grains that make up the final image can thus record the required detail provided their mean diameter is less than about 10 μm. Such grains would arise from silver halide crystals that were originally 1–2 μm across, as there is much clumping of grains during development. In fact, crystal sizes go up to tens of micrometers in diameter, the largest crystals being most sensitive to light. Now, a hologram is a photographic record of interference fringes that have a separation of less than 1.5 wavelengths. This implies that the mean crystal diameter in a holographic emulsion needs to be about 60 nm; for a reflection hologram, with its fringe spacing of only one half of the wavelength, it needs to be 25 nm or even less. The sensitivity of a silver halide crystal goes down very rapidly as its size decreases: the *ISO Index* (arithmetic) of reflection holographic material is about 0.01, as against 200 for an average amateur film.

Emulsions designed specifically for holography are shown in the table below. Agfa-Gevaert and Ilford emulsions are available world-wide; Eastman Kodak emulsions are available in the United Kingdom only if large quantities are ordered.

Manufacturer	*Designation*	*Sensitivity*	*Purpose*	*Film or Plate*
Agfa-Gevaert	8E75HD	Red/blue	Reflection	Both
	8E56	Green/blue	Reflection	Both
	10E75	Red/blue	Transmission	Both
	10E56	Green/blue	Transmission	Both
Eastman Kodak	SO–424	Green/blue	Transmission	Film
	SO–343	Green/blue	Reflection	Film
	SO–173	Red	Both	Film
	120–02	Red	Both	Plate
	SO–253	All	Transmission	Film
	649F	All	Reflection	Plate
Ilford	SP673	Red/blue	Both	Both
	SP672	Green/blue	Both	Both

Holographic plates of extremely fine grain (about 10 nm mean crystal diameter) are made in the Soviet Union, but are not obtainable outside. They do not travel well, and in the Soviet Union are usually coated shortly before use. Emulsions are also produced in Romania and China, but no information is available at present.

Sensitometric Requirements for Holographic Emulsions

In photography, the most important technological pointer to the quality of the photographic image is the H & D curve (called after Ferdinand Hurter and Vero Driffield, who first devised it). It shows the relationship between the density of the processed negative (or transparency) and the exposure which produced it for given development conditions. As visual perception is approximately proportional to the *logarithm* of the stimulus, both scales are *logarithmic scales*, and the part of the curve that shows a *linear relationship* between density and log exposure represents the useful range of luminances (or subject contrast). Unfortunately, this curve is of little use in holography, as it is the relationship between amplitude transmittance and exposure that has to be linear. Also, the photographic definition of contrast is inappropriate, and a quantity called *modulation* is used instead. These concepts are discussed in more detail in Appendix 2.

In practice there is some latitude in exposure for a bright reconstruction. This is because the fringes in the average hologram are so close together that their second and third harmonics are spatial frequencies that are too high to diffract visible light at all, because they represent fringes that are closer together than the wavelength of the light.

Beam Intensity Ratio

This is the ratio of the intensities of the reference and object beams, measured straight up the beams, usually without taking polarization into account. It is important, as it has a direct bearing on the fringe modulation, which is related to the brightness, signal-to-noise ratio and diffraction efficiency of the final image. The subject is dealt with in more detail on pp. 47–8 and p. 143.

Constituents of a Developer

The progress of holography as a display medium was held up for a number of years by preconceived ideas about processing technique derived from practices in photography. The traditional constituents of a photographic developer are as follows:

1 *Developing agent*. This is usually, though not invariably, based on a substituted benzene molecule containing amino (NH_2), hydroxyl (OH), methyl (CH_3), etc groups. A developing agent is a reducing agent that will reduce the silver halide crystals in the emulsion if and only if they bear a latent image. It is often advantageous to use two developing agents rather than one, as some combinations, eg

phenidone and hydroquinone, form a superadditive combination (their activity together is greater than either alone). Most developing agents operate efficiently only in alkaline solution (ie at a pH greater than 7). It is therefore necessary to add:

2 *Alkali.* The most popular alkali is sodium carbonate, a substance which maintains a large reservoir of OH^- (hydroxyl ions) when in aqueous solution. These neutralize the H^+ (hydrogen ions) produced in the reduction process, which would otherwise lower the pH to unacceptable levels.

These two components are all that are necessary to develop the emulsion. However, in photographic practice it is usual to keep the developing solution exposed to the air for fairly long periods of time. As the developing agents react with atmospheric oxygen and lose their reducing power, it is necessary to add a further reducing agent which is not a developing agent to protect the solution from oxidizing in the atmosphere:

3 *Preservative.* This is usually sodium sulphite, added in fairly large quantities. In a coarse-grain emulsion such as an everyday photographic emulsion it actually helps the development process by forming complexes with the silver compounds.

Finally, to prevent the developer from attacking unexposed crystals it is usual to add:

4 *Restrainer.* This may be potassium bromide or an organic antifoggant such as benzotriazole, or both.

In proprietary developers there may also be other additives, the most common being wetting agent. Altogether, a photographic developer is quite a complicated brew, and thousands of papers have been written attempting to explain exactly what goes on in it when an emulsion is being developed. One thing is agreed: many of the known developing agents form soluble complexes with silver halides, and sulfite ions play a large part in this. In photography the silver halide crystals are more than 100 times as large as those in holographic emulsions, and very little is lost into the solution by this action. In holographic emulsions the crystals are so tiny that a small amount of sulfite, or even the presence of chloride ions (Cl^-) in the water used to make up the solution, can have a considerable effect on the developed grains. This is because the relative surface area of a crystal increases as its diameter decreases. If a holographic emulsion containing the same amount of silver halide as a photographic emulsion has a mean crystal diameter one-hundredth as large, the total silver halide surface area in the holographic emulsion will be one thousand times as great.

One important effect is associated with what is often called (not too accurately) *physical development.* This term was originally coined to describe a method of development involving a solution containing silver nitrate, which deposited metallic silver on the latent image specks; it could be used before or after fixing. As this process became obsolete, the term had its meaning adjusted so that it came to mean the action of a developer containing a high concentration of complexing substances which dissolved the smallest, unexposed silver halide crystals and deposited the reduced silver on the developing grains, which acted as seeding centers. Of these substances the most important is sodium sulfite; others include thiocyanate and thiosulfate. Developers with a strong physical action have their uses in holography, particularly in the making of transmission master holograms, where their tendency to produce smooth rather than whiskered silver grains improves the signal-to-noise ratio, and the scavenging of the unused silver halide means that fixation actually removes little silver. However, in reflection holography such developers may produce images that are not very bright, especially if an inappropriate bleach is used, and it is preferable to use a developer that contains nothing but the developing agent(s) and alkali. Some developing agents (in particular pyrogallol) are used in *tanning* developers; the oxidation products promote cross-linking in the molecules of the gelatin so that it resists shrinking even when material has been lost from the emulsion (see p. 435).

Thus, although some of the ingredients may be the same, holographic developers should not in general have the same formulae as photographic developers; and even when they do, the action of the developer may be quite different. In holographic developers containing no sulfite, it is important to exclude air, as otherwise the solution will quickly oxidize and become useless. The easiest way to do this is to float another tray the same size as the developing tray on the developer solution. This can be kept in position even while a film or plate is developing, though as developing times are short this is not usually necessary. A selection of developer formulae suitable for the different types of hologram is given in Appendix 10.

Bleaches

When a holographic emulsion is developed, all the crystals of silver halide which bear a latent image are converted into opaque grains of silver which replicate the pattern formed by the interference of the object and reference beams. If we can turn the silver into a transparent substance of high refractive index no light will be absorbed; if we get everything right, all the light will go to form the holographic image and we shall get a considerable improvement in diffraction efficiency. This will happen, as explained on pp. 46–7, if the transparent substance delays light waves passing through it by up to one half-wavelength. This is then equivalent to creating a second hologram with its fringes interleaved with those of the first. Silver bromide has all the characteristics required for this substance. In addition, silver bromide is the predominant form of silver halide present in holographic emulsions. There are various bleach techniques available for obtaining fringes in silver bromide:

1 Rehalogenation after fixation The emulsion is developed in a more or less traditional developer such as D-19, and fixed, removing the undeveloped silver halide. The developed fringes are then converted back into silver bromide by means of an *oxidizing agent* such as ferric nitrate with potassium bromide. This method has to some extent fallen out of favor recently, as some of the original silver halide is lost from the emulsion and is wasted. However, the technique gives very clean images in transmission master holograms; but as it is the largest grains that remain, while the smaller ones are removed, this technique is less suited to the resolution demands of reflection holograms.

2 Reversal bleach This is so called after an analogous process used in photography for the production of direct positive transparencies on negative film. In the *reversal bleach* technique the fringes are developed as in *1* above, but instead of fixing away the undeveloped silver halide it is the silver image that is removed. This takes two chemical stages, both in the same bleach bath. First, the silver image is oxidized using, say, potassium dichromate; this *oxidation* process results in a silver compound which is soluble in sulfuric acid or sodium hydrogen sulfate. The undeveloped silver bromide is already virtually fully oxidized and is almost completely immune to this action. As it occupies the space between the original fringes it is a 'positive', in the photographic sense (hence the 'reversal' tag), but it is a replica of them, and with a finer grain structure, as it consists of the crystals that were too small to record a latent image. It is thus particularly suitable for reflection holograms. However, as this process plainly removes a great deal of material from the emulsion, it is necessary when using reversal bleach baths to use a tanning developer such as pyrogallol. Even so, it is difficult to avoid some change in the reconstruction wavelength, and this method is therefore unsuitable for workers wishing to use such a hologram as a laser-illuminated master hologram for transfer purposes.

3 Physical Bleach The best method of obtaining a hologram which will reconstruct in the original laser light and at the same angle of incidence to the beam as that at which it was originally made, is to use a developer that has neither a physical nor a tanning action, and to use a *physical bleach* solution. The method was first proposed by Benton (1, 2), and his account of the mechanism still remains one of the best.

Most bleaches work to some extent in a physical manner. After development and fixation, a rehalogenation bleach appears to cause the enhancement of the fringe contrast by dissolving silver in the low-density areas and re-depositing silver bromide in high-density areas. However, if the emulsion is *not* fixed after development, the deposition of silver halide appears to take place in the reverse direction, ie onto the undeveloped silver bromide. It is interesting to note that this happens only in aqueous solutions of bleach; if a dry bleach such as bromine vapor or a solution of bromine in methanol is used, no fringes at all can be detected with an electron microscope. Water thus plays an important part in physical bleaching. If a non-physical developer is used without fixation, followed by an aqueous rehalogenating bleach, we obtain strong positive fringes in the spaces between the original negative fringes (the terms 'negative' and 'positive' being used in the photographic sense). Plainly it is desirable to use a process that does not lose valuable fringe-forming material from the emulsion; so this principle forms the basis for the third method.

After a straightforward development in a solution having neither physical nor tanning effects, a bleach is used that absorbs silver bromide to the unchanged silver bromide crystals that remain in the emulsion. This results in a high-resolution image that is free from scatter. If this method is used in conjunction with pre-swelling techniques, very bright transfer images in yellow, green and even blue are possible. This is the basis for one of the pseudocolor techniques for reflection holograms.

It should be noted that Ilford materials show approximately 10% shrinkage of the emulsion when processed in this manner, and if they are to be used for reflection master holograms they should be processed somewhat differently (see Appendix 10). It should also be noted that Ilford materials show *reciprocity failure* at long exposures, and should not be used for exposures of more than a few seconds.

4 Total bleach If the emulsion can be made sufficiently rigid, the silver remaining in it after fixation can be removed by reversal bleaching. This technique is known as 'total bleach' or, somewhat incongruously, as 'silver-halide gelatin'. The fringe patterns now consist only of gelatin and air spaces. The technique has been successfully used in some laboratories, though workers with whom the author is acquainted have had little success. The optical principles of this method are shared with *dichromated-gelatin* holograms, which are dealt with in Chapter 19.

Colloidal-silver Processes

When fine-grain photographic emulsions are developed in solutions containing large quantities of thiocyanates or thiosulfates, much of the liberated silver forms soluble complexes which are eventually deposited in the emulsion in the form of *colloidal silver*. This substance, familiar to older photographers as *dichroic fog*, can form fringes which are powerfully reflecting for green light. Exposure and processing need careful control, but the results can be exceedingly bright. Development and fixation are carried out as in photographic processing; there is no bleach. Details of typical colloidal-silver processes are given in Appendix 10.

Other Processes

Processes that do not employ silver chemistry are attractive because (a) they are comparatively cheap, (b) they can give relief images which can be used to make replicas by mechanical processes and (c) they are grainless. The universal disadvantage is their low sensitivity to light, in particular to red light.

1 Dichromated-gelatin (DCG) Dichromated-gelatin holograms use a layer of gelatin sensitized to light by a dichromate, which catalyzes cross-linking in the presence of light. DCG holograms are usually of the Denisyuk type, and have extremely high diffraction efficiency. The method of production is discussed in full on pp. 274–7.

2 Photoresists These are substances which become either insoluble (negative) or soluble (positive) in certain common solvents after exposure to light. Along with photopolymers (see below) they are used as the precursors of stampers for producing embossed holograms.

3 Photopolymers These also have potential in embossing technology, and are the subject of active research. The fringes are formed by a change in refractive index of the exposed regions when processed, and they can (unlike photo resists) be used for reflection holograms. Images produced by holograms on photopolymer material can show a very high signal-to-noise ratio (300:1). They are the most light-sensitive of the non-silver materials, and can be sensitized to red light.

4 Photothermoplastic materials These are made for special applications in industrial research. The base material carries a conductive substrate, upon which is coated a material which becomes electrically conductive on exposure to light. On this is a layer of transparent thermoplastic material. The photoconductive layer is charged to a high potential, and this leaks away when exposed to bright fringes, leaving the fringe pattern as a variation of electrical charge. A brief electric current through the conductive coating heats the thermoplastic material, which deforms under the influence of the electric charge to replicate the fringes as a raised pattern. After examination, with or without photographic recording, the material can be regenerated by the application of a powerful electric current, and the cycle repeated.

References

1 Benton S A (1974) Intra-emulsion diffusion-transfer processing of volume dielectric holograms. *Journal of the Optical Society of America*, **64**, 1393A.
2 Benton S A (1980) Production of volume dielectric holograms. United States Patent No 4217405.

PART
II

PRACTICAL DISPLAY HOLOGRAPHY

CHAPTER

7

Making Your First Hologram

The White Rabbit put on his spectacles. 'Where shall I begin, please your Majesty?' he asked.
'Begin at the beginning,' the King said, very gravely, 'and go on till you come to the end: then stop.'
                LEWIS CARROLL, *Alice in Wonderland*

Making holograms is not in the least difficult. You can produce successful images with very little equipment and even less experience. You need a laser, of course, and that is the most expensive item. However, if you have any connection with an educational establishment, you will very likely be able to obtain access to a 0.5 or 1 mW HeNe laser, and this will produce perfectly good holograms up to 4×5 in size.

The simplest type of hologram to make is the Denisyuk or single-beam type. This set-up is capable of producing good results even in the absence of isolation from vibration. What is more, you can replay a Denisyuk hologram using any small white-light source, even a desk lamp.

Although it has been stressed that holography and photography belong to completely different worlds, they do have one thing in common: you can spend almost any amount of money on equipment. For your first holograms, just as for your first photographs, you would be well advised to keep to the simplest and cheapest equipment. Then, as you gain experience, you can upgrade it a little at a time. Some of the essentials such as darkroom equipment are the same as you would use for black-and-white photography; you can buy them from any photographic supplier. Other items are a little more unusual. A list of basic requirements is given in the Box.

Equipment and materials for making a Denisyuk hologram

Optical set-up
    Laser, 0.5–5 mW HeNe (see text)
    Concave lens, focal length approximately -6 mm; or $\times 40$ microscope objective
    Paving slab, 600 mm square
    Scooter inner tube

4 mm float glass, 150 mm square, or two pieces of anti-reflection glass, for film holder
Blu-tack (see footnote on p. 82) or hot-glue gun

Materials and processing
4 × 5 in film suitable for reflection holograms (see text)
Glass plate, at least 125 × 150 mm in size
Sink with running water
Four 4 × 5 in processing trays
Scales for weighing chemicals
Darkroom *safelight* (see text)
Minutes and seconds timer
Darkroom thermometer
Squeegee (the rubber-bladed type sold for window cleaning)
Processing chemicals and bottles (see text)

The Laser

As discussed in Chapter 3, several kinds of laser are suitable for holography, but the one that is used universally by amateur holographers, and, indeed, by many professionals (at least until they get their hands on an ion laser), is the HeNe laser. The very small lasers used for demonstrations in school physics departments, and in well-heeled conference halls as pointers, are perfectly satisfactory for making small single-beam holograms up to 4 × 5 in size, provided you are prepared to put up with rather long exposure times. If you are not too sure whether you want to take up holography seriously, you would be well advised to see if you can obtain the use of one of these for your initial experiments. However, if you do intend buying your own laser, you should choose one that is linearly polarized and as large as you are eventually going to need. The best value for money is in the 5–10 mW range; 5 mW is the lowest power suitable for serious work. If you are thinking about working regularly at 8 × 10 in, with the occasional larger hologram, 10 mW is the lower limit and 15 mW would be better. The coherence length of a laser beam goes down as the length of the optical cavity (and the power) goes up; by 25 mW the coherence length may be as little as 10 cm as against 30 cm or more for a 1 mW tube.

Laser Safety

With the increasing popularity of laser shows, people have become familiar with the sight of powerful laser beams, and you may be disappointed to find that the 5 mW laser you have just bought produces a beam that resembles that of a pocket torch rather than a searchlight. Don't be fooled. The energy of a laser beam is very concentrated: your laser beam is intrinsically brighter than the sun. So don't under any circumstances stare into the unspread beam even of a 1 mW laser. Just to catch a stray reflection off a metal surface can be an eye-watering experience that leaves an after-image lasting for several minutes. Watch out for any visitors, too. Make sure nobody can walk into the room and look into the beam either accidentally or deliberately, and ensure that nobody can operate the laser when you are out of the room. The power supply for even a small laser operates at several thousand volts, so never operate a laser with the cover off either the tube or the power supply. There are both international and national regulations for the operation of lasers; they are rather more stringent in the United States than in the United Kingdom. The main recommendations governing laser safety as they apply to holographers are given in Appendix 1.

The Need for Stability

When you start making holograms on your own, you will get one or two failures (everybody does). The most common problem is that the image has a large black patch on it, or it may disappear altogether at certain angles, or appear as if seen through a narrow slit. In all these cases either the object or the film has moved during the exposure. It doesn't have to move far: any disturbance of the optical paths of more than one-tenth of a wavelength (about 6×10^{-8} m) will seriously affect the visibility of the image, and a movement of just one half-wavelength (3×10^{-7} m) will wipe it out altogether. The best way to ensure complete stability is to place all the optical components on a heavy 300 mm (2 ft) square paving slab sitting on a fairly slackly-inflated scooter inner tube (Fig 7.1).

The Beam Expander

The purpose of the *beam expander* (beam spreader) is to increase the diameter of the laser beam so that it fills the film area. Using a concave lens of focal length

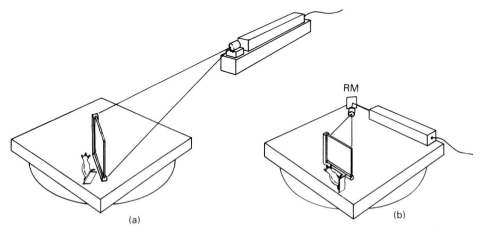

FIG 7.1 *To ensure stability for a single-beam (Denisyuk) hologram you only need a paving slab sitting on a slackly-inflated inner tube* (a). *If the laser is installed on the slab itself you will need a relay mirror* RM (b).

−6 mm to expand the laser beam, a throw of about a meter will produce a beam that will cover a 4 × 5 in film. A ×40 microscope objective will produce the same effect, and if used as part of a spatial filtering system (see pp. 99–100) is capable of producing a beam that is free from whorls and striations. However, a simple concave lens will produce quite satisfactory holograms. Affix the lens to the output port of the laser with Blu-tack*. It is not necessary to put the laser itself on the slab: it is only the optical paths after the beamsplitter (in this case the film itself) that have to be stable.

Film Support

You can use either film or plates. The latter are simpler to use, but are much more expensive than film. If you use film you will need to support it between two pieces of glass, preferably of the anti-reflection type. Hinge the two pieces of glass together, using masking tape. Along the opposite edge to the hinge, and inside the sandwich, lay a second strip of masking tape, and a third strip along one of the adjacent edges. These are to act as guides when you are loading the film (Fig 7.2).

If you use glass plates which do not have anti-reflection coating, your holograms will have fine patterns of lines all over them. These are known as *Newton's fringes*. They may not seem to matter as they do not obscure the image, but as you become more critical you will become less tolerant of artifacts of this type. Anti-reflection glass is one answer, but a better one is the use of *index-matching* fluid. You can find out more about the use of this on pp. 106–7.

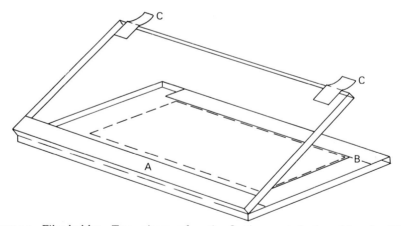

FIG 7.2 *Film-holder. Two pieces of anti-reflection-coated glass hinged with masking tape outside and inside (A). An L of masking tape (B) on the upper side of the lower plate locates the film. Flaps of masking tape (C) hold the sandwich shut.*

---

* Blu-tack (Superstuff in the United States) is like modeling clay, but is much more adhesive. Originally produced for affixing posters to walls, it has found a permanent place in optics labs for quick experimental set-ups.

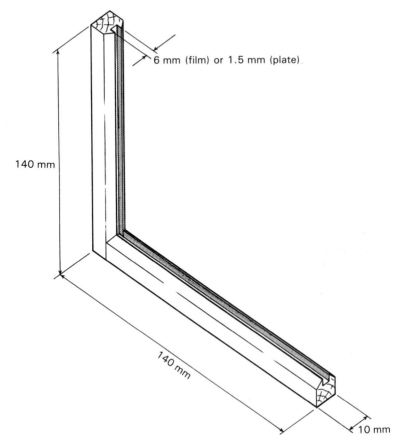

FIG 7.3 *L-shaped frame for holding plates. Made from wood or metal, sprayed black. The 45° chamfer is to eliminate shadow of the frame on the plate.*

The simplest kind of support for the glass is an L-shaped frame with two sawcuts into which the glass fits. You can make this from wood, or, if you have access to metal-working facilities, from square-section brass, sprayed matt black (Fig 7.3). At a pinch you can hold the plate in position simply by pushing it into a large lump of modeling clay (Fig 7.6).

You will also need some sort of support for your subject matter. A few wooden or metal blocks of different sizes are invaluable, especially when they are backed up by a supply of Blu-tack and cellulose tape. If you need to fix components more firmly, use a hot glue gun. Spray all your components matt black, including the paving slab.

Film

You need to use the type of film that is suitable for reflection holograms (see Table on p. 69). The availability of materials from the various manufacturers

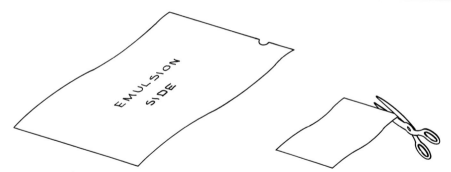

FIG 7.4 *When the film is oriented (in 'portrait' format) with the notch in the top right-hand corner, the emulsion is towards you. Nick the same corner with scissors when you cut up films.*

differs somewhat from one country to another. Films are more expensive than amateur black-and-white films; they cost about the same as professional color film. Plates are about three times as expensive as film. Of course, if you want to make test strips, film is a good deal easier than plates to cut up in the dark; but plates can be fitted directly into the plate-holder without bothering about anti-reflection glass. In order to avoid tedious repetition, the word 'film' will be used throughout this section, even though you may have opted for plates (in the later sections dealing with more complex set-ups, the opposite will be assumed).

However, with both films and plates, it is important to be able to identify the emulsion side. For cut films there is a universal identification code, a notch near one of the corners of each sheet of film. If you hold the film in 'portrait' format, then if this notch is at the upper right-hand corner, the emulsion will be facing you (Fig 7.4). When you cut up film for test strips or to make small holograms, always keep the cut pieces in the box emulsion side down. In this way not only will you be sure about which side is the emulsion side in the absence of a notch, but you will avoid touching the emulsion with your fingers (something you must never do with unexposed emulsion). An added safeguard is to make a nick in the top right-hand corner of all the smaller pieces of cut film with scissors.

To identify the emulsion side of a plate, place the extreme corner of the plate between your lips; the emulsion side will feel slightly sticky. Be careful when you do this, as the edges of the plates are sometimes very sharp. Plates were at one time packed back to back, but are now packed with edge separators, with all the plates facing the same way, so once you have identified the emulsion side of one plate you can mark the container for future reference. Keep any cut pieces facing the same way.

Safelight

Holographic emulsions are not very sensitive to ordinary light, and some workers do not bother about a safelight at all; they simply allow enough stray light to enter the lab to enable them to see their way around. This is reasonable as long as you are producing only simple holograms, but once you begin more complicated set-ups which require multiple exposures and pre-exposure baths you will certainly need a safelight. There is at present no satisfactory safelight material available for red-sensitive holographic emulsions, though the Wratten Series 2 red (orthochromatic) safelight filter is suitable for green-sensitive emulsions designed for use with Ar^+ lasers. When you begin more serious work you will find that you need to examine your films very close to the safelight, and you should test any light you use as described on pp. 105–6. However, provided you do not hold the film less than a meter or so away, a Wratten OB (printing) safelight filter to which you have added a cyan (blue-green) stage gel, which you can obtain from theatrical suppliers, will do, as will a Wratten Type 3 (panchromatic) safelight filter used with a 60 W lamp bulb. It is useful to have a safelight even for straightforward work, as you can use it to judge the density of a developing film by holding it up and looking at the safelight through it; you will quite quickly become capable of assessing the level of darkening.

Polarization of the laser beam

As explained on pp. 33–4, laser light is linearly polarized, ie the light waves oscillate in one plane only along the direction of propagation. If you are going to minimize internal reflections in the film-holder glass (an important factor in diffraction efficiency) you should have the plane of polarization such that both the incident and reflected rays lie in it. This is called p-polarization. The simplest way of checking the plane of polarization is to use Polaroid sunglasses. Place one of the lenses in the beam and rotate the sunglasses until the beam is cut off. If this occurs when the glasses are horizontal (ie the way they would be if you were wearing them) the axis of polarization is horizontal too. In the set-up described in the text you need the plane of polarization to be horizontal, and to achieve this you may need to turn the laser on its side. If the laser is randomly polarized, follow the direction of the axis of polarization by rotating the axis of your sunglasses over a period of a minute or so, and set up the laser midway between the two extremes.

Choosing a Suitable Subject

An important factor affecting the brightness of the final image is the beam ratio (see pp. 47–8). With multi-beam set-ups it is possible to control the beam ratio by means of a variable beamsplitter, but for the Denisyuk configuration we are using here this is not possible. We simply have to ensure that the object is as bright as possible, and as close as possible to the emulsion. Metallic objects and white objects make the best subject matter. It is also a good idea to use a light background such as a piece of metal or acrylic sheet sprayed with matt white paint. This should also be as near as possible to the emulsion.

Layout of Components

Set up the laser approximately 1 m from the center of the paving slab and switch it on. Place the plate-holder at the center of the slab at about $45°$ to the beam, with

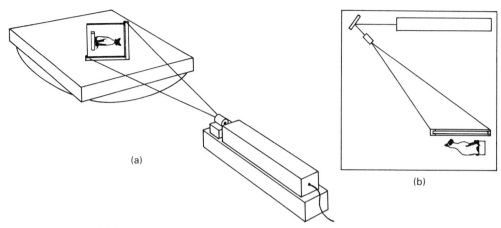

FIG 7.5 (a) *Laser set up at same height as the paving slab, beam expanded with concave lens or microscope objective to cover film area. The film is set at 45° to beam. The object is mounted on its side, top-lit.* (b) *Alternative set-up with the laser on the table.*

the vertical part nearer to the laser to prevent light from entering the edge of the glass plate. Put a piece of card the same size as your glass plate into the holder, and align the laser beam so that the spot of light is in the center of the card. Now fix your concave lens to the output port of the laser, using Blu-tack. If you are using a ×40 microscope objective it will need additional support. The arrangement is shown in Fig 7.5.

FIG 7.6 *Set-up for making a Denisyuk hologram on a paving slab, using a relay mirror and a microscope objective. The laser is a 1 mW HeNe laser as used in school physics departments for teaching purposes.*

You will probably see a number of dark swirls and patches on the card. These are caused by specks of dust and optical inadequacies in the beam. You can get rid of some of them by cleaning the lens with tissue. If any bad ones remain stubbornly, rotate the lens until you have minimized them. As mentioned earlier, on pp. 99–104, there is a set of instructions for using a device called a spatial filter, which eliminates this problem.

If you are short of bench space you can place the laser on the paving slab (Fig 7.6) and use a small piece of front-surface mirror to relay the beam diagonally across the slab. Now set up the subject matter on its side, on the side of the plate-holder opposite to the laser beam and as close as possible to the holder, oriented so that it is illuminated from 'above'. (Note that whenever terms such as 'above', 'left side' etc, appear in quotation marks, they refer to beams as seen from the point of view of the *displayed* hologram.)

Final Preparations

Now set out the processing trays. If you are unfamiliar with black-and-white processing in photography, you should refer back to Chapter 6, which explains the principles. Now, although it is possible to obtain reflection holograms using conventional photographic processing techniques, it is not easy to get bright images by this method. A better method is to follow conventional development by a reversal bleach. This is a robust technique which will produce reasonably good images with quite large errors in exposure or development time. Again, if this is unfamiliar ground you are referred back to Chapter 6. You can use any proprietary general-purpose developer, but if you are buying your processing chemicals especially for holography, a good developer to use is Kodak LX-24. This is an X-ray developer, and at the time of writing is available only in large quantities. Its predecessor, DX-80, is still around, and is available in one-litre containers. Either of these should be used diluted 1 : 1. The best developer for reversal-bleach processing is pyro-metol (see Appendix 10), but this has to be mixed immediately before use and requires a higher processing temperature (23–28°C). You will have to make up the dichromate bleach yourself (see Appendix 10). Both the bleach and developing solutions (except pyro-metol) can be poured back into the bottle after use, and re-used more or less indefinitely. You will also need a tray for washing the processed hologram, a squeegee and a glass plate for removing the surface water from the emulsion and, if you are impatient to see the results, a portable hair-dryer.

Loading the Film

Remove the white card from the holder, put a black card over the laser output port and switch off the main light. If you are using plates, place the plate in the slot in the holder and squeeze in two small pieces of Blu-tack at the outer edges to hold it firmly. If you are using film, open the glass sandwich, clean it carefully with tissue, and position the film, emulsion side down, against the masking tape guide strips. Close the glass sandwich and place it in the film-holder slot with the hinge outermost, using two pieces of Blu-tack as described above. The emulsion should be facing the subject.

Exposure

After loading the film, wait for at least 3 min before making an exposure. Use the black card as a shutter. The exposure time will be around 12–14 s for a 0.5–1 mW laser, and about 4–8 s for a 5 mW laser, depending on how wide the beam is spread. These suggested exposure times are very tentative, as the powers of lasers as given are often as much as 50% inaccurate. To avoid unnecessary waste of expensive film, it is a good idea to sacrifice just one piece of film in making a test strip. This involves exposing part of the film for 8 s, part for 24 s and the remainder for 72 s (for a 1 mW laser). Hold the black card blocking off the beam, then uncover the film completely while you count 8 s, then cover one-third of it while you continue counting to 24 s, then cover a further one-third while you complete the count to 72 s, then block off the light completely and remove the film. It does not matter if you don't hold the card absolutely steady during the exposure. Using a wax pencil, mark on the back of the film the area that received the longest exposure, or nick the corner of the film. If you do not have a seconds timer you can time seconds fairly accurately by saying 'One little second…two little seconds…' and so on; or you can listen to the ticking of the timer if the hands are not luminous. It is unimportant whether the 'seconds' are accurate as long as they are consistent.

Processing

Have the developing and bleach baths ready, with enough solution to cover the film (at least 100 ml for a 4 × 5 in tray). If possible, the solutions should be at approximately 20°C (warmer if possible for pyro-metol developer). If the solutions are colder than this the processing times will be longer. Place the wash tray in the sink, preferably with a drop or two of photographic wetting agent or ordinary washing-up liquid to promote uniform wetting.

1. Immerse the film in the wash tray, emulsion up. Agitate the film vigorously, and brush your fingers gently over the emulsion to dislodge any air bubbles. This agitation is particularly important if you have used index-matching fluid, as it is oily, and if not removed completely it can interfere with the action of the developer. After a minute or so lift the film out of the tray by one corner, rinse it under the cold-water tap, and examine it by the light of the safelight to make sure that there are no greasy streaks on the emulsion.
2. Start the timer and immerse the film in the developer, emulsion up. Do this in one clean sweep. Rock the tray during development constantly in all directions. If you are using pyro-metol developer the development time will be short ($1\frac{1}{2}$ min at

20°C to under 1 min at 25°C). With other developers the time of development will need to be longer, up to 3 min. The exact duration of development is governed by the darkness of the emulsion: the final density should be about 2.0 (quite dark but not totally opaque). You can judge this by briefly taking the film out of the developer and looking at the safelight through it. After a while you will become expert at judging density, but to begin with you will need to become familiar with the visual appearance of a density of 2.0. A good method for acquiring this experience is given in the Box on pp. 108–9.

If you have made a triple-exposure test strip as suggested above, one of the exposures will probably be close to the desired density after the appropriate development time. If the film is overall dark or overall light, never mind; continue the processing, and in the light of your judgment of the final image you can make a fresh test strip giving a new range of exposures. By the way, if you have not been able to fix up a safelight at this stage, you can get by if you leave the door open a crack, and examine the hologram by the light coming through it (the light will not affect the hologram at this stage).

3 When development is complete rinse the film in water from the tap, then immerse the film in the bleach solution, emulsion up. You can now turn the lights on. Rock the tray until the film has been completely clear for about a minute.

4 Wash the film for at least 5 min in running water. When you first put the film into the wash, brush the emulsion gently all over with your fingers as you did in the pre-wash. This is because the bleach bath sometimes leaves a deposit on the emulsion surface. Empty and re-fill the tray with water several times during the washing period.

5 Give a final rinse in de-ionized water to which you have added a few drops of acetic acid. This is not essential, but helps in the prevention of *printout* on subsequent exposure to bright light.

6 Lay the film, emulsion up, on a sheet of glass and remove all the surface water with a squeegee. Take care not to leave any streaks on the surface, as these will give rise to discolored marks. Hang the film up and allow it to dry. If you are in a hurry to see the result, leave the film for about 2 min, then complete the drying with a portable hair-dryer. By the way, this somewhat cavalier treatment is not recommended for the high-quality master holograms you will be making later (see Chapter 12).

If you want a sneak preview before drying the film, you can hold the just-bleached hologram up to a small-source lamp and look through it as though it were a transmission hologram. You should see a spurious transmission image in a bluish color. The fact that you can see this image (which has parallax) does not of itself guarantee that there will be a bright reflection image, but it is encouraging. If there is no spurious transmission image, you can be fairly certain that there will be no reflection image either, so at least you have been forewarned.

Assessing the Test

As the surface emulsion begins to dry and the hologram catches the light, you will see the holographic image begin to appear, dark red at first. Enjoy this moment; you

FIG 7.7 *The appearance of a processed test strip with the correct exposure given to the center strip. As a rule the difference in brightness in the images will not be as marked as this: actual exposures were* 4 s, 20 s *and* 200 s *(right to left).*

will never again experience it in quite the same way. The image will strengthen steadily as the successive layers of the emulsion dry out. When the emulsion appears to be completely dry, take your hologram and examine the image under a spotlight, with the emulsion down and the top of the hologram towards the light. Vary the angle of incidence until you have found the position that gives you the brightest image. If you have made a triple test exposure as suggested above, you may find that the image varies in hue between the exposed sections; you will also find that one of the strips gives a brighter image than the other two. If this is the middle strip, you need experiment no further; you have found the correct exposure. If the brightest image is the upper or lower exposure, make a further test strip, making three more exposures, starting at the upper or lower end of the previous three as the case may be. Fig 7.7 shows a typical test strip.

Once you have found the optimum exposure, make two more holograms with the same exposure. Develop them together, back to back, turning them over every 20 s or so. Take one out of the developer at $3\frac{1}{2}$ min and the other at $5\frac{1}{2}$ min. Compare the finished holograms with those from the previous test; the brightest image represents the optimum combination of exposure and development time. As long as you stick to the same combination of configuration, exposure time and development time, you can be sure of good results.

Changing the Image Color

The process described above tends to give a greenish-yellow image. If you have been using an ordinary amateur film developer you can get a stronger image by using the pyro-metol developer described in Appendix 10; this solution oxidizes very rapidly, and you need to protect it from the atmosphere by floating another processing tray on its surface all the time you are not actually developing a film. It also stains anything it touches, particularly fingers. You can get film tongs with sponge-covered ends from photographic dealers, and these will save your fingers from staining. As mentioned above, the development time for pyro-metol is much shorter than for ordinary developers, and it works better at higher temperatures: 75–90 s at 25°C (77°F) is about right. You will probably have to adjust the exposure downwards. Shorter exposures and lower final densities give red or orange images; longer exposures and higher final densities give yellow or green images. Chapter 18 explains how you can control the color of the final image within narrow limits by controlled swelling of the emulsion before exposure.

The Pseudoscopic Image

As we saw earlier (pp. 43–4) there is a real image associated with every virtual image, and in the case of a single-beam reflection hologram you can find it by flipping the hologram (ie turning it over about a horizontal axis) under the replay light. The image will now be inverted, but you can correct this by turning the hologram through 180° in its own plane (we have called this action 'rotating' it) and moving the replay light so that the hologram is illuminated from 'below'. You will then see the real image standing out in front of the hologram (Fig 7.8), and it exhibits the weird phenomenon of pseudoscopy: the parts of the image that appear to be at the back throw shadows on the parts that appear to be at the front, and if you move to one side the image appears to turn so that you see more of the other side. The illusion of movement is heightened by closing one eye; with images of human faces the impression is very powerful.

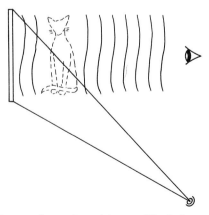

FIG 7.8 *Viewing the pseudoscopic real image. The hologram is spun (ie rotated about a vertical axis); the replay beam must now come from below. If the hologram is merely flipped in the original beam the image will be inverted.*

An 'Orthoscopic' Real Image

We saw earlier (pp. 49–51) that one way of converting a pseudoscopic real image into an orthoscopic real image is to use it as the object for a second hologram, which is again flipped for viewing; this technique is discussed in Chapter 13. However, there is a simpler way of producing a real image that has all the appearance of being orthoscopic. If you have ever looked at the inside of a mask of a face you may have noticed a strange illusion: it does not look hollow but convex, just like a real face; moreover, the illusion exhibits all the strange behavior of the pseudoscopic image we have just been discussing. It is, as it were, a pseudoscopic object. If you own a small porcelain or terracotta head or bust, or a set of Scottish chessmen with kings and queens and little warrior pawns, you have the ideal model. Get a package of self-curing mould material (obtainable from large educational stationers, model shops and dental suppliers), and paint the object with the material according to the instructions. Make sure that there is a really good build-up of mould material, so that the final mould is resilient; then, when it is thoroughly cured, strip it off and cut it cleanly into front and back halves. Keep the back half: you can use it later to make a separate hologram of the back of the object, and you can mount the two holograms back to back to give a 360° hologram (see p. 135). Spray the inside of the mould with aluminum spray paint, and paint the cut edges matt black. You may find

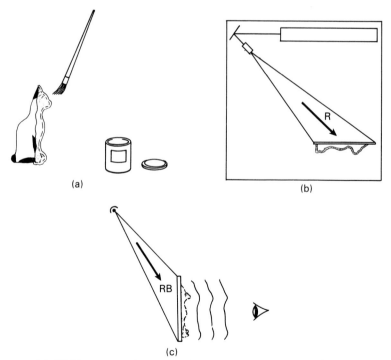

FIG 7.9 (a) *Making a pseudoscopic object from moulding material.* (b) *Table geometry: illumination/reference beam* (R) *from 'below'.* (c) *Viewing the real image: replay beam* (RB) *from above.*

MAKING YOUR FIRST HOLOGRAM

FIG 7.10 *One of the author's earliest holograms (left). This little Scottish chessman is actually half of a hollow mould (right); when the image is viewed flipped the perspective appears normal and the image is in front of the film plane.*

that the mould opens out a little when you spray it, but this is not important. If you are using dental-type moulding material, pour it into a shallow rectangular tub of the kind used for some brands of margarine, and push the model into the soft dough to slightly less than the halfway mark. When the material has set, remove the model and clean up the edges of the mould before spraying the inside and painting the edges. If you feel that a metallic finish is inappropriate, you can use glossy or matt white spray instead. Set up the pseudoscopic object in contact with the emulsion side of the film for exposing; alternatively, attach it to the glass with a few spots of adhesive and locate the film on the other side with the emulsion facing the object. Set up the film-holder and object so that the illumination comes from 'below'. This is necessary because the hologram will be displayed flipped; the replay beam will then come from above, and the lighting will look natural (Figs 7.9 and 7.10).

Displaying Your Holograms

Reflection holograms look best when they have been sprayed on the back with black paint. Don't use cheap cellulose spray that smells of peardrops or acetone, as it will almost certainly contain solvents which will attack the fringe structure and cause the image to appear unsharp. Reliable brands are Krylon and Spectra, which use toluene as a solvent. Gloss and matt finish are equally satisfactory. It is also possible to blacken the back of the hologram chemically; a method is given in Appendix 10. Mount film holograms behind glass or acrylic sheet or in a photo-frame; plates need only a clip frame. Hang the hologram on the wall with a spotlight shining on it from about 45° elevation. Try out the angles first with a friend holding the light; it may need to be moved around a little in order to get the brightest result. You will sometimes get a sharper image if you use a gel of the same color as the image in front of the light. You can use any decorative spotlight as illuminant, but if you are buying one especially for the purpose, look at several models. Hold your hand between the lamp and the wall, and see how sharp the shadow is. The one that gives the sharpest shadow is the one to buy. At the time of writing the 'best buy' is the Wotan Minispot.

CHAPTER

8

---- ✸ ----

Single-beam Techniques

'The method employed I would gladly explain
While I have it so clear in my head.'
LEWIS CARROLL, *The Hunting of the Snark*

Chapter 7 showed how to make a simple Denisyuk hologram. This chapter shows how to make quite sophisticated types of image with single-beam configurations using a frame made from the kind of slotted strip sold for building shelving and workbenches. One of the difficulties associated with the Denisyuk configuration is that of obtaining sufficient intensity in the object beam, and we shall look at some simple ways of increasing the illumination on the subject matter.

Optical Configurations

In the method described in Chapter 7 the optical set-up is on its side. Although this is the simplest method it does have the disadvantage that the subject matter is off balance, and unless it is mounted securely it may move and wipe out the holographic fringes. If the glass plate holding the film is tilted forward rather than sideways, and held in position by a sloping frame, the subject matter can be rested against it and prevented from sliding off by a plinth (Fig 8.1).

This arrangement is fine for solid objects, but does not work well for holograms of jewelry or collections of objects, as they simply slide down. One solution is for the glass plate and film to be held horizontally over the subject matter. This means that the reference beam must come from above (Fig 8.2).

This is also a good arrangement for subject matter consisting of highly-reflective material which must not be allowed to make contact with the emulsion

FIG 8.1 *A more stable configuration than mounting the object sideways. The plate (or film-holder) and object are supported on a frame tilted forward at* 45°.

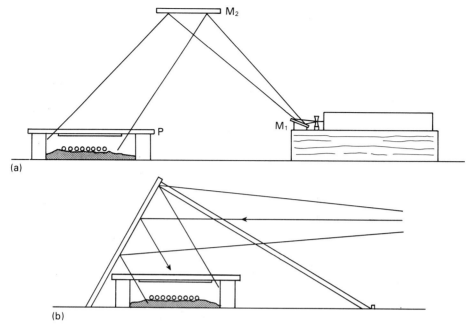

FIG 8.2 *A more stable set-up for jewelry, etc. In* (a) M_1 *and* M_2 *are front-surface mirrors. The film is mounted emulsion side down with index-matching fluid, beneath the glass plate* P. *Only this part of the set-up requires isolation.* (b) *shows a simpler arrangement using a single large front-surface mirror.*

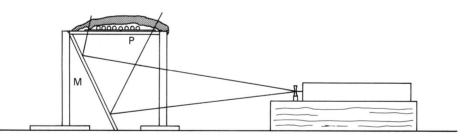

FIG 8.3 *An alternative set-up with the beam from below. The mirror* M *is at 22.5° to the vertical. The film is mounted emulsion side up on the glass plate* P, *and the subject matter rests on it, with the background material lying on top.*

because of the risk of *burn-out* (gross over-exposure which destroys the fringes locally) (see p. 126). Another solution is to have the object lying on the film horizontally, with the beam coming from below (Fig 8.3).

Construction of a Frame System

It is possible to produce good holograms with any of these configurations using crude stands and holders, if you are prepared to put up with long settling times and a proportion of wipe-outs. Indeed, blocks and Blu-tack (the holographer's equivalent of string and sealing wax) are the best way to go about an experimental set-up. But once you are satisfied with the arrangement you need something more permanent, something that will not move under any circumstances. In the case of single-beam holograms one excellent solution to the problem of steadiness is a rigid frame that holds all the components. The frame described below is capable of producing all types of single-beam holograms, both one-step and transfer.

The frame is made from heavy-duty L-section slotted strip ('Dexion'), held together by square-shouldered bolts specially designed for the material. You can get it from builders' suppliers and large ironmongery (hardware) stores; the square shanks of the bolts prevent them from rotating in the slots when they are being tightened. A good basic measurement for the frame dimensions is 300 mm (1 ft) square; this will give you plenty of room for an 8×10 in hologram in 'landscape' or 'portrait' format. You will need a minimum of eight pieces 300 mm long, and four pieces 320 mm long. The supplier will probably have a guillotine designed for this material, and will cut it to length for you for a small extra charge; this will save you a good deal of labor with a hacksaw. You will also need a minimum of 32 bolts and washers, 12 wing-nuts and 20 plain hexagonal nuts. While you are about it, however, it is worthwhile investing in another four 300 mm pieces and another 12 bolts, washers and wing-nuts; these will enable you to add extensions for various purposes. You will also need a minimum of two pieces of 4 mm *float glass* 275×300 mm (11×12 in) in size, which you can get from any glazier; ask the glazier to stone the edges, to make them safe to handle in the dark. The smaller dimension is to enable the glass plate to be inserted from above when there are two separate supports for glasses (eg in making transfer holograms with the frame). Again, three or four such pieces will increase the versatility of the set-up. Finally, you will need a piece of front-surface mirror 300×324 mm (12×12.75 in) in size; you can obtain this from one of the addresses listed in Appendix 12.

The basic construction of the frame is shown in Fig 8.4; the assembled frame appears in Fig 8.5. Note carefully which parts of the frame go inside others, or you may not be able to get all the slots lined up. When you have completed the assembly, but before you tighten the nuts, place the framework on your paving slab and adjust it until it stands firm and level. Tighten the lowest horizontal frame, making sure nothing goes out of alignment. Now slide one of the glass plates onto the middle frame, and adjust it until the glass does not rock at all. Tighten the wing-nuts, and repeat the operation for the uppermost frame. Now insert the front-surface mirror, holding it with handkerchief tissues so that you do not touch the surface (wipe off any accidental fingermarks immediately with a tissue), fit the $33°$ stops and adjust these to the correct angle with a protractor. Now move the mirror back as far as it will go and do the same for the $22.5°$ stops. Mark the positions of both stops so that if you need to take them out to modify the frame you can put them back in the same places. Remove the mirror and glass, and spray the entire frame matt black. (You may prefer to spray the components before assembling them, but take care not to overdo the spray on the bolts, or you may find it difficult to get the nuts onto them.) Once the frame is complete, refit the mirror, and slide the glass plate into the middle frame, with its smaller dimension at right angles to what will be the laser beam axis (this is why you need slots cut in the vertical framework). The frame

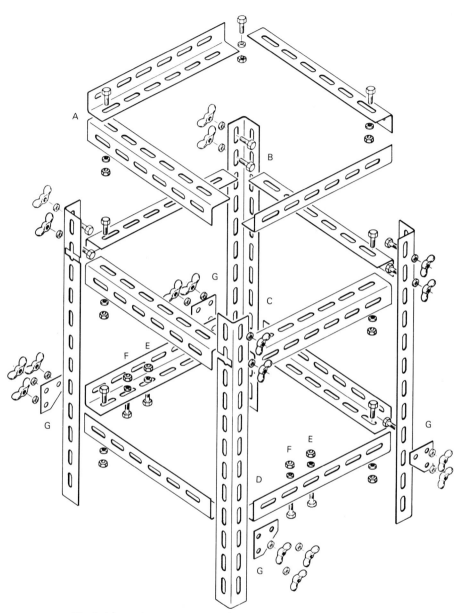

FIG 8.4 *Single-beam frame. The horizontal pieces are 300 mm long and are bolted together first, hand tight, with the nuts on the underside. They are then bolted to the 120 mm legs with the nuts outside. The frame is lined up with a set square and the nuts tightened, a little at a time. The upper frame A is used only for transfers and is removed when not required. Concealed corners B, C and D are pinned as at A. Locations at E, E and F, F are mirror stops for 33° and 22.5° inclinations respectively (nuts uppermost). Four gusset pieces G hold the frame square while adjustments are being made to the upper horizontal pieces.*

SINGLE-BEAM TECHNIQUES 99

FIG 8.5 *The assembled frame. When set up for transfers, as here, two longitudinal pieces can be used to support the lower glass plate and two lateral pieces for the upper glass plate.*

is now ready to use for making reflection and transmission masters and image-plane transfer holograms without any further modification.

Spatial Filtering

In Chapter 7 it was noted that there is a way of getting rid of dark swirls in the illumination patch of the expanded laser beam. The method is called *spatial filtering*. The principles underlying the method are given in the Box.

Spatial filtering

You have already learnt (pp. 8–10) that a diffracted beam of light contains information about whatever it was that caused the diffraction, coded in terms of the spatial frequencies that describe the object. When a beam of laser light is focused by a lens, the electric field at the principal focal plane is the *optical transform* of the field of the object, and it forms a pattern at the focal plane in which the highest spatial frequencies (representing the finest detail) form spots farthest away from the optical center of the pattern. The swirls in an unfiltered beam are caused by the interference of parts of the beam that have been diffracted by dust and optical imperfections in the laser and the microscope objective. At the focus of the objective, and only there, these diffracted beams have separated into discrete spots, the spot at the optical center being the optical transform of the undiffracted beam, narrowed down to a 'waist', the diameter of which can be calculated. If you position a pinhole of the right size at the focal point, only the 'clean' beam will be transmitted. The diameter of the pinhole is important, as if it is too large some of the 'rubbish' will get through and spoil the uniformity of the beam, and if it is too small not all the clean beam will be transmitted. The minimum permissible size of pinhole is given in terms of the wavelength by the approximate formula

$$D = \frac{0.6\lambda}{Md}$$

where D is the required pinhole diameter in micrometers, M is the magnification of the objective, λ is the wavelength of the laser light in nanometers and d is the diameter of the beam in millimeters. Where focal length is given instead of magnification, the magnification can be found from the approximate formula

$$M = \frac{250}{f}$$

where f is the focal length in millimeters. For various magnifications used in holography, the figures for a beam diameter of 1.5 mm are as follows:

Objective magnification	10	20	40	60	100
Focal length (mm)	25	12.5	6.3	4.2	2.5
Minimum pinhole diameter (μm)	25	13	7	5	3
Size commonly used (μm)	50	20–25	10–15	7–10	5

NOTE: These figures are for a HeNe laser with a beam diameter of 1–1.5 mm. For an Ar^+ laser, λ is smaller and d is larger, and the pinhole diameter should be reduced by 20–30%. This also applies to HeNe lasers with outputs greater than about 20 mW, which have larger beam diameters and thus smaller focal 'waists' than lasers of lower power.

A useful money-saving tip is to buy pinholes of non-standard sizes (eg 9 or 11 μm rather than 10 μm). These are often as little as one-third of the usual price.

The figures for minimum pinhole size have been rounded up to whole numbers. Actual pinhole sizes used tend to be 50–100% larger than the calculated minimum; this does not affect their effectiveness and makes alignment easier.

Even when you use a high-quality spatial filter made by a precision optics manufacturer, lining it up can be quite tricky, and if you have to learn to do this with a homemade spatial filter you may be in for a good deal of frustration and waste of valuable time. When I made my first holograms in the early 1970s the cheapest spatial filter available cost more than £500 without objective, pinhole or mount, more even than the laser! I made my own pinhole from a milktop and a specially-sharpened needle point, mounted it on a washer and fixed it to the objective with a ring of modeling clay. The first time I tried, I got it aligned in three minutes. The second time it took

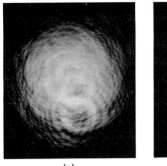  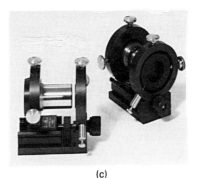

(a) (b) (c)

FIG 8.6 (a) *A 'dirty' expanded laser beam;* (b) *a beam cleaned up by a spatial filter;* (c) *a spatial filter designed by the National Physical Laboratory and manufactured by Scie-Mechs plc.*

four hours. The third time I gave up, and thereafter made my holograms with a dirty beam. Since then, one or two quite good designs for do-it-yourself spatial filters have been published, eg McNair (1) and Brooks (2). However, such contrivances, however ingenious, are no substitute for a properly-engineered spatial filter holder (Fig 8.6), and as prices of these have now come down considerably, the decision to acquire one demands much less heart-searching. Suppliers are listed in Appendix 12.

Aligning a spatial filter A spatial filter holder usually has x, y, z *movements* (lateral, vertical and axial respectively) on one component holder and x, y movements only on the other; it does not usually matter which one you pick for either component, though for single-beam work you will find it easier if the z-adjustment screw is on the side away from the laser.

Start by raising the laser to a level where the horizontal beam strikes the mirror about halfway up and goes centrally through the horizontal glass plate. Failing a proper optical bench you can use almost anything rigid to support it, such as a wooden box. A better support is bricks; you will probably need three at each end of the laser. Make sure they do not rock (Fig 8.7).

Each time you set up a spatial filter you need to go through the routine given below. When printed out in full it looks very tedious, but in fact it is quite quick once you have had a little practice.

1 Set up the spatial filter holder, without its pinhole or objective, so that the beam passes as closely as possible through the center of the two rings. You can check the alignment by putting a piece of card behind each of the rings in turn. If the holder is not horizontal, put *shims* under the base until it is as near horizontal as you can judge.
2 Fit the microscope objective. A $\times 40$ objective will produce a 300 mm diameter beam at a distance of about 1.2 m. If you have room to spare, use a $\times 20$ objective at about 2.4 m. Adjust the x and y settings of the objective until the disk of light covers the frame aperture. You can often judge the centering better by looking at the reflection of the beam on the ceiling. Examine the patches of light reflected

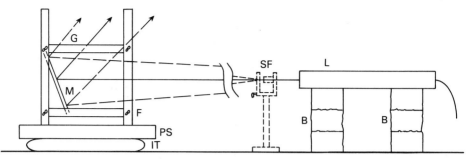

FIG 8.7 *Initial setting up of a single-beam frame. The frame F sits on the paving slab PS, which rests on the inner tube IT. The laser L is mounted on bricks BB so that the beam strikes the mirror M and is reflected through the center of the glass plate G which will carry the film. After this alignment the spatial filter SF is inserted into the beam.*

back by the lens surfaces of the objective towards the laser output port and adjust the holder until they are all concentric with the port. The microscope objective is now correctly aligned. If you are using a triangular optical bench and saddle (see Box) to support the spatial filter you need not fix the bench in position as it is heavy enough to remain in position even if accidentally knocked; but if your base is less massive, it is a good idea to fix it in position with a squirt of hot glue.

Care of pinholes When you are not actually using it, always keep your spatial filter covered with a plastic bag to keep out the dust. Never put an unmounted pinhole down on a bench top, no matter how clean you may think the surface is, and on no account allow any fluid to make contact with the pinhole. If the diameter of the pinhole is correct for your laser and microscope objective (see Table of pinhole sizes and setting-up routine above), but does not allow a clean beam through when it is in optimum alignment, then the pinhole is probably dirty. This is not a common occurrence, but it can happen. To check this, set up the pinhole in the unexpanded laser beam, and examine the *Airy diffraction pattern* (you will have to darken the room to see this properly). The central disk, and the rings around it, should be circular and uniform. If they are not, do not use the pinhole until it has been cleaned. The only satisfactory way of cleaning a pinhole is by ultrasonic cleaning. If you do not have access to such a facility you can try a cotton bud moistened with acetone (flammable!), carbon tetrachloride (toxic!) or audiotape head cleaner (safe but less powerful). Use a gentle rotary action, a fresh bud for each side, and finish with a dry bud. But don't be too optimistic about the result.

By the way, if you get your pinholes mixed up and need to check their diameters, you can also do this by placing them in the unexpanded laser beam. The diameter of the *Airy disk* produced in each case is inversely proportional to the diameter of the pinhole, ie the bigger the disk the smaller the pinhole. The exact formula is $d = 2.44 x/D$, where d is the pinhole diameter, λ is the wavelength of the laser light, x is the distance of the pinhole from the screen and D is the diameter of the Airy disk. (All measurements are in meters.)

> **Triangular benches**
>
> Sooner or later in practical holography you are going to find a need for what is called a triangular (optical) bench. This is a heavy steel bar, in cross-section an equilateral triangle with a deep channel milled along both upper sides. It supports mounts called saddles, which hold vertical pins on which are mounted optical components. Triangular benches can be obtained in various lengths, the most useful of which is 300 mm; but if you decide to mount your laser and spatial filter on the same bench, you will need a 450 or 600 mm one. Triangular benches are deliberately made heavy so that a casual knock will not disturb them.

3 With the pinhole mount well clear of the objective, fit the pinhole and adjust the z-axis control until the pinhole is 2–3 mm clear of the objective. Adjust the x- and y-controls until the pinhole is approximately centered. Now bring your eye down to the level of the pinhole and observe it from a distance of 20 cm or so. You should see a weakish red point of light at the pinhole. Keep looking at the pinhole, and move the x-control to see if you can make the point of light any brighter. Then do the same with the y-control. Repeat this a few times, getting the point of light brighter each time. Then, quite suddenly, you will see a bright red flash. This is what you have been seeking. (It is not dangerous, as the beam is very much attenuated.)

4 Now you can begin the alignment proper. Switch off the lights and pick up a piece of white card. Hold it about 10 cm away from the pinhole. You will see a small disk of red light on the card, surrounded by faint rings (the Airy diffraction pattern).

5 Adjust the x- and y-controls, observing the white card, until the Airy disk is at maximum brightness. Now cautiously operate the z-control to bring the pinhole nearer to the objective. The spot will get larger and brighter; then at some point it will begin to wander off-center and become fainter again.

6 Adjust the x- and y-controls until the disk is once again centralized and bright, and the rings round it even and symmetrical.

7 Repeat actions 5 and 6 in turn until the patch of light suddenly becomes very much brighter, large and uniform, and the rings disappear. At the exact focal point the patch does not move to one side or the other when the x-control is adjusted, but spreads out horizontally and disappears, behaving similarly in a vertical direction for the y-control.

NOTE: If adjustment of the x- or y-control results in movement of the patch in the same direction, you have not yet reached the principal focus of the objective. If adjustment of the x- or y-controls results in movement of the patch in the opposite direction, you have passed it.

8 Check that the disk of light is still accurately centered on the frame. If it is slightly off-center you can re-align it by moving the laser. Do this very carefully; usually a movement of less than a millimeter will be enough. Provided you aligned the objective accurately in the first place the patch will not disappear, but you may

need to make a small *xy* adjustment to bring it back to full intensity. In general you need to move only the front end of the laser and if it is free-standing you can lever it with a pencil. Check the patch of light on the ceiling while you do this, and ensure that it is centered on the shadow of the frame.

If no amount of adjustment produces a completely even patch of light, clean the laser mirror and/or the glass surfaces of the microscope objective with a lens tissue wrapped round a matchstick (be very gentle when you do this). If the patch of light is still uneven, replace the pinhole with one of smaller diameter. If, on the other hand, the patch of light is even, but is less bright than it should be and surrounded by rings even at the point of best focus, the pinhole is too small.

Making an Electrically-operated Shutter

Some workers do not bother about a shutter, but simply use a black card over the laser output port. However, for large open-table set-ups it is preferable to make the exposure by remote control; making one is worthwhile even for single-beam arrangements. All you need is an old DC voltmeter, 0–10 V or thereabouts, a 6 V battery of the type that has press-fastener terminals, a connector to fit it, some flex and a switch. Remove the back and the scale from the voltmeter and glue a disk of

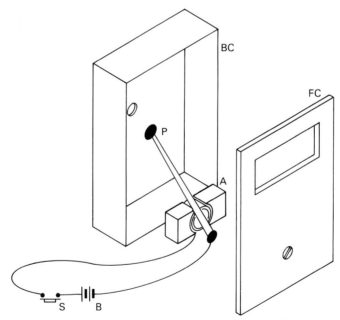

FIG 8.8 *An electromagnetic shutter made from an old voltmeter. The glass is removed from the front cover FC, and the back cover has a hole drilled to align with the zero position of the pointer P, which bears a disk of black card. When the armature A is energized via the battery B and push switch S, the pointer moves out of the way of the light beam.*

black card about 7 mm in diameter to the needle in the plane of its movement. Counterbalance this weight by a spot of Blu-tack at the other end of the needle. (Fig 8.8).

Drill a 5 mm diameter hole in the back of the meter casing in such a position that the black disk covers it when no voltage is applied. Connect the voltmeter in series with the battery and switch, and fix it on a block so that the laser beam passes centrally through the hole. When you operate the switch the disk swings out of the way, allowing the laser beam to pass through. The device should be free-standing, so that you can move it out of the way when you are setting up the table.

If you want your exposures to be timed more accurately than is possible by counting seconds you can make up a more sophisticated circuit using an electronic timer of the type sold for photographic enlargers. Instead of a battery you will need a low-voltage DC source of the type sold for operating portable radios from the mains. This plugs into the timer output and operates the voltmeter; you operate the shutter from the timer remote control. Don't be tempted to use a mains voltmeter as a shutter; even if you are a qualified electrician, it is much safer to work with low voltages with any apparatus you are going to handle in darkness.

Safelights

The use of a safelight was mentioned on pp. 84–5; as you come to make more complicated types of hologram you will find an increasing need for a reliable safelight that is bright enough to allow you to inspect glass and emulsion surfaces without fogging the film. You can buy safelight lamps from photographic suppliers: the most useful size takes 8×10 in safelight filter glasses.

Safelights

It was suggested earlier that a Wratten OB printing safelight filter with a cyan stage gel over it would make a satisfactory safelight. However, if you are going to do serious work you will need to examine the emulsion surface, both unexposed and exposed, quite close to the safelight, so it is important that it should be really safe. A combination of gels that has been found satisfactory is two thicknesses of cyan gel plus two thicknesses of straw (deep yellow), with a 40 W lamp. An alternative is a double thickness of Cinemoid no 39 (tricolor green). It is not possible to be completely specific, as there is considerable variation between the products of different manufacturers, and even a single make is often not consistent from one batch to the next. When you go to the theatrical supplier to buy your gels, take a diffraction grating with you (if you don't possess one, you can easily make one holographically: see pp. 230–1). Test the gels by looking at a bright small light source, preferably a filament lamp, holding the gels in front of it and examining the spectrum through the diffraction grating. With the right combination of gels you will see only the green part of the spectrum, no red and no blue. For a final test, back in the lab, take a piece of 4×5 in holographic film, put it about 60 cm from the safelight and place four coins on it. Cover up the section with the first coin after $2\frac{1}{2}$ min, the next after

5 min, the next after $7\frac{1}{2}$ min, and remove the film completely after 10 min. Develop it normally, and fix it; do not bleach. Place the processed film on a white surface and examine it for outlines of the coins. If any are visible you will have an idea of the 'safe' time of exposure at that distance. At other distances the safe time will be in proportion to the square of the distance. Thus if the safe time at 60 cm was 5 min, then at 120 cm it will be 20 min and at 180 cm it will be 45 min. If there is visible fogging at $2\frac{1}{2}$ min, the safelight will not be adequate for its purpose. You could add more gels, checking the effect with your diffraction grating. Alternatively, you can reduce the wattage of your bulb to 25 W.

When you come to make pseudocolor reflection holograms or pulse-laser portraits, you will need a brighter safelight that is nevertheless effective. Such a safelight is described on p. 239.

Index-matching

If you used a sandwich of ordinary glass for your first holograms, you will have noticed a closely-spaced pattern of lines much like the contour lines of an Ordnance Survey map (Newton's fringes), which is caused by interference of light waves reflected between the various surfaces. This does not happen with anti-reflection glasses, but these have the disadvantage that they disturb the polarization of the beams, weakening the image. Newton's fringes can be totally suppressed if all the spaces between the film and the glass plates are filled with fluid of a suitable refractive index. The technique is known as index-matching, mentioned earlier on p. 82. There are a number of fluids that fit the bill to a first approximation*: those that have most often been used in the past include glycerol, liquid paraffin (mineral oil), carbon tetrachloride, xylene and trichloroethylene; the last is probably the best. All of these have serious practical disadvantages, and a cheaper and safer alternative is ordinary white spirit (turps substitute) which is sold for thinning gloss paint. In the United States this is sold under the name of 'Thin-X'. If you use this you need only a single sheet of glass, as the surface tension of the liquid holds the film in close contact with the glass, and even if left overnight it will not peel away or shift.

Take the sheet of glass and clean it thoroughly. The slightest speck of dirt between the film and the glass will cause a black disk or doughnut in the plane of the hologram on replay. The most common cause of this, by the way, is dandruff – so either wash your hair before undertaking a holography session or wear protective headdress; or at least avoid bending over the glass. Examine the glass by reflected light to ensure it is completely free from foreign matter. Now take up 1–2 ml of the fluid with a dropper and deposit it in the center of the plate. Put out the main light; get out a film, and lay it emulsion side up on the glass, bowing it so that the center of the film touches the index-matching fluid first. Now, using a blade or roller squeegee, squeeze out all the surplus fluid, working from the center outwards. Follow this action with a careful wipe with a pad of tissues to remove all

---

* Ideally, this should be the geometric mean of the refractive indexes of the film and the glass. In practice it is difficult to find a liquid that is chemically and physically acceptable and has exactly the right refractive index.

traces of fluid from the surface of the emulsion. Check that there are no streaks by examining the reflection of the safelight on the emulsion. Now slide the glass plate with the film uppermost into the middle tray and lay the subject matter on it. Allow at least three minutes settling time.

Exposure and Preparation for Processing

Make a test exposure as described on p. 88, using a black card. With Agfa-Gevaert 8E75HD film and a 15 mW laser, the beam covering a 300 mm square, the exposure should be about 8 s. Other powers of laser will require proportional changes of exposure time. The same emulsion on Agfa-Gevaert plates may demand up to three times the exposure required by film. Ilford materials are similar (but see p. 74). During the exposure keep as far away as you can from the frame so that you do not induce any warm air currents round it, and keep quiet. At the end of the exposure close the shutter (or put the black card over the laser port again) and remove the subject matter. Lift up a corner of the film with your fingernail, remove it from the glass and put the film in a lightproof box such as an old film box.

If your processing facilities are in a room separate from the lab, you can use the settling-down period for getting the processing solutions ready. You will need processing trays for developer, bleach, and final acid rinse and a larger tray for washing, as well as a dish to use as a floating lid for the developer. For some processes you will need a tray for a fixing bath. Label all the trays to be used for solutions on the outside with a waterproof pen, so that they do not get used for anything else. This is very important in holography, much more so than in amateur photography. If you intend using different types of bleach you need one tray for rehalogenating bleaches and another for reversal bleaches.

The developer and bleach formulae differ, depend on whether you want a red (master) hologram or a yellow or green (final) hologram. These formulae are given in Appendix 10. The ascorbic acid and DX-80 (or LX-24) developers keep fairly well (the proprietary developers can be poured back into the bottle after use and used repeatedly), but the pyro-metol developer lasts for only 10–15 min unless you use a floating lid, which may extend its life to several hours (see p. 72). The bleach baths can all be used repeatedly until the bleach time becomes inconveniently long, provided they do not become contaminated with developer from inadequate rinsing of the film between baths. You will also need some wetting agent. Kodak Photo-Flo is the best-known of these, and is often recommended in American texts; but if it gets into the developer it can cause fogging, and if it is used in the final wash it appears to be a factor contributing to print-out. Ilford 'Ilfotol' wetting agent, however, seems to have a clean bill of health. Ordinary washing-up fluid for washing dishes is as good as anything marketed for the photographer, and is a good deal cheaper.

Processing

Process the film according to the instructions given on pp. 88–9. For the highest-quality results it is important to give the optimum exposure and processing. The suggested density of 2.0 is a guide only; the correct density for a given combination of emulsion and processing chemistry is best found by trial and error. Two methods of assessing density which are accurate enough for this purpose are given in the Box.

Estimating density

If you measure the transmittance of a piece of grey film, then take the reciprocal of this figure and then take the logarithm of the result, you will have the measurement of its *photographic density*. This sounds pretty complicated, though it is easy enough if you have a pocket calculator. *Densitometers*, which are devices used in photographic sensitometry for such measurements, read directly in densities. The concept of photographic density may seem far removed from reality, but we do perceive things in a logarithmic way, and tend to rate subjects in terms of darkness rather than lightness, so in fact 'density' is a fairly close match to what we perceive. Nevertheless, in order to appreciate what a density of, say, 2.0 looks like, one has first to look at a piece of film that has a density of 2.0 in order to be able to recall its appearance. There are two ways of obtaining a series of calibrated densities:

1 Buy a set of Kodak ND (neutral-density) filters. These form a progression; 0.3, 0.6, 0.9, etc up to 1.5. By putting them together you can obtain higher densities by simple addition: thus two ND filters of value respectively 0.3 and 0.9 when superimposed give a density of 1.2. The important densities in holography are 0.7 (for transmission masters that are not bleached); 1.5 (for transmission holograms that are to be bleached); and 2.0, 2.5, 3.0 (for various processes for reflection holograms). Using combinations of ND filters you can get densities of 0.6, 1.5, 2.1, 2.4 and 3.0, which are near enough to work from. You can cut the filters into quarters and stick them with cellulose tape to your safelight along the edge, to use as a comparison with the film you are developing.

2 Buy a Kodak Step Tablet No 3. This is a staircase of equal density increments from 0 to about 3 in steps of about 0.15. Cut it in half lengthwise and place one strip over the other. All the marked densities will be doubled, so that it now runs in steps of about 0.3 from 0 to about 6. Again, you can tape this to your safelight to act as a comparison; this strip will allow you to match your density much more accurately. You can get step tablets through any Kodak agent, though you will probably have to order them specially.

You can measure density with an ordinary exposure meter, too, as nearly all meter scales are basically logarithmic. The way you do it is to aim the meter at a white surface illuminated brightly enough to give the meter a reading near the top of its scale. Note the reading; then, without moving the meter, interpose the piece of film you wish to measure. Count the number of blocks on the scale that the meter needle falls. Each whole block represents a density increment of 0.3, and each one-third of a block represents a density increment of 0.1; so if the reading falls from, say, 9 to 6, ie three blocks, the density of the film is 0.9. This also works for parts of a block: if the reading falls from 9 to $3\frac{1}{3}$, ie a fall of $5\frac{2}{3}$ blocks, the density is 1.7. You simply multiply the number of blocks by 3 and divide by 10.

> If you have a camera with semi-automatic exposure control (the type where you have to rotate the exposure time or aperture setting to bring a needle into a central position), you can use the same method: each successive exposure setting (or f/no setting) represents a density increment of 0.3, in the same way.

Assisted Single-beam Configurations

There is probably no such thing as a true single-beam hologram, as it is almost impossible to avoid stray scattered illumination of the subject matter. In any case, such lighting is less than desirable, as the sensitizing dyes in the emulsion absorb light and affect the beam ratio adversely, and without stray light the shadows would be totally blank. The simplest way of feeding some light into the shadows is to have a matt white background. Matt white acrylic material is good; metal sprayed with matt white paint is even better, as it is more rigid. In the frame, the background can be supported in the upper tray, which should be situated for this purpose as close to the subject matter as possible. If you use a white background of this type you need to use a film that is large enough for the shadow of its edge to be well out of the way of the image area. In this connection it may be necessary to use the larger mirror angle of 33° rather than 22.5°.

In the simple frame system described above, a good deal of light can fall on the subject matter directly from the laser, and this can sometimes be used to enhance the level of intensity of the object beam. However, for some subjects lighting from 'below' can give undesirable effects, in which case you will need to card off this light (*carding-off* is simply blocking off unwanted laser light with a piece of black card or metal. When you begin using sophisticated set-ups you will need a great deal of black card.).

By fitting outriggers from the frame (Fig 8.9(a)) to support mirrors, you can catch light from the side that would otherwise be lost. You can use concave mirrors to produce spot highlights (combinations of plane mirrors and convex lenses will do the same job). The important thing is to prevent any of this light from falling directly onto the film. Jeff Blyth (3) has designed a 'sigma' configuration (Fig 8.9(c)) to be used with a frame designed by the graphics artist Ray Mumford. The 'sigma' is the Greek capital letter Σ, which represents the arrangement of the beams; in the system illustrated the reference beam comes from above. The subject matter needs to be distanced from the film so that the light beam from the lower mirror can be carded-off from the film while still illuminating the subject.

Single-beam Real-image Techniques

Using the frame with the light coming from 'below', you can make pseudoscopic object holograms very easily. You need to make a hollow mould of your subject matter first, as described on pp. 92–3, and to rest it on the film oriented so that the light strikes the object from 'below'. When you flip the completed hologram and illuminate it from above, the image will appear to be solid, standing out in front of the hologram. With the frame described above you can make real-image holograms of life-size masks. In order to avoid blank shadows in the deeper images you should set the mirror to the 33° setting. It is also a good idea to set the laser as far as possible from the frame, using a ×20 or even a ×10 objective to expand the beam. The reason for this is that for correct perspective the replay beam should

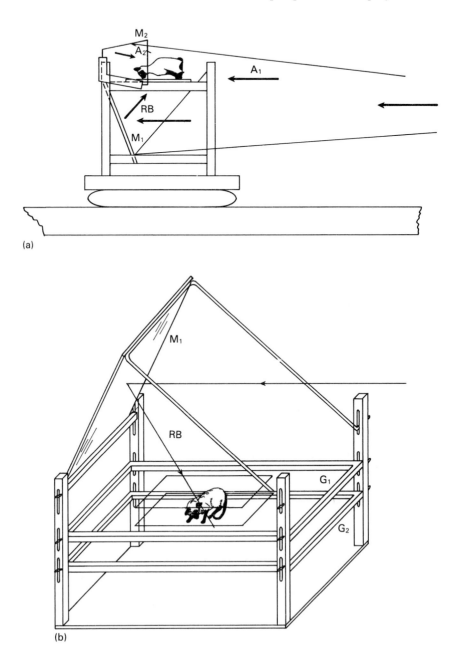

(a)

(b)

SINGLE-BEAM TECHNIQUES

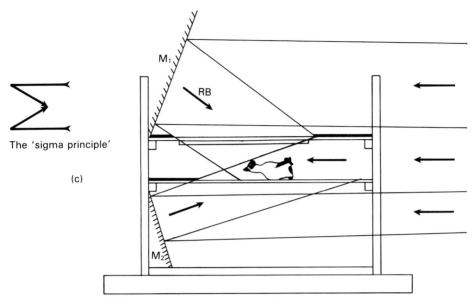

FIG 8.9 (a) *The overspill from the laser beam can be redirected to give extra illumination to the subject matter from below (A_1) and from the side via outrigger mirrors (A_2); only one mirror (M_2) is shown for clarity. The reference beam RB reaches the object via a mirror M_1 and provides Denisyuk-type illumination. (b) A modified frame designed by Ray Mumford. In this configuration the reference beam is from above via the mirror M_1. The film is affixed to the upper glass plate G_1 and the subject rests on the lower glass plate G_2. Slots are milled in all the uprights to allow adjustments. (c) Blyth's modification to the Mumford frame to give a 'sigma' illumination configuration. Other mirrors may be placed at the sides, but no light other than that from M_1 must reach the emulsion directly.*

be the conjugate of the reference beam; that is to say, if the reference beam is divergent, the replay beam should be convergent by the same amount. If both beams are divergent, the perspective and depth will be exaggerated, and faces will acquire grotesquely long noses (the reason for this is explained in Appendix 2). A long throw for both reference and replay beams reduces this effect.

Real-image holograms need not necessarily be made from subject matter consisting of hollow moulds. There are many objects which have inherently-ambiguous perspective; dental casts can be turned into dental impressions (see p. 91) and vice versa; spheres can become cups, and geodes can appear as crystalline globes. Once you have been bitten by the pseudoscopic-real-image bug you can begin to dream up your own surrealistic sets with perspective reversal combined with normal perspective in the same composite image. You can do this by multiple exposures in a single hologram, or by superimposing several holograms. The latter is simpler, and often gives brighter results; the main thing to remember is the order of assembly; the red image film should lie below the green or blue image film. This is because most processes leave a yellowish stain.

Multi-exposure Techniques

As previously explained, it is possible to have more than one image in a single hologram. One of the most impressive of these is a 360° hologram which shows an image of the front of the object from one side, and of the rear of the object from the other side. The technique for making such holograms is dealt with in Chapter 7.

The simplest multi-exposure technique is the real-image name-plate or logo. This comes out with the name or logo standing out in front of a background which can be an abstract design or a collection of objects. You can produce a real-image nameplate very simply, using the single-beam frame shown in Fig 8.7. Make up the name in white dry-transfer lettering on a piece of clear acrylic sheet at least 3 mm thick, and place it on two spacer pieces 6–8 mm thick (two pencils will do) on the glass above the film, so that when you look down at the letters oriented correctly for reading the beam is coming from 'above'. The film emulsion should be facing up. Make an exposure in the usual way. Now remove the nameplate and flip the film. For the second exposure, lay your background matter on the film, oriented so that the beam comes from 'above'. Make a second exposure then process the hologram in the usual way. You can use the same technique for a number of themes, but it is important that the real image should be brighter than the virtual image. Making sure that the emulsion is facing the real-image object helps, but you must choose a foreground subject that is as bright as possible, otherwise the background will show through.

There are many creative possibilities that do not involve deserting the single-beam format. Martin Richardson (4), who worked exclusively with such a system for several years, found it at first a constraint and later a challenge. The task is to choose (or construct) a model that will realize the image that is in the mind's eye. For example, we can prepare a grid of lines converging on a background in conventional draughtsman's perspective. A second element could be a white polyhedron. With the grid processed to a red image and the polyhedron to a green one, the latter mounted flipped and the former straight, the result will be a pseudoscopic polyhedron set against a grid which shows perspective but not parallax within itself, even though there is parallax relative to the polyhedron (Plate 7). This kind of perspective-juggling is unique to holography.

The Transfer Principle

As explained on p. 11, a hologram produces a reconstituted wavefront, and can record literally anything that can be seen, including the image produced by another hologram. We have already shown that if a hologram is flipped to produce a real image, this can be used as the object for a second or transfer hologram. The image can be situated behind, before, or across the plane of, the new hologram. Because this image is pseudoscopic, the transfer hologram has itself to be flipped for viewing, to make the image orthoscopic. If at the time of transfer the real image was between the two holograms, then on flipping it will be in front of the final hologram, ie it will be a real image, in correct perspective and standing out in front of the plate. Now, this may be very exciting – indeed, many exciting real images have been made

– but unless you have a very large master hologram there will not be much parallax (see Appendix 2). This is because you are looking at the image through the real image of the outline of the master, like a ghostly aperture in front of your face; and quite a small movement to one side or the other will bring the edge of the image of the master hologram sweeping across like a high-speed curtain, so that the image vanishes. In addition, deep real images in reflection holograms need very good spotlights – effectively point sources – and these may need to be filtered. Most holographers who work with simple set-ups produce image-plane holograms, in which the image is partly real and partly virtual. These images reconstruct well even under a poor light source, and can be given more parallax and depth than fully-real images. So, for the time being, we will settle for an image-plane hologram, made in the first instance from a reflection master hologram.

Making a reflection master hologram There is no essential difference in the making of a reflection master hologram from making any other single-beam reflection hologram, but it does have to give an image that is the same color as the original laser wavelength, and the image has to be of the highest possible quality. The subject matter should not be touching the film, to avoid the possibility of burn-out, and it is thus better to shoot reflection masters with an overhead beam. You can do this with the frame by simply turning it over (Fig 8.10). You need all the assistance you can get from side beams. Good single-beam masters need bright subject matter too.

You will need to process the film in one of the solutions given in Appendix 10 for reflection masters, otherwise the image will be the wrong color and will give only a weak image under laser illumination. You will also need to experiment with different exposures until you have found the one that gives the brightest image.

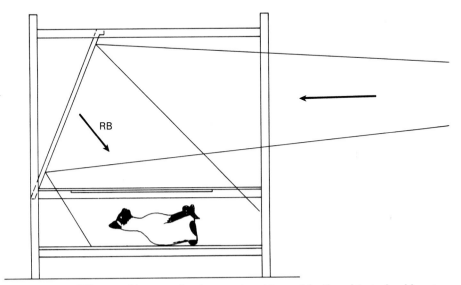

FIG 8.10 *When making a reflection master, if possible the object should not touch the emulsion. One way of achieving this is to invert the frame and shoot downwards.*

Making a reflection transfer hologram With the frame set up as in Fig 8.10 adjust the position of the upper plate of the two frames so that the lower surface of the upper glass plate (which will carry the transfer film) will intersect the real image. You will probably find you need to make final adjustments with the master hologram in position on the lower glass plate, flipped emulsion up, to produce the real image. The best plane for the transfer hologram is usually about one-third back from the front of the image. In adjusting the upper glass plate you need to remember that the image is back to front: the greater the separation of the plates, the more of the final image will be in front of the final hologram (Fig 8.11). (It may be more convenient to use the frame positioned as in Fig 8.7, in which case the master hologram is positioned emulsion down on the upper of the two frames).

One way of locating the image precisely is to stick a pin on the underside of the upper glass plate with a blob of Blu-tack, and watch the image as you move your head from side to side. The part that always remains stationary with respect to the pin is the part that is in the plane where the transfer hologram will be.

Now adjust the tilt of the mirror to give the brightest image from the master hologram. You may find that it needs to be at a slightly steeper angle of incidence, depending on your processing technique, but this is not important. When you are satisfied, affix the master hologram to the upper surface of the lower glass plate with index-matching fluid, taking all the usual precautions. In particular, be very careful not to scratch the emulsion when squeegeeing, as if you do, there will be an image of the scratch in the final hologram. Now position the upper glass plate on its frame support. Lay a piece of waste film, or a card cut to the exact size of the film, on the glass, and adjust its position until the shadow of the edges of this film or card coincides with the edges of the master hologram. Mark the position with a wax pencil on the upper surface.

NOTE: If the separation between the two glasses is more than a few centimeters this offset may be so large that you will need to use a larger piece of film in order to get the image central in the final hologram, and to trim away the upper part after processing (Fig 8.12).

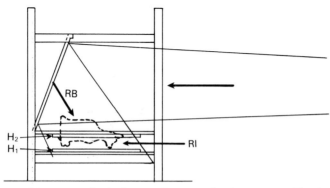

FIG 8.11 *Making a transfer hologram from a reflection master. The frames are spaced so that the real image* RI *produced by* H_1 *is cut by the plane of* H_2. *The two films should be emulsion side up.*

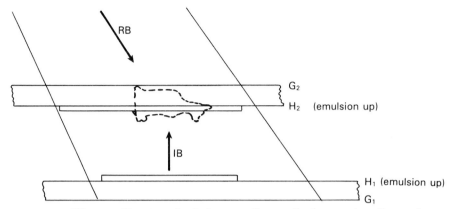

FIG 8.12 *When making transfer holograms from reflection masters it may be necessary to use a larger piece of film for H_2 to avoid a shadow of its edge falling on H_1. Notice that H_1 is on top of the glass plate G_1 and H_2 is underneath G_2.*

Now remove the upper glass plate, turn it over and affix the unexposed transfer film to it, emulsion side down, using the wax pencil marks as a guide. Remember that as these are on the opposite side you will have to offset the film slightly to allow for the obliquity of the beam. Cut off the laser beam and place the glass plate in position with the film underneath, so that both emulsions face the same way. Allow the system at least 3 min to settle down, and then make the exposure. The exposure time will be similar to that used for making the master hologram.

This time you can process the hologram in one of the solutions that give a yellow or green image. If you do this, the image will appear brighter, simply because you eyes are more sensitive to yellow and green than to red. As shorter wavelengths are diffracted less than longer ones, the parts of the image that are out of plane will be farther away from the hologram when it is reconstructed, giving a slight exaggeration of the perspective; this is not usually of any importance. The effect is aggravated if both reference and replay beams are strongly diverging, which is why it was recommended (p. 110) that if there is room you should use a $\times 20$ rather than a $\times 40$ objective as a beam expander. The reason for this is explained in Appendix 2. Films with a polyester base sometimes produce transfer images that show a number of black patches in the plane of the film, even though there has been no movement during the exposure. This is due to birefringence of the base material. For methods of dealing with this problem see p. 124.

An alternative method of obtaining a yellow or green image, or even a blue one, is to pre-swell the transfer-hologram emulsion with triethanolamine before processing in the solutions used for master reflection holograms; this often gives brighter, less noisy results. The technique is described in Appendix 10.

You can also use the pre-swelling technique to produce multiple-exposure images in different colors. For this you need to remove the transfer film between exposures to successive masters, pre-swelling by a calculated amount each time to give the desired color. The technique is described in full in Chapter 18. With the techniques described so far, it is not really feasible to produce more than two images on one film, and in general you will find you get better results by superposing several holograms in the same frame for viewing.

Transmission Master Holograms

You can use the same frame for making single-beam transmission master holograms, and transfer holograms of both full-aperture and rainbow type. To make a master hologram you need a reference beam from underneath, and two glasses; the lower glass bears the subject matter and the upper one the holographic film (Fig 8.13). Because of the strong side and back lighting, with little or no illumination from the front, the system works best with subject matter that is translucent (Plate 8).

You can use a faster emulsion such as Kodak SO-253 or Agfa-Gevaert 10E75 for transmission holograms, as the fringe structure is coarser than for reflection holograms. Process the film according to one of the methods given for transmission holograms in Appendix 10. You can view a transmission hologram by laser light, and this gives the best and sharpest image; but you can also view it by quasi-monochromatic light such as that emitted by a low-pressure sodium lamp or, indeed, by any point-source light fitted with a narrow-band filter. As with reflection holograms, the use of light of shorter wavelengths for replaying a laser transmission hologram results in some distortion of perspective.

Most of the early creative holograms were laser transmission holograms, and a few of these (see Plates 9 and 10) are striking enough to have survived the takeover by white-light-viewable images.

Making the transmission master hologram With Benton's invention of the white-light-viewable or 'rainbow' transmission hologram (pp. 50–1) the laser-viewable hologram rapidly lost its pre-eminence in creative and display holography. The main disadvantage of the rainbow hologram is that it has no vertical parallax,

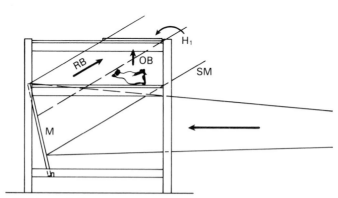

FIG 8.13 *Making a full-aperture transmission master* (H_1). *The subject matter SM needs to be far enough away from H_1 not to cast a shadow on it, and the mirror M needs to be at an angle of only about* $15°$ *to the beam for the same reason.*

as this has been eliminated and replaced by a holographically-generated diffraction grating. The so-called full-aperture transfer hologram (pp. 185–7) does have vertical parallax; it gives an almost achromatic but very shallow image. If you choose your subject matter carefully so that its total depth does not exceed about 25 mm, you can produce a compromise between these two techniques which we may call a wide-slit transmission transfer hologram.

The master hologram can be as wide as you like; indeed, the wider it is, the more parallax you get. It should be about 25 mm deep. You can cut a 25 mm wide strip from an 8 × 10 in film using a guillotine or a knife and a metal template, or you can use 35 mm holographic film, which is somewhat cheaper. The advantage of the narrow format is that you have a more extensive space for the subject matter than you would have with a full aperture master (Fig 8.14). The wide slit gives a better beam ratio than either the full aperture or a narrow slit.

When making a transfer you will probably find that a larger angle of incidence such as 60° is better than 45° for the reference/reconstruction beam, as it reduces the likelihood of the shadow of the edge of the master hologram falling across the final hologram. It is a good idea, therefore, to use this greater angle for the master.

Making a transmission transfer hologram The set-up for the transfer is much the same as for a reflection transfer except that the geometry is different: the master hologram is still below the transfer film, but this time the light comes from below (Fig 8.15).

Find the position of the master hologram and the upper glass plate so that the real image lies across the plane of the transfer hologram (you can use the pin method described earlier) and the shadow of the master hologram is well clear of the transfer hologram area. Affix the master hologram to the upper surface of the lower glass plate in its correct position using index-matching fluid, and mask off its upper and lower edges, or the 35 mm perforations, with black strips. The strip on the side nearer to the mirror should not be so wide as to prevent light from reaching the far

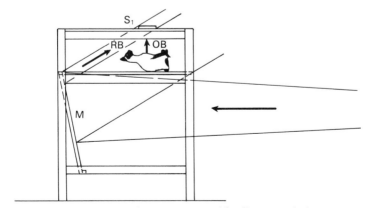

FIG 8.14 *Making a restricted-aperture or wide-slit transmission master (S_1), which can be up to 25 mm wide if the object is shallow.*

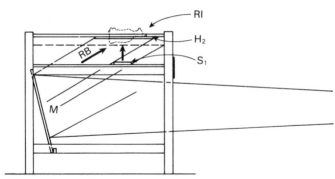

FIG 8.15 *A single-beam slit transmission transfer. The master slit S_1 gives rise to a real image in the plane of the transfer hologram H_2, and the upper part of the beam acts as the reference beam for H_2.*

edge of the transfer film. If there is not much room you can omit the masking, though if you do so it is sometimes possible to see the image of the edges of perforations when viewing the final image.

With wide-slit holograms the real image produced by the master hologram must be close to the plane of the final hologram at all points. You can achieve this without going to the trouble of checking parallax with a pin, if you simply lower the upper glass plate as used for the mastering by a little more than half the depth of the subject matter. The transfer hologram should have its emulsion facing in the same direction as that of the master hologram, and should be affixed to the underside of the glass plate as with reflection transfer holograms. The final image will be almost achromatic and somewhat brighter than that of a full-aperture transfer hologram. It should be viewable under almost any lighting conditions.

Deeper images in transmission transfer holograms In Chapter 13 a modification to the rainbow transfer technique is described, using a wide slit and a plano-convex cylindrical lens. You can use this method with the single-beam configuration too, and it is capable of giving bright results with deeper images. You can also get deep images by narrowing the slit to about 6 mm width using two opaque strips (metal rules do very well). However, you may find the beam intensity ratio so high that the image produced by the final hologram is unacceptably weak. The solution is to attenuate the lower part of the beam, ie the part forming the reference beam for the transfer hologram. The simplest way to do this is to insert a Kodak ND 0.9 neutral-density filter in the upper part of the laser beam at some point before it reaches the mirror.

When making a slit transmission master the subject matter should be at least 10 cm from the film. It is also advisable to use a ×40 objective to expand the laser beam rather than a ×20 objective, otherwise it may not be possible to see the entire image when the final hologram is viewed from a comfortable distance. The reason for this is explained on pp. 188–9.

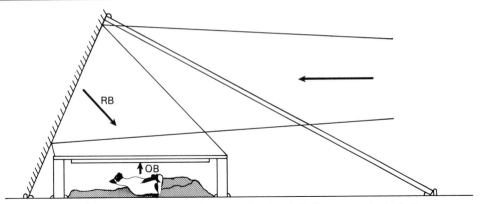

FIG 8.16 *A very simple set-up for large Denisyuk holograms. A single large front-surface mirror is supported by a long stick; the glass plate is laid on blocks. Everything is fixed to a large isolated slab with hot glue.*

Further Frame Techniques

It is a simple matter to make Denisyuk holograms on a frame of this type simply by laying the object on the film, but there are types of subject where the areas in contact with the film may reflect enough light to overload the emulsion and cause local burn-out of the fringes. In such cases you can usually get an acceptable hologram by separating the subject matter from the film by the thickness of the glass: you simply affix the film to the underside of the glass with the emulsion towards the glass and place the subject matter on top of the glass. An alternative, already suggested for reflection masters, is to turn the whole frame over so that the beam falls on the film and subject matter from above. You can then lay the subject matter on a suitable surface such as white-sprayed metal or other background material – even black velvet if you feel daring and the subject is very bright. You can, of course, then make the subject–film separation as large as you like. Indeed, it is possible to dispense with the frame altogether, for the occasional hologram that is larger than your frame. Using wooden or metal blocks, stays, a large enough glass plate and mirror, and plenty of hot glue, it is possible to make holograms up to 30×40 cm by this method (Fig 8.16 and Plate 7).

You can also make transfers of these large holograms, by simply building another story onto the glass plate with more blocks and a second plate, with black velvet at the bottom. You can even combine transfers and real objects in a single exposure.

Overhead Reference Beam

Strictly speaking, the reflection configurations described above are all overhead reference beam set-ups. However, this term is usually taken to mean configurations where the beam travels downwards from a considerable height. It is possible to shoot an object straight down a flight of steps (Fig 8.17), though such conditions

FIG 8.17 *Shooting a Denisyuk hologram down a flight of steps with an overhead reference beam. The subject matter is the original architect's model of the Albert Memorial. Photograph by Michael Langford.*

are uncommon. The more usual configuration is with the beam directed upwards towards a horizontal mirror, which then reflects it down at 45° or so into the hologram area. This has two advantages: it gives a long beam throw in a confined space; and it permits both horizontal and vertical arrangements. You can use a cast-iron base with a tall pillar and clamp, known as a burette stand, to hold the elevated mirror, or, for greater rigidity, a heavy-duty camera tripod or theodolite stand fitted with a suitable clamp at the top. It is just possible to operate this configuration on a 600 mm square paving slab (see Fig 8.18), but if you are able to get hold of a larger slab you will have more room to work in.

The overhead reference beam set-up lends itself particularly well to the introduction of supplementary subject lighting using one or more beamsplitters without

SINGLE BEAM TECHNIQUES

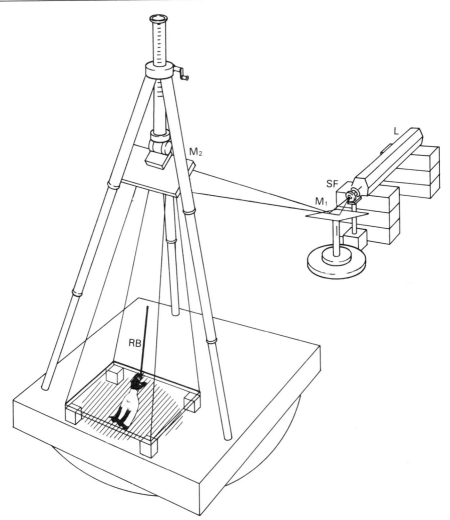

FIG 8.18 *Overhead reference beam configuration using a camera tripod. Light leaving the laser L passes through the spatial filter SF and is deflected at an upward angle by the relay mirror M_1 onto the large reference mirror M_2 (which may be a collimating mirror). This directs the reference beam downwards at an appropriate angle onto the film and subject matter. M_2 is mounted on a camera tripod with the central pillar inverted.*

disturbing the essential Denisyuk configuration. Such an arrangement is described in detail in Chapter 9.

Interferograms Using Frame Set-ups

You can make several different types of interferogram using the Denisyuk principle; the techniques are described in Chapter 22.

Copying Holograms

All single-beam holograms can be duplicated by contact copying. This process must not be confused with contact printing as applied to photography; it is simply a single-beam transfer holographic technique in which the master and the transfer holograms are so close together that there is effectively no change in the position of the image. It is important that a reflection master should reconstruct in the right color to produce a strong image when replayed by laser light; if the original image is orange or yellow rather than red it may be necessary to use an angle of incidence close to 0° or to swell the master hologram with triethanolamine or sorbitol (see Appendix 10) so that it replays in red. The master and transfer films should be emulsion to back with the laser beam going through the transfer film first in the case of a reflection master (but see p. 124). Both films should be affixed using index-matching fluid. You may find it easier to affix the films to opposite sides of the glass plate. Recently it has been found possible to make high-quality contact copies using a laser scanning device rather than a beam expander. Such devices are used in laser display equipment and can be programmed to produce a uniform rectangle of any required size. The exposure is thus completely uniform, and no light is wasted. A compromise between linear scanning (which can sometimes produce *raster* lines) and a stationary expanded beam, is a beam spread one-dimensionally, moved at a constant speed across the master hologram by a motor drive. This technique works well with Ilford materials which are unsuitable for long exposures.

Nick Phillips (5) has found that very bright copies can be made using a quasi-monochromatic light source based on filtered white light, provided contact between the emulsions is sufficiently intimate. Indeed, copies (of a kind) have been produced using unfiltered white light from a commercial spotlight. However, if you try this method, you are recommended to use a very good point source and a narrow-band (5 mm bandwidth) dielectric filter.

To copy transmission holograms, including rainbow holograms, the procedure is very similar. Set up the master hologram (ie the hologram to be copied) so that it replays the brightest possible image with a laser beam. This time the transfer hologram should be on the side farther from the laser, so that the beam goes through the master hologram first. To make a silver halide copy of an embossed hologram, illuminate the hologram with a laser and find the angle of incidence at which the image is brightest; as with a laser-illuminated rainbow hologram you will be able to see the image only at one particular angle.

Now place the film, emulsion side down, on the master hologram. Do not use index-matching fluid. Hold the film in place with an anti-reflection-coated glass plate, heavily weighted down, and make the exposure. This technique actually produces two images, one reflection and one transmission; the latter will be much stronger than the former, especially if you use an emulsion intended for transmission holograms.

Mounting and Displaying

1 Reflection holograms The general instructions given on pp. 93–4 apply. If you can't find a suitable spray paint to blacken the back of the hologram, don't just use any old spray; the solvents may damage the fringe structure (p. 127). Settle for one of the more expensive, slower-drying types. Blyth (6) has worked out a method of chemical blackening which needs a little skill but works very well; it is described in Appendix 10. Another alternative is to make the transfer with the emulsion side towards the master hologram. You will then be spraying the base side, and there will be no danger of damaging the fringe structure. You can protect the emulsion side by polishing it with a high-quality wax polish such as Simoniz (the non-abrasive type) or one of the colorless waxes sold for antique furniture; or you can mount the hologram behind glass. A good type of display frame, especially for images using more than one layer of film, is the sunk-mount type, which grips films firmly but holds them away from the glass. Other methods of display are described in Chapter 21.

2 Transmission holograms The only really satisfactory way to display a laser-viewable hologram is with a laser, and this usually means the construction of a darkened box (see p. 299). White-light transmission holograms are best sandwiched between two sheets of acrylic material and either framed with U-channel clips (the type used for holding press releases etc together) mitered at the ends, or drilled and fixed near the four corners with pop rivets or decorative screws and nuts.

Transmission holograms need to be illuminated from behind, and thus need their own alcove or recess; alternatively, you can mount them on a mirror (not necessarily a front-surface one) and illuminate them from the front. Benton has developed a *Fresnel prism mirror* for this purpose; it is described on pp. 295–8. Transmission holograms are much less easy to display than reflection holograms. The whole topic of holographic displays is dealt with in detail in Chapter 21.

Trouble-shooting

'By-the bye, what became of the baby?' said the Cat. 'I'd nearly forgotten to ask.'
'It turned into a pig,' Alice answered.
                    LEWIS CARROLL, *Alice in Wonderland*

In some respects holography resembles photography: a silver-halide emulsion is exposed to light, and a latent image is formed and then developed. But there, as we have said before, the resemblance ends. This is nowhere more true than in the diagnosis of faults. An experienced photographer can look at a negative – sometimes even at a print from a negative – and instantly appreciate that it has been under-exposed, or over-exposed, or incorrectly processed, or that it is out of focus, or that the camera moved during the exposure, and so on, often at a glance. This is not necessarily so for a hologram. Indeed, over such troubles as noisy images or light scatter, by no means all experienced holographers would even agree as to the cause in any particular hologram. Let us look at the faults most commonly seen in single-beam holograms and try to sort out some of the possible causes.

1 No sign of any image This, easily explicable in photography, can be baffling in holography. There are certain questions the answers to which may furnish a clue. Did the hologram go black in the developer? If you could still see the safelight through the hologram, the cause is not gross over-exposure but almost certainly movement of the object or a gross change in one of the optical paths (rare in a single-beam set-up). If it is a reflection hologram, hold it up to the light as if it were a transmission hologram. Can you see a ghostly greyish image? If so, then there was *out-of-plane* movement by the film or the object; this destroys the reflection image but has little effect on the spurious transmission image.

2 Just a flicker of an image from some viewpoints Movement of the object relative to the film. The visible bit represents the part that remained stationary or moved only parallel to the direction in which parts are fleetingly visible. You can sometimes work out, from the image that does exist, in what precise way the object was unstable. One of the most common faults of this type is where the object has rolled on the film; the image will be a thin sliver, as though seen through a narrow slit (Fig 8.19(a)).

3 Part of the subject matter is missing If the subject matter consists of an assembly of objects and one of these undergoes gross displacement during the exposure, the image-forming fringes corresponding to that object will be totally destroyed, and there will be a three-dimensional 'black hole' in the shape of the object. The appearance of this artifact is so striking that once you have seen it you will readily recognize the symptom the next time round (Fig 8.19(b)).

4 Dark patches in the plane of the image The dark parts represent parts of the subject matter that have vibrated or distorted during the exposure. In a reflection hologram the movement was out of plane; in a transmission hologram (less common) the movement was in plane. This fault is most common in holograms of flexible or floppy objects.

5 Dark patches in the hologram plane (masters) If you can see round them, they are not caused by image movement but are very large Newton's fringes. They are usually disk- or doughnut-shaped. They are caused by very small foreign particles, usually tiny flakes of skin, trapped between the film and the glass plate. If they occur on only one of the transfer holograms, the fault was on the transfer not the master hologram (Fig 8.19(c)).

6 Large dark patches in the plane of the transfer hologram Almost certainly caused by birefringence in the film base, which results in rotation of the plane of polarization of the transmitted beam. As a result, in certain regions interference between the two beams is weak. Try changing the format from the landscape to portrait or vice versa, or change to material with a thin or an acetate base, or position the films emulsion to emulsion (Fig 8.19(b)).

7 Image is feeble but not noisy (By 'noisy' is meant that unwanted uniform light is swamping the image.) Usually, both under-exposure and over-exposure produce similar results. Did you need to leave the film in the developer for a very long time? Or did you have to snatch it? If neither, then probably the beam intensity ratio was well outside the optimum. Was the subject dark? Try adding a light background.

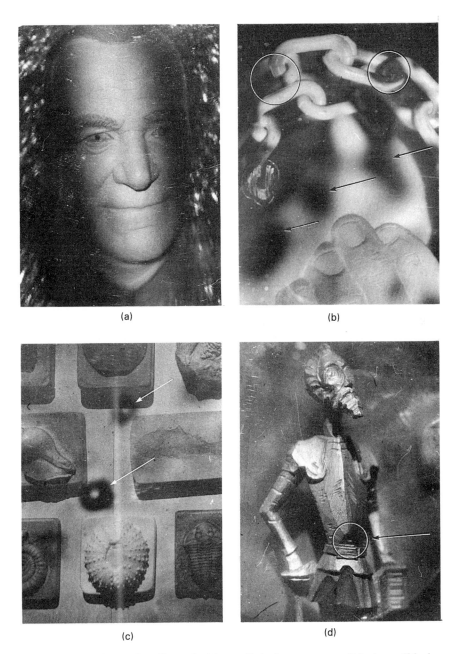

FIG 8.19 (a) *Shows the effect of object roll during exposure;* (b) *shows 'black holes' (ringed) where part of the object moved; these are in the plane of the image. The dark smudges (arrowed) are in the plane of the hologram and are caused by birefringence in the film base (see text).* (c) *shows dark doughnuts (arrowed) which are in the plane of the hologram, caused by a small particle trapped between the film and the glass plate. There is also a vertical streak which is a drying mark resulting from poor squeegeeing.* (d) *(ringed) Shows an example of burn-out from local gross over-exposure. The black smudge is in the plane of the hologram and does not show up well in this photograph. Holograms supplied by friends of the author who prefer to remain anonymous.*

8 *Parts of the image are hidden behind grey-black patches* This is known as 'burn-out'. The areas in question have directed too much light towards a small area of the emulsion. This has totally blackened the emulsion at this point and swamped the fringes. Such points are usually where a shiny object has been in contact with the emulsion. Try again, putting the film on the side of the glass opposite to the subject. If this still does not work, cut the exposure by 30% and/or increase the angle of incidence of the reference beam (Fig 8.19(d)).

9 *The image is surrounded by a bluish haze* In a reflection master hologram this is nothing to worry about. If your processing solutions were uncontaminated by traces of chlorides it is probably an idiosyncrasy of the particular emulsion batch. If it happens with a yellow transfer, put an amber filter over the display light.

10 *There are tiny criss-cross fringes all over the plane of the hologram* Either you are not using index-matching fluid, or you have been sold plain glass instead of anti-reflection glass. Take it back and complain.

11 *The image is there, but doesn't have any real guts* Provided you did everything else right, and conditions were absolutely stable, and all your holograms are exactly the same, ten to one you made up your solutions with tap water containing chlorides. Make up fresh solutions using de-ionized water, and use it for the final wash, too.

12 *Brown or blue spots appear on the film after a day or two* Tap water again, this time iron rust. Try re-bleaching, and wash in several changes of de-ionized water. Sometimes dark specks are caused by particles of undissolved chemical in the processing baths; these usually become apparent almost at once.

13 *Hologram goes dark after a week or two* This is called 'print-out' and is caused by re-sensitization of the bleached image to light. Certain types of final wetting agent, in particular Kodak Photo-Flo, have been blamed for this; residual traces of triethanolamine or sorbitol swelling agent can certainly cause it; but the most likely cause is your own slipshod processing. Don't skimp the washing. A final rinse in de-ionized water containing a few drops of acetic acid or Kodak Stopbath helps to prevent print-out. Try re-bleaching.

14 *There is scum on the hologram* This usually comes from the bleach bath. It will wash away, given a long enough washing time, but as it is on the surface you can remove it by gently wiping the emulsion with a cellulose sponge or with your fingers while the film is still in the wash. If you notice it only after the hologram is dry, re-wash the hologram in de-ionized water to which you have added a little wetting agent or washing-up fluid.

15 *There is a deep yellow stain all over the hologram* This is not usually a problem except with deep green reflection images. It is caused by the oxidation products of the developing agent (usually pyrogallol) and can be eliminated by the stain remover given in Appendix 10.

16 *The white background doesn't show* It is so far back from the hologram plane that the distance there and back from the film is beyond the coherence length of the laser beam.

*17 **Your assisted side-lighting isn't having much effect*** Same as above: the total round trip of the light compared with the direct distance to the film is greater than the coherence length of the laser beam. Try to shorten the optical path length of the route via the side mirrors.

*18 **The image is generally below par, though exposed and processed correctly*** The most common cause of poor image quality is very slight movement of the fringes caused by tiny shifts in the optical paths during a long exposure. Try doubling the settling time; turn off all heaters; use a remote exposure control. If none of these improves the image, the film may be faulty; send the film back to the manufacturer in its original box with a specimen hologram and a full description of the symptoms.

*19 **The image is redder and slightly unsharp in the deeper parts** (reflection holograms)* The solvents in the black spray paint you used have attacked and weakened the fringes. Use a spray paint that is not based on acetone and similar harsh solvents. If you are already using a satisfactory spray, you may not have given the hologram sufficient time to fully dry out, or you may have used a hair dryer on an emulsion that was still wet, resulting in *reticulation*.

References

1. McNair D (1983) *How to make holograms.* Tab Books Inc, pp 213–25.
2. Brooks L D (1979) A design for an inexpensive spatial filter. *Holosphere*, **8**, 3.
3. Blyth J (1986) Simplified single-beam holography. *Photographic Journal*, **126**, 448–9.
4. Richardson M (1985) Horror Holograms. *Royal Photographic Society Holography Group Newsletter* **2**, 8–9.
5. Phillips N J and Van der Werf R A J (1985) The creation of efficient Lippmann layers on ultrafine-grain silver halide materials using non-laser sources. *Journal of Photographic Science*, **33**, 22–8.
6. Blyth J (1985) Notes on processing holograms with solvent bleach. *Proceedings of the International Symposium on Display Holography*, **2**, 325–32.

CHAPTER

9

360° Holograms

'You may look in front of you, and on both sides, if you like,' said the Sheep; 'but you can't look *all* round you – unless you've got eyes in the back of your head.'

LEWIS CARROLL, *Through the Looking Glass*

There are a number of ways of making holograms with 360° of parallax, ie holograms you can walk right round and see an image the whole time. With the exception of the various sorts of holographic stereogram, which are a hybrid of holography and photography and are dealt with separately in Chapter 17, all the usual types of 360° holograms can be made using single-beam techniques.

Cylindrical Holograms

These are basically single-beam transmission holograms, but owing to the very large angle of incidence of the reference beam (between 65° and 85°) the Bragg condition is very powerful, and it is possible to obtain a reasonably-good image with white light reconstruction, using an amber filter. The film is inserted emulsion inwards inside an opaque roll of cardboard or plastic piping, which should be sprayed with matt black inside. For 8 × 10 in film (cut into two pieces each 4 × 10 in) you need an inside diameter of around 80 mm (3 in); not larger, as this would leave a gap in the image. The offcut should be about 204 mm (8 in) long, which will enable you to make use of the full width of the film if you want to make a hologram of, say, a full-length statuette. For more ambitious work you can go up to 190 mm (7.5 in) inside diameter × 250 mm (10 in) height, which will take 50 × 60 cm film cut down to half-width.

If you use an overhead reference beam (see p. 121) you can easily modify it to shoot straight down into the cylinder (Fig 9.1).

It is important to choose a subject that looks good when it is illuminated from directly above. Translucent objects such as shells, corals and glass figures all work well. The light reflected from the cylinder walls helps to illuminate the shadows. You may like to try placing your subject on an aluminum-sprayed plate, or on a mirror, which will provide an image of the reflection of the subject as well as extra illumination. It is worth noting that this type of illumination gives a very large depth, as the

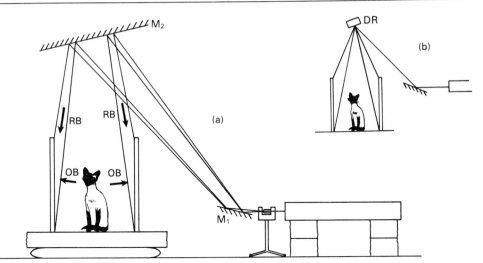

FIG 9.1 *A cylindrical transmission hologram. (a) The light path is the same as in* FIG 8.18, *with the angles adjusted to give a vertical cone of light.* RB *and* OB *are the reference and object beams respectively. (b) Alternative lighting using a diffuse reflector* DR.

optical path difference between the reference and object beams is very small at any point. This is generally true of single-beam transmission holograms.

If your table does not allow you to set up an overhead reference beam you can use a direct horizontal beam as in Chapter 8; you will need to set up the cylinder and subject matter on their side. Fix the cylinder in position with hot glue and the subject matter with contact adhesive (Fig 9.2).

An alternative lighting arrangement is also shown in Fig 9.2. The base is aluminum-sprayed and the unspread laser beam is fired straight past the subject at the base. This reflects the light without depolarizing it, and behaves much like a short-focus beam expander. This set-up reads like an optical horror story, but, oddly enough, it works very well (Fig 9.3(a)). If you feel that this arrangement

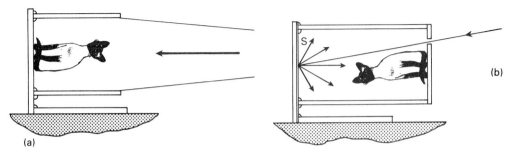

FIG 9.2 *A cylindrical hologram shot on its side, using a supporting frame. Lighting in (a) as in* FIG 9.1. *In (b) the base of the cylinder is towards the laser, the beam entering unspread through a small hole. The light is spread by being reflected from an aluminum-sprayed surface* S.

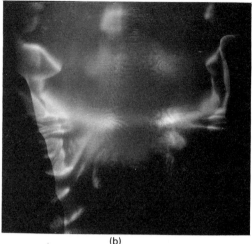

FIG 9.3 (a) *Two cylindrical holograms made as described on p. 129, shown with the original statuette, and* (b) *opened out for display. Holograms by Robin Horrex.*

is inherently inefficient (which theoretically it is), you can opt for a mirror with a short-focus concave lens resting on it.

The best way to display a cylindrical hologram is on a pillar, using an amber-filtered spotlight directly above. If your cylindrical hologram has been made with the illumination from below, you can place the replay light inside the pillar, close to the position of the original 'source'. You can get a somewhat surrealistic effect by opening out the hologram and lighting it as if it were an ordinary flat white-light transmission hologram, with or without a filter (Fig 9.3(b)). You can get equally interesting distortions by displaying the pseudoscopic real image either flat or cylindrically.

Conical Holograms

If you place a cone of holographic film over an object and expose it to laser light from above, you will get a 360° reflection hologram of the object. Again, you can use an overhead reference beam or simply shoot horizontally into the set-up laid on its side. Fig 9.4 shows the second of these arrangements.

To make the cone you will need a semicircular template with a diameter equal to the longer dimension of the film you are going to use. 8×10 in is the smallest useful size, giving a 60° cone 110 mm (4.3 in) high and 127 mm (5 in) in diameter, or, if you use the maximum area of the plate (*B* in Fig 9.4), a cone of approximately 90°, 1.75 in height and 3.5 in diameter. You will need a circular template with a 110° sector cut out for this. You will also need a sharp knife and a device that will punch out a semicircular notch at the center. A leather punch is best, but you can use an ordinary paper punch if you align the film carefully. The reason for the notch is that the film will not bend round to a perfect point, but will go into a sharp fold at the

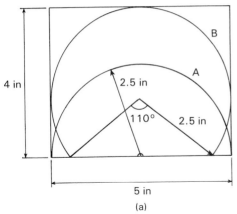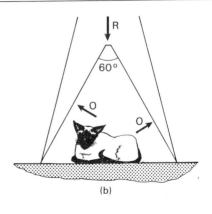

FIG 9.4 *Conical Denisyuk hologram. A semicircle marked out on 4 × 5 in or 8 × 10 in film will give a cone with a vertex angle of 60° (a). By using a template with a 110° sector removed, a broader cone with a vertex angle of 90° will be obtained. (b) The exposing beam does not need to be precisely vertical, and the set-up may be placed on its side.*

top unless it is notched in this manner. Put away the leftover piece of film to use for test exposures for other work. Roll the film into a cone, emulsion side in. Overtighten it a little at first, then let it relax. Butt-join the edges, or better, overlap them a little, and secure them with a narrow strip of cellulose tape (Fig 9.5). Put the cone away in a lightproof box.

Next, take a heavy wooden or metal base, preferably sprayed matt white, and mark a circle on it in pencil, the diameter of your cone. Affix a length of putty sealing strip to this circle (you can use Blu-tack or modeling clay instead) and make sure it is adhering thoroughly all round. If you are using the sideways set-up, affix the subject matter to the base with contact adhesive. Set up the subject in the expanded laser beam, close the laser shutter, and affix the film cone to the ring of putty sealing strip, with the join opposite the least important part of the subject matter. Allow at least 15 min for the film to settle, then make the exposure.

Remove the cellulose tape and open the film out for processing. After the processing, when the film is dry, rebuild the cone and affix it to a black plastic or cardboard base. Illuminate it with a white spotlight from directly above. You can also get interesting abstract effects by turning the cone inside-out and illuminating it from below, or by opening it out and displaying it flat. If you intend doing this, you will need to think well ahead and choose appropriate subject matter.

NOTE: Films with a polyester base can show dark bands when used for either conical or cylindrical holograms. This is because stress causes the base material to become birefringent and to rotate the plane of polarization of the object beam relative to the reference beam so that they do not form high-contrast fringes. The effect can be minimized by omitting the spatial filter from the optical set-up and instead reflecting the undiverged laser beam off a white diffusing reflector directly over the cone or cylinder. You can make a good diffuser out of an aspirin tablet or a Vitamin C tablet (which of the two you have handy depends on the nature of your particular hypochondria).

You can make cylindrical Denisyuk holograms, too, by wrapping the film into a

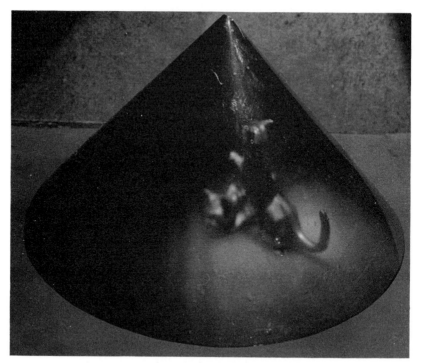

FIG 9.5 *A conical Denisyuk hologram.*

cylinder rather than a cone. This type of hologram requires two reference beams, one for the front of the subject and one for the back. Three reference beams respectively at 0°, 120° and 240° give a better result still.

360° Holograms with Plates

If you work with plates you can produce some interesting variants on cylindrical and conical holograms. The equivalent of the cylindrical hologram with plates is the cuboidal hologram, using four plates set up in a square. Because of the weight of the plates, the arrangement with an overhead beam is the only practicable one. The plates can be held together at the top by clips (see Fig 9.6), which also serve the purpose of blocking off light from entering the plates through the edges. The lower edges can be set in putty sealing strip or Blu-tack as described above for conical holograms.

It makes an interesting variation to close the top of the cuboid with another plate. This gives a Denisyuk reflection hologram; but its image is continuous with the sides. The processing is different, of course, and you may need to give the plate a separate exposure; also, you will need to be careful with the filtering of your eventual replay beam, if you are going to get a good color match between the two types of hologram. Once your tests have shown good results, cut the top plate square with

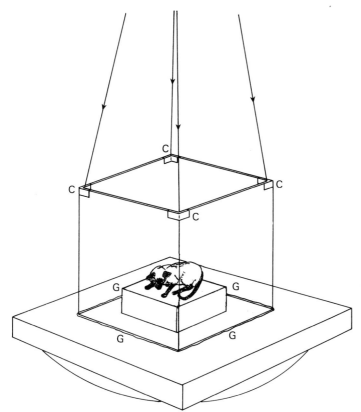

FIG 9.6 *Cuboidal transmission hologram. The plates are fixed at the base with hot glue lines GGGG and at the top corner with clips CCCC. The top edges of the plates must be protected with black paint or opague PVC strip cut to width. A further plate laid on top produces a Denisyuk cover.*

a glass-cutter (see pp. 174–5) and cement all the edges of the holograms with transparent adhesive, or use clips. You can have more than four sides, but you will then have to cut down the width of the plates. The most economic number would be eight, with four plates cut into halves along the longer direction.

The glass-plate version of the conical hologram is the tetrahedral or pyramidal hologram. In order to cut the plates to the right shape you will need to make a cutting template (see Fig 9.7).

Three 4 × 5 in plates will give you a tetrahedron with an altitude of 98 mm (3.9 in) and four will give you a square pyramid with an altitude of 94 mm (3.7 in), less than the height of the tetrahedron but with a greater base area. The tetrahedron and pyramid are intrinsically self-supporting, but it is still worthwhile using the putty sealing strip and putting a small spot of transparent adhesive at the top. Incidentally, you can also use film to make tetrahedral and pyramidal holograms, but the unsupported film is liable to vibration, and absolute quiet is necessary. In this type of hologram it is impossible to avoid shadows of the edges of the plates appearing

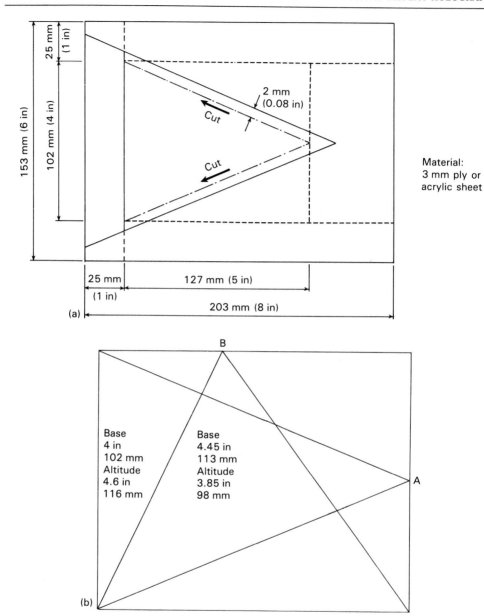

FIG 9.7 (a) *Template for a pyramid made from 4 × 5 in plates. Make the baseboard first and glue the end and side strips in position, using an old plate as a guide (the right-hand side is left open). Mark the position of the cut on the top board, allowing a 2 mm rebate for the shoulder of the glass cutter. Draw the cutter in the direction indicated and make both scores before removing the plate and snapping off the waste (see pp. 174–6). (b) Shows the two possible ways of cutting the triangles from a 4 × 5 mm sheet, drawn to scale. B gives a pyramid with a slightly larger base and less altitude than A. The template is made similarly.*

in the image, so they need to be carefully positioned with respect to the subject matter. Fig 9.7 also shows an alternative way of cutting a plate to give a somewhat wider base, with slightly lower altitude.

360° Flat Holograms

On p. 92 it is noted that it is possible to make a 360° hologram of a 'pseudoscopic' object by making a real-image hologram of the front view of the object on one side of the film and one of the rear view of the object on the other side, the images being in register. You can make the two exposures on a single film, exposing as though the holograms were entirely separate (which, indeed, they are), or you can obtain a somewhat brighter result by making two separate holograms and mounting them back to back with a layer of black material between them. This also eliminates the possibility of cross-talk, which can otherwise make both images visible from either side. You can achieve a similar result with any object (not a pseudoscopic object) if you mount it on a plinth which can be rotated through 180°. First you make master holograms of both front and rear views, on separate films, then transfer holograms in which the hologram plane contains the centre of rotation of the subject matter. When making two exposures on a single film in this way, the second exposure needs to be about 30% greater than the first for the two images to be equally bright.

You can display such holograms so that they can be walked round, with a spotlight on each side. Such holograms are also amusing as conversation pieces which you can hang on the wall by a cord, and turn round to show the rear view. This type of hologram does, however, have a serious potential use in instructional media, particularly in architecture and the plastic arts, and in such disciplines as biology and stereochemistry.

A very important branch of 360° holography is the field of holographic stereograms. These are very different from the type of hologram we have been dealing with so far, and are discussed in detail in Chapter 17.

CHAPTER

10

Introducing Further Beams

'I gave her one, they gave him two,
You gave us three or more.'
LEWIS CARROLL, *Alice in Wonderland*

In this chapter the idea of a controllable beam intensity ratio is introduced. Without entirely abandoning the Denisyuk approach, extra illumination of the subject from the sides and possibly the back is introduced, using part of the laser beam that has been selected from the reference beam by means of one or more beamsplitters.

When do we Need Extra Light?

As has been explained, the beam intensity ratio is an important factor in the eventual brightness of the image. If the subject matter is deep, there are two serious problems with the Denisyuk configuration. The first is that the image depth is limited to half the coherence length of the laser beam; beyond this distance the image fades rapidly, and at roughly twice the distance it will have disappeared completely. The second is that *illuminance* falls off as the square of the distance from the light source (which in this case is effectively the object), so that if the nearest part of the subject matter is, say, 5 mm from the film, and the farthest part is, say, 50 mm from the film, the beam ratio for the front part will be approximately 3 : 1 and that for the back part as low as 300 : 1. Assistance from side reflectors and a light background can help, but any single-beam hologram is working close to the limiting beam ratio for a bright image. So we need to split the beam in two using a beamsplitter, and this serves the dual purpose of attenuating the reference beam and supplying light to illuminate the subject matter from the side. This gives better control over the level of illumination of the subject matter, and at the same time solves the problems of the inverse-square law and of coherence length. We can also begin to think of lighting in a more creative way.

Plain-glass Beamsplitters

A beamsplitter is simply a partially-reflecting mirror. There are several different types of beamsplitter, some operating by simple partial reflection and some by quite different physical principles; those most frequently used are described on pp. 138–40. You can use a piece of plain glass as a beamsplitter; indeed, for all of the set-ups described in this chapter and for many of the more complicated ones discussed in later chapters, this is all you need. The glass should be as thick as possible, 6 mm or even more if you can get it. You must ask for float glass, which, although quite cheap, is of full optical quality. Any glazier will let you have some pieces about 25 × 36 mm for a few pence. The reason for the thickness is that a second beam is reflected from the rear (or 'second') surface of the glass, and if you allow these beams to overlap when expanded they will produce an interference pattern of alternate dark and light bars which is large enough to be obtrusive in the final image. You therefore need to have the two beams sufficiently well separated for you to be able to card off the unwanted secondary reflection before you expand the beam.

A variant on the plain-glass beamsplitter is the 3° prism. This is also of optical-quality glass, but is slightly wedge-shaped, as its name implies; the first- and second-surface reflections separate quite quickly, so that the second-surface reflection is easier to card off. It is unimportant, optically speaking, whether the wedge is horizontal or vertical, but carding-off is usually easier if the thick end is at one side rather than at the top or the bottom (Fig 10.1).

Such 3° wedges are more expensive than plain glass, but are more convenient in use. A Fresnel biprism (see pp. 8–9) is a suitable substitute, if one is available; you simply use one half of it as the prism.

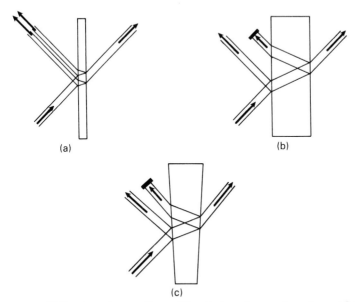

FIG 10.1 (a) *Thin plane beamsplitter. The first and second surface reflections overlap and will cause a broad interference pattern when expanded.* (b) *With a thick plane beamsplitter the second surface reflection can be blocked off.* (c) *Use of a 3° prism sends the first and second surface reflections in different directions, facilitating blocking off.*

Beam ratios in plain glass beamsplitters

The amount of light reflected from a plain glass–air interface at normal incidence (ie perpendicular) is about 4%, and this is the same whether the interface is air–glass or glass–air. For unpolarized light this ratio remains constant up to about 50° incidence. However, for linearly-polarized light the amount reflected depends on the angle at which the beam is polarized. For *p*-polarization (ie in the plane containing the incident ray and the normal) the proportion of reflected to transmitted light begins to fall at about 20° incidence, until at approximately 56° (the Brewster angle) it is zero, ie no light is reflected at all. Thereafter it rises rapidly to 100% at 90° incidence. Light polarized in the *s*-direction (perpendicular to the *p*-direction) is reflected in a proportion which rises gradually from 0.1 at 45° to about 0.18 at 60° and 0.33 at 72°, after which it also rises rapidly towards 100% at 90°. Fig 10.2 shows the intensity of reflected light as a function of the angle of incidence for both *p*- and *s*-polarized beams.

If we want the beam ratio to remain fairly constant regardless of the angle of incidence we need to have *p*-polarization and an angle of incidence of around 20°. The transmittance (allowing for first and second surface reflections) is roughly $0.95 \times 0.95 = 0.90$; the reflectance at the first surface is about 0.05, giving an 18:1 beam intensity ratio. This is rather large for most purposes. The second option is

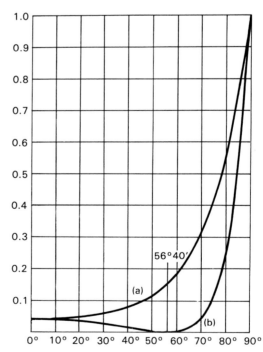

FIG 10.2 *Reflectance at a glass surface for* (a) *an s-polarized beam,* (b) *a p-polarized beam.*

to allow the ratio to vary as linearly as possible. Light polarized in the *s*-direction (oblique to the plane of the glass and orthogonal to the *p*-direction) gives a range of beam intensity ratios that vary from 10 : 1 at 40° incidence through 5 : 1 at 55° incidence to 1 : 1 at 78° incidence. The amount of transmitted light wasted due to the second-surface reflection varies from 8% at 40° incidence through 13% at 55° incidence to 46% at 75° incidence. For medium angles incidence this compares favorably with the losses in aluminized beamsplitters, which can be as large as 30%.

Metallized Beamsplitters

Apart from plain glass, the most commonly-used type of beamsplitter is the metallized type. These have the advantage over plain glass ones that the beam intensity ratio is independent of the angle of incidence or direction of polarization of the beam. The most useful ratios are 1 : 1 and 9 : 1. The second-surface reflection can usually be ignored; in the more expensive beamsplitters it is suppressed by dielectric coating.

More useful than the fixed-ratio beamsplitter is the variable-ratio beamsplitter. This is much more expensive, but it simplifies the balancing of intensities to such an extent that it is worth the extra expense. The metallizing is in the form of a broad ring on a rotatable disk, and the reflectance ratio is calibrated. A somewhat less expensive version is a metallized straight strip which you have to calibrate yourself. The metallizing has traditionally been aluminum, which has a better resistance to oxidation than silver. However, it has a comparatively low reflectance, and the metal titanium is gaining ground. This not only has a higher reflectance than aluminum, but it is much harder, and it will stand up to the batterings of high-power lasers.

Other types of beamsplitter

A more efficient alternative to the metallized beamsplitter is the dielectric-coated beamsplitter. This is a leaky interference mirror, and wastes no light at all. The wavelength needs to be specified, of course, and it operates correctly only over a fairly small range of angles. Variable dielectric beamsplitters are also available. You can also make interference beamsplitters holographically, either as diffraction gratings or as interference mirrors (see pp. 230–3).

The best and (of course) most expensive variable beamsplitter is the variable-polarization cube. This is one of the few types of beamsplitter that can handle an undiverged pulse-laser beam without damage. It is described in detail on pp. 237–8.

Other types of beamsplitter include split cubes, which vary the amount of light transmitted by varying the (very small) separation between two 45° prisms. The beamsplitter makes use of the *evanescent wave* which appears when total internal reflection occurs. There is an electromagnetic field just outside the surface which can be picked up by a surface placed sufficiently close; the closeness determines the amount that is picked up. A popular way of avoiding the problem of second-surface reflection in photography is the pellicular beamsplitter, which is an extremely thin partially reflecting membrane; unfortunately, it picks up the slightest vibration, and therefore cannot be used in holography. A further type is the dichroic beamsplitter, which is another type of dielectric beamsplitter: it reflects a broad band of wavelengths including green and blue, and transmits red; it has applications in color holography (see p. 262).

A Larger Platform

As suggested in the previous chapter, the overhead reference beam arrangement lends itself to the Denisyuk configuration with assisted split beams. However, as soon as you introduce a beamsplitter into the system there is an important new requirement, that of the stability of the beams. Hitherto, the only stability requirement has been that the subject matter and the holographic film remain perfectly stationary with respect to one another during the exposure. In fact, under reasonably quiet conditions it is possible to make Denisyuk holograms without any special antivibration precautions. However, as soon as a beamsplitter is introduced, there are two quite separate optical paths within the system, and not only do these need to be equal in length to within 1 cm or so, but if either of the paths changes during the exposure by as much as one-quarter of a wavelength the image-forming fringes will be wiped out. Such changes in optical path can be the result of a small vibration on one of the mirrors, a heavy footfall in the vicinity of the table, or even a current of warm air.

If you are not going to be continually frustrated by cramped optical conditions you will need something larger in the way of tables. For the layouts described in this chapter the minimum useful size is about 0.6×1.2 m (2×4 ft). There are many sources of suitable slabs: a visit to a builder's or stonemason's yard may be profitable; a prowl round your local junkshop or auction room may turn up an old billiards or pool table – if you are lucky, a full-size one with a slate base – or, if you have a taste for the macabre, you may come across a cache of discarded headstones in a churchyard that is being refurbished. Cast-iron presses from cloth factories, engineers' marking-out tables, armor-plate from a naval breaker's yard, all have been pressed into use as holographic tables. Many holographers turn out superb multi-beam work on optical tables made from material found in the most unlikely places. What is more, if you have ever laid a concrete path you will have no difficulty in casting an excellent table from concrete by yourself. The stability requirements for large holograms involving heavy optical equipment and long beam throws are discussed in the next two chapters.

Additional Equipment

In order to make holograms with extra beams introduced via beamsplitters you will need the following further equipment:

    3 front-surface mirrors (20 mm diam) and mounts
    Plain glass or 3° wedge beamsplitter

1 : 1 metallized beamsplitter
Heavy-duty camera tripod with reversible column or similar stand
Mounts for all optical components
8 of each metal or hardwood blocks, thicknesses 1, 2, 4, 8, 16, 32 mm
Black velvet or other background material
Concave lens, focal length approximately -6 mm, or semi-diffuse reflector
Professional-quality photographic exposure meter, eg Gossen Lunasix*

The steering mirrors can be ordinary pieces of front-surface mirror secured to posts by Blu-tack; but you will find adjustments much easier if you invest in proper steering mirrors in mounts with x and y adjustments. The beamsplitter also needs an adjustable support. You can obtain all these items from the suppliers listed in Appendix 12. The best way to mount the overhead mirror is to glue it (with a hot-glue gun) to a metal plate which has a $\frac{1}{4}$ in Whitworth tapped hole for the tripod screw (Fig 10.3).

It is worthwhile getting the spacer blocks made from mild steel by a local company if you do not have your own metal-working facilities. They will prove useful throughout your holographic career, no matter how advanced your lab may

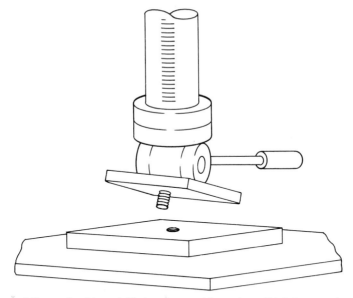

FIG 10.3 *Mirror glued to a drilled and tapped baseplate which fixes to the tripod screw on an inverted column.*

---

* If you have some knowledge of electronics you can build your own photometer using a silicon photodiode, an amplifier and a microammeter with switchable shunt resistors for changing the scale. If you do not already possess a sufficiently-sensitive exposure meter this will save you a good deal of money. Ron Graham (1) has designed a simple power-density meter based on a cadmium sulfide (CdS) photoconducting element. Designs for do-it-yourself power-density meters also appear in amateur electronics periodicals from time to time.

become. If you have access to a good carpentry shop, hardwood or opaque acrylic material is very nearly as good.

The diffuse reflector can simply be a metal can lid sprayed with aluminum paint. This provides a more widely-spread and less harsh and uneven beam than a concave lens, and avoids the need for a mirror. With a larger table there will be room to put the laser on the table itself; this has the advantage that any movement of the table does not misalign the beam.

Setting up with an Overhead Reference Beam

The procedure for setting up is as follows:

1 Set up the tripod in the middle of the table with two legs on a line parallel to its long side, the large mirror horizontal and about 600 mm (2ft) above the table (Fig 10.4).

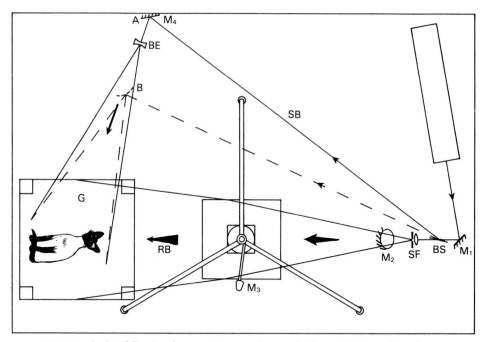

FIG 10.4 *Assisted Denisyuk geometry on a larger platform. M_1 is a relay mirror, BS the beamsplitter, SF the spatial filter and M_2 a further relay mirror directing the beam up to the reference mirror M_3 on a tripod. RB is the reference beam. The glass plate G is supported on shims with the subject matter below. The secondary beam SB provides extra object illumination at table level via the relay mirror M_4 and beam expander BE. The optical path length from the beamsplitter to the plate must be the same for both beams, and light from SB must not fall directly on the film. The beamsplitter ratio should be 1 : 1 if possible; otherwise use plain glass. A shows the position of a mirror/concave lens beam spreader; B the position of a diffuse reflector (beam indicated by broken lines).*

2 Set up the blocks to support the glass plate at the required height above the subject matter, with the center of the arrangement about 45° off the center of the large relay mirror when seen from the side.
3 Set up the laser and relay mirror as shown. The laser is horizontal, at an angle of about 25° to the edge of the table. Align the relay mirror so that the laser spot is in the center of the overhead mirror and is approximately centered on the glass plate.
4 Set up the beamsplitter for a relay mirror/concave lens (position A) or diffusing reflector (position B). Make sure you leave enough room for the spatial filter. Adjust the components so that the subject is suitably illuminated. You need not card off the secondary reflected beam if you are using a diffusing reflector. Ensure that none of the beam illuminating the subject can reach the film directly; card off as necessary.
5 Readjust the reference-beam relay mirror, which will have been affected by insertion of the beamsplitter.
6 Insert the spatial filter between the beamsplitter and the second relay mirror, and align it (see pp. 101–4). If the final beam is not uniform over the whole area of the glass plate, adjust the relay mirror and overhead mirror as necessary.
7 Check that the optical path lengths of the two beams from the beamsplitter to the film are the same to within 10 mm or so, using a piece of string.

Checking the Beam Intensity Ratio

This arrangement will probably give you a beam intensity ratio that is close to the optimum. However, if you have a sensitive photographic exposure meter or photometer, you can check the ratio as follows:

1 With the meter cell open (as for reflected-light readings) take a reading straight down the object beam, ie from directly above. Note the reading.
2 Card off the side beam and point the meter straight at the overhead mirror from the glass plate. Note the reading.
3 Compare the two readings. With a photographic exposure meter, work out the beam ratio by counting the number of blocks difference between the readings: each block represents a doubling of the intensity. Thus two blocks difference represents a beam ratio of 4:1. With a photometer giving direct readings you simply divide one reading by the other. If the ratio is outside the range 1.5:1–4:1 for a reflection hologram, bring the concave lens or diffuse reflector nearer to the subject (but keep the optical path length the same).

With the laser installed on the table as in Fig 10.4, the mirror M_1 is useful, as you can adjust it through small angles as necessary to obtain precise centering of the reference beam without disturbing the overhead mirror. When you come to more complicated set-ups you will find that a relay mirror early in the system, equipped with fine x- and y-adjustments, is a great time saver.

Double-beam Object Illumination

By using a second beamsplitter you can illuminate the subject with a second beam which can be brought in from the other side, if your table is large enough. In this way you can use a lens on one side to give modeling and a diffuse reflector on the other to fill in the shadows. The principles are discussed in more detail in Chapter 11. A number of different lighting effects are discussed on p. 154. Fuller discussions of the aesthetics of lighting are to be found in books on photography, painting and sculpture. Nevertheless, these principles apply to holograms as much as to any other creative form. As a general rule, the second illuminating beam is derived from the object-illuminating beam with a 1 : 1 beamsplitter. However, if you try this and find that the overall level of object illumination has been reduced to the point where the beam intensity ratio is unacceptably high (over about 4 : 1), leave the object-illuminating beam as it is, and instead derive the second illuminating beam direct from the main laser beam using a 1 : 1 beamsplitter.

Fiber-optics Light Guides

Fiber optics has become a popular method of guiding illumination in photo-macrography, where tiny objects demand bright and cool lighting at short distances. This is exactly the sort of thing that is needed in making holograms, and now that fiber-optics light guides are readily available it is worthwhile exploiting their uses. If the object illuminating beam is fed into a light guide of the correct length, it can be used to illuminate the subject matter from any angle, once the path lengths have been matched, without the necessity of re-matching paths by moving mirrors around. In addition, there are devices obtainable which split the beam into two or more beams. However, you cannot use these light guides for the reference beam, as the internal reflections destroy the polarization and spatial coherence of the beam. In the object beam this is no disadvantage: it also helps to reduce laser speckle.

NOTE: Fiber-optics light guides are known as 'coherent' if they preserve spatial relationships from one end of the bundle of fibers to the other. This simply means that they can be used for visual observation and photographic imagery; it has nothing to do with coherent light. The fibers themselves have a diameter many times the wavelength of light, and are known as *multimode optical fibers*. Fibers which have a diameter comparable with the wavelength of light are called *monomode optical fibers*, and these preserve optical coherence. The use of monomode fibers in holography is discussed in Appendix 8.

Polarization Considerations: Use of a Half-wave Plate

It was shown on pp. 138–9 how the direction of polarization affects the reflectance of a plain-glass beamsplitter, and it was suggested that Denisyuk holograms were best made with p-polarized light as this more or less eliminates internal reflections within the film and the supporting plate. It is important that the reference and object beams should be polarized in the same plane, otherwise they will not form strong interference fringes; if the planes of polarization are orthogonal (ie at right angles to one another) no fringes at all will be formed. Now, a totally diffusing object, such as one made of unpainted plaster or sprayed with matt white paint, de-polarizes any light falling on it regardless of its initial polarization. Such light does, however, form fringes with a reference beam. The reason for this is easy to understand if we think of 'unpolarized' light as being made up of two coincident beams of equal intensity that are polarized in mutually perpendicular directions: thus half of the beam energy produces useful fringes, while the other half does nothing except add to the fog level of the emulsion. But very few objects are in fact totally diffusing, particularly metallic ones. In most cases the reflected light has quite a strong polarization, and this is not necessarily in the same direction as that of the beam illuminating the object. So in order to get the best alignment of polarization angles you may need to rotate the plane of polarization of the reference beam.

To do this a further piece of optical equipment is required; it is called a *half-wave plate*. This is made of a substance (usually a thin sheet of quartz or mica) which has two refractive indices at the same time, ie it is birefringent. If a plate of this substance is made with a thickness such that when a beam of light passes through it the two emergent wavefronts are precisely one half-wavelength out of phase, it can be used to rotate the plane of polarization of a laser beam by any desired amount simply by rotating the plate. A half-wave plate placed in the reference beam can be used to align the plane of polarization to match that of the object beam. There are several methods of lining up a half-wave plate, and they all involve the use of a polarizing filter of the type used in color photography for eliminating reflections or darkening the sky. You can use Polaroid sunglasses if you do not have one of these. Position the half-wave plate in the reference beam, then card it off. Examine the subject matter from the plane of the hologram, through the polarizing filter, and rotate the filter until it looks brightest. Note the orientation of the filter. Now card off the object beam instead. Put a piece of white card in the hologram plane and hold the filter in front of it, in the same orientation, so that it casts a shadow on the card. Rotate the half-wave plate until the light passing through the filter is at its brightest. You now have the optimum conditions for producing fringes. You may find the method easier if you look for the darkest, rather than the brightest, orientation, as this is easy to establish positively, especially for the reference beam (which disappears altogether at the correct orientation).

You can use an ordinary polarizing filter in the reference beam if you do not want to go to the expense of buying a half-wave plate, but there will be some loss of light.

Reference

1 Graham R W (1982) A simple power density meter for holography. *Journal of Photographic Science*, **30**, 102–4.

CHAPTER
11

Deep-image Reflection Holograms

'Well, I should like to be a *little* larger, Sir, if you wouldn't mind,' said Alice: 'three inches is such a wretched height to be...'
LEWIS CARROLL, *Alice in Wonderland*

So far we have been considering holograms in which the main illumination of the object also does duty as the reference beam. This restricts the lighting to the front and above; also, the subject matter needs to be close to the film, otherwise the edge of the film casts a shadow on it. The alternative is to separate the subject matter from the film by a distance sufficient to allow the reference beam to miss the subject altogether, and to provide separate illumination for the subject matter. There are two advantages to this: first, the lighting can be much more attractive and imaginative; and secondly, the subject matter can be much deeper, as the optical paths can be better matched. On the debit side, you will need more optical equipment and a larger table, and stability requirements will be much more stringent.

Types of Optical Table

The first thing you have to decide is the kind of optical table you are going to have. Up to this point the 'tables' we have been discussing have been decidedly ad hoc. Tables made specifically for holography are made of steel with a honeycomb base, and usually have the surface pierced with a grid of threaded holes into which the various optical components can be screwed. These tables are very expensive, and many holographers opt for engineers' marking-out tables, which are made of cast iron, and can often be found going quite cheap, though transport and installation can be expensive unless you have access to a truck, a dozen burly friends and a good insurance company in case of accident. One of the advantages of a steel table is that you can use magnetic bases to hold your components. If you have had some practice at laying a concrete path you can cast your own table from concrete, and face it with steel plate if you want to use magnetic bases. However, if you are anticipating moving out of your lab at any time, you will want a table you can dismantle; many holographers favor the sand-table system (see below), though others view the presence of such quantities of sand with something of the pessimism of the Walrus and the Carpenter. A sand table is simply a large box filled with sand, into which the optical components, mounted on posts, are pushed. The sand is a good damping medium, and its heavy weight provides the inertia necessary to prevent vibration. The main objections are the long settling times that are necessary with complicated set-ups using long beam throws, and the incompatibility of sand with delicate optical surfaces.

Choosing a Suitable Site

The ideal site is a basement with a concrete floor, preferably well removed from anything that produces vibrations or low-frequency sound, such as town traffic. A stone-floored garage is the next best thing. Don't choose a room with a wooden floor, as the slight flexure when you move about on it, or even when someone moves around outside the room, will cause sufficient distortion of even a concrete table to spoil your hologram. Block off any draughts, and any sources of convection currents such as radiators and freezers. If you are unable to do this, and have to share your lab with such equipment, install a heavy curtain right round the table. This is also essential if the environment is noisy. You would be well advised in such a case to line the walls with sound-deadening material; if you can't afford acoustic paneling, eggboxes are just as good.

Building a Sand Table

Various designs of sand table have appeared in the literature, sometimes with extremely detailed instructions for building (1, 2, 3, 4). The simplest type of table is a tension table, a design attributed to Lloyd Cross and described in detail by Unterseher *et al* (2). Its construction is shown in Fig 11.1. The material is heavy chipboard (particle board); the threaded rod can be bought in lengths of about 2 m. Five lengths will be sufficient for a table 1.25×2.5 m (4×8 ft) in size and 300 mm (1 ft) deep. It is not necessary to glue the box, as it is held together by the tension of the rods, hence its name. However, it is very important to mark out and drill the holes accurately, otherwise the tensions will not be at right angles and the box will collapse while you are building it.

Assemble the sides first, then the top and bottom, with the nuts finger-tight. Then go round and round, giving each nut half a turn at a time with a spanner. Do not allow any thread to remain projecting on the underside; indeed, it is a good idea to countersink both washers and nuts so that they project only a little. This is because you are going to mount the box on inner tubes, and any external rough edges might damage them.

The simplest form of stand for the table is six concrete blocks on end; you can obtain these from builders' suppliers. Alternatively you can build a Dexion frame as described by Saxby (3). Stand the blocks or frame on squares of carpet; this doesn't help the damping, but it does cure any possible trouble from unevenness of the floor. Don't buy expensive carpet tiles: you can usually get squares of carpet samples of discontinued lines very cheaply from home furnishers. Place further pieces of carpet on top of the blocks or frame, then lay the under-board on these.

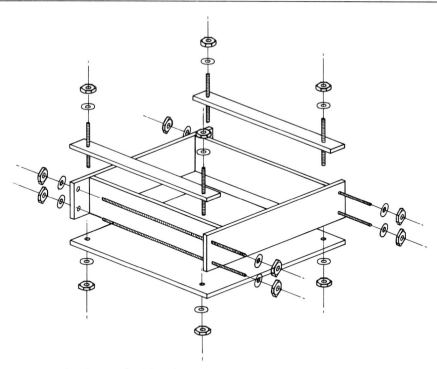

FIG 11.1 *Tension sand table (after Unterseher et al (2)). The material is heavy-duty chipboard (particle board).*

Next come the inner tubes. Scooter tubes are the best, six of them, about $3\frac{1}{2} \times 10$ in size. Choose the ones with long cranked valves. Inflate them so that they are firm but not hard, place them on the under-board with the valves lined up with the holes and lay the table on top. As the joins in the table have not been sealed, it is a good idea to line the box with thin polythene sheet before you begin to fill it with sand.

The sand can be any kind, even sea sand you have collected yourself, as long as it has been washed and is thoroughly dry. However, ordinary sea sand or builders' sand tends to become dusty after a time, and it is better to put up with the extra expense and buy coarse white silica sand, which comes in bags rather than by the truckload. In this way you will avoid having it dumped in your driveway in the middle of a rainstorm, and will be able to carry it to the lab clean and dry. For a table of the size suggested you will need about 1.5 m^3 (40 ft^3), or around 1.5 tonnes weight. You will probably find you have too much to begin with, but as it settles down you will need to add more to top it up.

Once the sand is in the box you can check the pressure in the inner tubes. Rock the box: it should rock just twice, over a period of 1–2 s. If it rocks more than twice, the tubes are too hard; if it rocks only once, put in some more air. Check that the height of the box above the under-board is the same all round, and adjust the tube pressures as necessary.

If you want a more permanent table, or one larger than the one described above, you can make a sand table on a concrete base. There is a very detailed

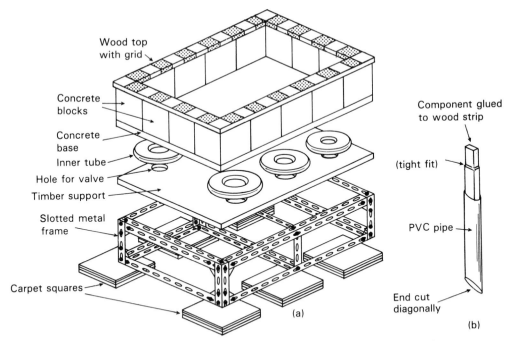

FIG 11.2 (a) *Concrete-based sand table (after McNair (1)), supported on a frame built from heavy-duty angled strip (after Saxby (3)).* (b) *Optical component inserted into a piece of* 38 mm *plastic piping.*

account of the building of such a table in McNair (1). If you can find an old billiards or pool table with a slate base, this is ideal, and there will be no need for you to go to the trouble of casting a concrete base; you can simply build the side walls onto it, using concrete blocks and ready-mixed mortar from the local do-it-yourself shop. Fig 11.2 shows such a table, supported on a Dexion frame. The dimensions of the frame are 900 × 1500 mm (3 × 5 ft) × 300 mm (1 ft) high. For this you will need four pieces 1500 mm long, six pieces 900 mm long, six pieces 300 mm long, 24 gusset plates and 80 bolts and nuts, ie one gusset plate and three bolts and nuts for each corner and eight bolts and nuts for the center stays.

Supporting the Optical Components

Small optical components such as mirrors, beamsplitters and reflectors are best supported on 38 mm (1.5 in) diameter plastic piping, which you can buy from do-it-yourself suppliers. Saw the end that is to stick into the sand at an angle of 45° (make two at a time by taking a double length and sawing it in half at 45°), and clean up the rough edges with a sharp knife. The more expensive optical components are usually supplied on 16.7 mm pins intended for use with optical benches; so it is a good idea to make some plugs which are a tight fit in the pipes and are bored to take pins. Cruder components such as reflectors and diffuse reflectors can be held by bulldog clips screwed to 15 × 36 mm (5/8 × 1½ in) battens with the edges rounded off. You can fit these into the pipes by putting your foot on the pipe to make it slightly oval, and inserting the batten; when you release the pressure the batten will be held firmly.

Steadiness Checks

The finest holographic set-up in the world is useless if there are any disturbances present. There are two checks for steadiness that you need to carry out before you can start making holograms. One is very simple; the other is more subtle.

The simplest test uses no more than a laser beam and a saucer of water. Place the saucer in the middle of the table and direct the laser beam (expanded) so that it is reflected onto the ceiling. Any stray vibration is immediately obvious. This may seem a crude test, but it is remarkably sensitive. Try walking round the table, tapping it, leaning on it; see what you can get away with. The saucer test checks the effect on the table of fairly sudden movements, as well as continuous vibration; it is not sensitive to very slow movements such as you get from changes in temperature or humidity, or from the gradual settling of components. For this you need a more sophisticated optical test, using an arrangement known as a *Michelson interferometer* (Fig 11.3).

The method is as follows:

1. Switch on the laser and adjust the relay mirror so that the beam is across the table at an angle of roughly 45°.
2. Place a mirror at the end of the beam throw so that it sends the beam back along the same path. The two spots at the relay mirror should be alongside one another, almost but not quite coinciding.
3. Place a beamsplitter vertically in the center of the beam so as to reflect it at approximately 45°. A piece of ordinary float glass will do.
4. Place a second mirror in the reflected beam, the same distance (to within a few millimeters) from the beamsplitter as the first mirror, so that the light is reflected back along its original path, as with the first mirror.
5. You should now see two spots on the opposite wall, close together. Adjust the second mirror and the beamsplitter in turn until the spots from the two mirrors exactly coincide both at the relay mirror and at the wall. If your wall is dark, hang a piece of white paper on it so that the spots are at the center.
6. Place a lens in the two overlapping beams so that they spread into a patch large enough for you to see detail from a couple of meters distance.

The detail you will see is a set of dark and light fringes, which to begin with will be moving around unsteadily, perhaps quite fast, so that you may not see them at first. There will also be some stationary fringes, caused by double reflections in the beamsplitter, which you should ignore. The moving fringes are a very sensitive detector of movement, and the way they move can tell you a great deal about the nature of any instability present. If they are skittering, there is vibration present. Listen hard for any noise that may be causing it (usually thumps and rumbles). If the fringes move around in an irregular kind of way, but keep coming back to rest in their original position, there are convection currents in the optical path, and you

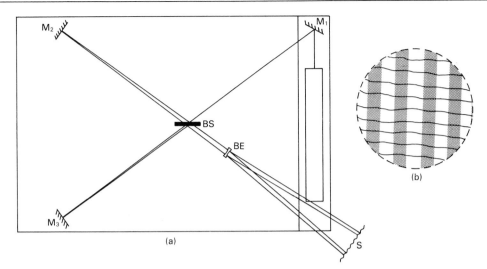

FIG 11.3 (a) *Set-up for Michelson interferometer steadiness test. The relay mirror* M_1 *directs the beam onto the beamsplitter* BS. *The two beams are reflected back to the beamsplitter along the same paths and combine to form interference fringes at the screen* S, *the pattern being expanded by the lens* BE. (b) *An impression of the pattern seen. The broad vertical bars are the genuine fringes. Note that the bars are often skewed rather than vertical: this does not matter.*

will need curtaining (but try holding your breath first). If there is a steady slow movement of the fringes in one direction, this is due to the table settling down, and is quite normal. Make a note of the time it takes for the fringes to become stationary to within a quarter of a fringe movement in 2 min; this will be the minimum settling time for any exposure. This is easier if you mark the position of one of the fringes with a black felt pen.

While you are about it, do some tests to ascertain how much it takes to disturb the fringes. Lean on the table. Shout at the mirrors. Breathe into one of the beams. Wave a piece of card close to a beam. Jump up and down on the floor at various points near the table. Get a friend to walk around outside the room and make various kinds of disturbance such as whistling, talking loudly, rattling teacups, etc. While you are watching the fringes and making notes, bear in mind that a movement during the exposure of as little as one-tenth of a fringe will adversely affect the quality of the holographic image, and movement of half a fringe will destroy it completely.

NOTE: You can also use the interferometer to assess the coherence length of your laser beam. Starting with the optical path lengths exactly equal, move one of the mirrors along the beam until you can no longer see the fringes clearly. You have now moved the mirror through a distance equal to one-quarter of the total coherence length (not one-half as you might think, because the light is reflected from the mirrors and is therefore traveling double the distance you have moved the mirror). The useful depth of object space is about half the total coherence length. For a 5–10 mW HeNe laser the useful object space will usually be about 200–300 mm (8–12 in); it may be somewhat less for higher-powered lasers.

It is advisable to repeat the saucer and interferometer tests at intervals of a few weeks, in case conditions have changed. Also carry the tests out before any session when you are producing master holograms, or where the geometry involves long beam throws.

Optical Set-ups for Reflection Holograms

A photographer arranging a display of objects for an advertisement, or an art director preparing a TV commercial, or a window dresser arranging a display, all know that one of the secrets of success is the lighting. Holography is no different. The Denisyuk format allows only a very limited range of lighting angles, and these are not always appropriate to the subject matter. A wealth of lighting variations is possible for any subject. There are plenty of excellent books on both photographic and display lighting, and you would be well advised to study some of the better ones, if you have little or no background in this aspect of display work. Lighting for display holography differs in only one way from lighting for any other type of display: it has to be derived from a single light source, with all the beam paths the same length (the last part can be waived if you are using a high-powered laser equipped with an etalon (see pp. 36–7). With this restriction in mind we will begin to look at some basic lighting set-ups for deep-image reflection holograms.

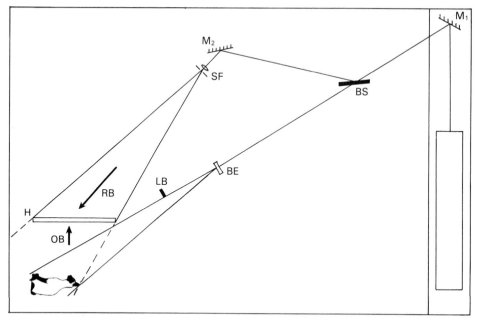

FIG 11.4 *Basic two-beam layout. The object is illuminated by light passing directly through the beamsplitter* BS *and the beam expanding lens* BE. *The reference beam* RB *is derived from the beam reflected at the beamsplitter via a relay mirror* M_2 *and a spatial filter* SF. *This light provides extra object illumination. Stray light is carded off from the emulsion by the light block.* LB. BS *is plain glass or a* $3°$ *wedge.*

Basic Two-beam Layouts

Two basic layouts are used for making reflection holograms: in the first, the whole set-up is on its side, so that the reference beam travels parallel to the surface of the table; in the second, the reference beam comes from above, so that the subject matter is in its normal upright position. The first of these is easier to set up, optically speaking, but in most cases it is more difficult to support the subject matter. The set-up is also more limited from the point of view of illumination of the subject matter. The most basic two-beam layouts will be considered first. In each case the geometry depends on whether or not you possess a variable beamsplitter. The first set-up is for a plain glass or 3° prism beamsplitter with the object on its side. The beam must be *s*-polarized with respect to the beamsplitter, ie in this instance, vertically polarized (Fig 11.4).

This is a simple and reliable set-up for your first experiments in split-beam reflection holography. At the angle shown in Figure 11.4 the beamsplitter will be splitting the light in a ratio of about 5:1 in favor of the reference beam by the time the beams reach the hologram plane. A 9:1 metallized beamsplitter will give you

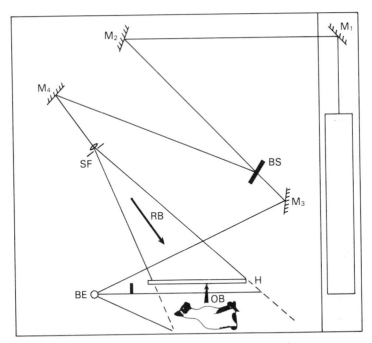

FIG 11.5 *Reflection hologram on a table of restricted area. The beam reaches the 1:1 metallized beamsplitter BS via the relay mirrors* M_1 *and* M_2 *and is split into the reference beam RB via the mirror* M_4 *and a spatial filter SF to illuminate the hologram H and provide auxiliary lighting. The other beam is directed by the mirror* M_3 *onto the beam expander BE. Alternatively,* M_4 *can be elevated to give an overhead reference beam; the subject matter is then placed in an upright position. BE can also be elevated. OB is the object beam.*

a final beam ratio of about 2 : 1, which is better, and you will be able to use it at a less steep angle, which will simplify path matching.

An alternative way, suitable for shorter tables, is to fold both beams with mirrors (Fig 11.5). You will probably have to expand the reference beam somewhat before it strikes the mirror, which will need to be large enough for this purpose.

The beam-spreading reflector can be of a number of types. Notice that in both of these set-ups some light will fall on the object through the film unless it is sufficiently far away from the object. This does not matter provided the shadow of the edge of the film does not fall obtrusively across the object.

This second layout is more flexible than the first, in that you can elevate the reflector so that the object is illuminated diagonally rather than straight ahead. By tilting the beamsplitter so as to throw the reference beam upwards onto an overhead mirror the arrangement can be converted into an overhead-beam configuration. The overhead mirror can be mounted on a camera tripod or other rigid mount, as described on p. 121. If you are working with a sand table you will need to place a platform on the sand for the tripod legs; this platform can be anchored in place with four wooden spikes, one at each corner.

Multi-beam Layouts

Some published layouts for reflection holograms show a second illuminating beam coming from the other side of the object. This is a poor and unrealistic kind of lighting and is suitable only for translucent objects, which often look well when illuminated from below. In general it is much better to have one of the beams horizontal and the other elevated. It is best to stand the subject matter upright or tilted forward; you can then illuminate it from both sides (Fig 11.6).

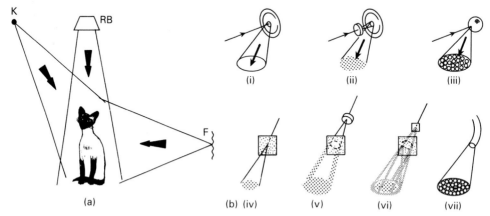

FIG 11.6 (a) *Lighting for modeling.* K *is the undiffused key light from above, side and front.* F *is the diffused fill-in light from the side and front.* RB *is the reference beam, falling on the emulsion from above and in front, and supplying extra fill-in light.* (b) (i) *and* (ii) *show a metal-sprayed can lid diffuser;* (iii) *is a ball-bearing producing a reticulated pattern;* (iv), (v) *and* (vi) *are transmission diffusers making use of ground glasses;* (vii) *is a fiber-optics bundle.*

In Fig 11.6 the beam coming from the left (K) is shown as directed light, from an elevated position. This is the *key light*. The beam from the right is more or less horizontal, and is called the *fill-in light*; it is in general of lower intensity than the key light, and it may be more or less diffused. The reference beam passing through the film provides additional fill-in light. This type of illumination is typical of the illumination used in the photography of objects, sculptures or people. Fig 11.7 shows a table geometry applying the principle.

In general, you should have a directed light source where you would use a photographic spotlight if you were taking a photograph of the subject matter, and various degrees of diffusion where you would use softer lighting. Because it remains fully polarized, a beam that is spread with a lens or a convex mirror produces a stronger effect than appears from a visual examination. If you want to see what the lighting contrast really is (as far as the holographic image is concerned) you need to look at the subject matter through a polarizing filter oriented to match the reference beam. Diffusers reduce laser speckle, as they lower the spatial coherence of the light. You can also use non-uniform reflectors to produce special lighting effects. Due to the crystalline structure of the steel used to make it, a large ball-bearing produces a strongly marbled pattern which can be attractive with certain subjects; if sprayed with aluminum paint it acts as a partial diffuser. A ball-bearing is basically a convex mirror, with a focal length equal to one-quarter of its diameter. A 25 mm (1 in) ball-bearing will spread the laser beam in roughly the same way as

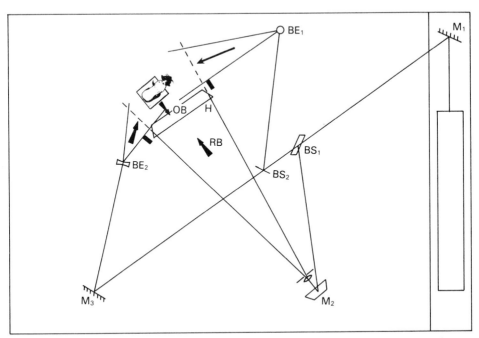

FIG 11.7 *Application of lighting of* FIG 11.6 *to a table. The symbols have the usual meanings.* BS_1 *and* BS_2 *are metallized and the ratio is approx* 1:1. M_2 *is elevated, and H and the subject matter are tilted forward.*

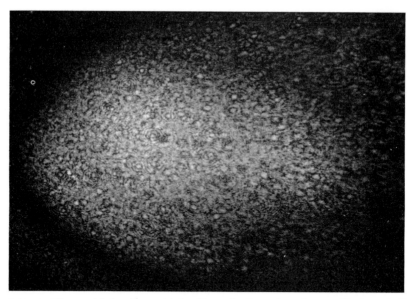

FIG 11.8 *The marbled pattern obtained by reflecting a laser beam off a large ball-bearing.*

a ×40 microscope objective (about 1 in 6, or 10°). The sort of pattern a ball-bearing produces is shown in Fig 11.8. The pattern can be extended horizontally or vertically by the use of a cylindrical lens in the beam. A somewhat similar pattern can be generated by a fiber-optics bundle; the general appearance is of greatly-enlarged laser speckle. Some types of specular and diffuse reflector are illustrated in Fig 11.6(b).

Backlighting, Shadowgrams and Black Holes

By taking one of the illumination beams behind the subject matter you can get a rim-light effect that is very effective in showing surface textures. If you take the light right round to the back of the subject you can produce three-dimensional shadowgrams (Fig 11.9). It is advisable to have enough diffusion to avoid throwing a sharp shadow on the hologram.

Another shadowgram of a kind is a so-called *black hole*. This usually unwanted phenomenon occurs whenever part of the subject matter moves during the exposure. If you try to make a hologram of living subject matter, say a flower, you will usually find that a part of it has moved during the exposure; the result looks as if that part of the subject had been painted coal black. You can see a kind of three-dimensional hole where the image should have been. The effect is particularly striking with parts of the human body such as hands, or the face in profile. If the subject stays as nearly stationary as possible during the exposure, the black hole will be photographically sharp and will have a dramatic appearance.

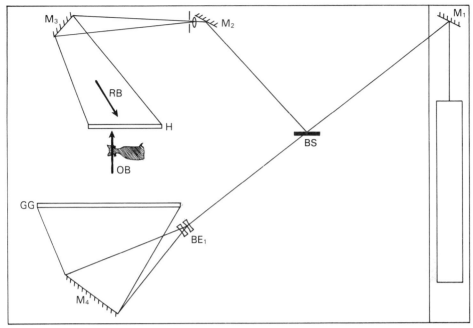

FIG 11.9 *Layout for a holographic shadowgram. GG is a ground glass plate, which must be large enough (out of the plane of the diagram) to suppress any direct shadows from the subject matter.*

Matching Beam Paths

With these more complicated set-ups, and even with more simple ones, it may be difficult to match all the optical path lengths. In such cases it may be preferable to forget all about path matching when first setting up, and to insert path-extenders as necessary. These are pairs of mirrors mounted on a sliding frame such as a length of triangular optical bench, forming a narrow Z in the beam and acting as a kind of optical sheepshank. In a sand-table arrangement they can be placed on wooden bases equipped with spikes.

Large Reflection Transfer Holograms

On a table of the size we have been discussing you can make reflection holograms up to 30 × 40 cm in size (you will probably have to fold the reference beam unless you use a ×60 objective to expand it). The best way of making transfer copies is to use a transfer frame (see pp. 164–5), but if you are making only the occasional copy of this size you may feel it is hardly worthwhile going to the trouble of making one, or to the expense of buying one. The simple answer is to use a frame with a large plane mirror, as described earlier (Fig 8.2, p. 96). To minimize distortion you should use as long a beam throw as possible, and have the image crossing the plane of the hologram rather than being wholly real or wholly virtual. If you want a very deep image without distortion you will need a collimated beam (see p. 166).

Color control   The importance of color control in both master and final hologram has been stressed in Chapter 8. Let us now look into the techniques a little more closely.

In the case of the master hologram, the important thing is that the emulsion should be exactly the same thickness before and after processing. We saw in Chapter 6 that there are several ways of achieving this. The most important are as follows:

1 Process the film in a developer with a strong physical action, fix, and use a rehalogenating bleach.
2 Process in a developer with a strong tanning action and use a reversal bleach.
3 Expose and process as in 2, but with the emulsion immersed in water throughout exposure, including the transferring process.
4 Process in a non-physical, non-tanning developer and use a bleach with a strong physical action, without fixing.

Of these, methods 3 and 4 give the brightest image, though as far as method 3 is concerned, it is not very convenient to have to keep the master hologram permanently underwater. As far as obtaining the correct reconstruction wavelength is concerned, method 4 is the most reliable. Formulae for all these techniques are given in Appendix 10.

A correctly-processed master hologram will give a bright image when it is placed back in a position identical with the one it occupied when it was made (but flipped). If it needs a replay beam at a smaller angle of incidence (ie more nearly perpendicular), then the emulsion has shrunk. The image may then show some change in apparent depth when replayed. It is possible to regain the correct angle by swelling in parabenzoquinone, triethanolamine or sorbitol (see Appendix 10), but every additional process adds to the noise, and it is better to avoid the necessity for such treatment.

The color of the final transfer hologram is usually modified to make it visually brighter, by shortening the replay wavelength to golden-yellow or green. This can be achieved in four different ways:

1 Develop in a developing solution which has a weak physical action, fix and rehalogenate.
2 Develop in a tanning developer containing a complexing agent such as sodium sulfite, and reversal-bleach.
3 Develop in a developer which generates colloidal silver, fix, then post-swell as necessary.
4 Pre-swell, expose, develop in a non-physical, non-tanning developer and use a physical rehalogenating bleach.

Of these, methods 2 and 4 seem to give the brightest results, particularly method 4. Method 3 seems to work well with some batches of Agfa-Gevaert plates, though it does not seem to be very successful with films.

Finally, there are methods of differential swelling used in making pseudocolor reflection holograms. These are described in Chapter 18.

Collimated reference beam   When the master hologram is played back to produce the real image that is to act as the object for the transfer hologram, the replay beam should, strictly speaking, be the conjugate of the original reference beam. This means that it should (for all the set-ups described so far) be a converging rather than a diverging beam. This is difficult to achieve in practice, as it necessarily involves quite large and well-corrected optics. Where correct perspective is important, as in images that are wholly real, the best solution is to collimate the master reference beam using a large lens or curved mirror. The conjugate beam is then also a collimated beam, and this can be achieved using the same lens or mirror. The techniques of producing a collimated reference beam are dealt with in Chapters 13 and 14.

References

1 McNair D (1983) *How to Make Holograms*, pp. 128–53. Tab Books Inc.
2 Unterseher F, Hansen J and Schlesinger R (1982) *The Holography Handbook*, pp. 23–43. Ross Books
3 Saxby G (1980) *Holograms: How to Make and Display Them*, pp. 73–8. Focal Press.
4 Dowbenko G (1978) *Homegrown Holography*, pp. 150–2. American Photographic Publishing Co.

CHAPTER

12

Transmission Master Holograms

'The question is,' said Humpty Dumpty, 'which is to be master – that's all'.
LEWIS CARROLL, *Through the Looking Glass*

When making holograms up to 8 × 10 in in size, reflection masters are convenient because they can be transferred on a simple frame with a Denisyuk lighting arrangement. For larger holograms, and for the highest-quality results, transmission masters offer a number of advantages:

1 The reference beam does not illuminate the object, so there is no danger of the shadow of the edge of the film falling on it.
2 Exposures can be shorter, as faster film can be used and developed density is lower.
3 There is usually more room for illumination.
4 The diffraction efficiency of the master hologram is no longer so important (though signal-to-noise ratio still is).
5 Transmission and reflection transfers can be made equally easily.
6 Master holograms can be made with a laser of one wavelength and transferred using a laser of a different wavelength, eg a pulse-laser master can be transferred using a HeNe laser.
7 The fringe contrast lost through out-of-plane movement is less than one-fifth of that lost through a similar movement in a reflection hologram.

Stability Requirements

It is generally agreed that the optical problems associated with making holograms go up roughly in proportion to the square of the linear dimension of the hologram; and as we are now considering holograms of 8 × 10 in upwards intended for transfers of the highest quality, we shall have to raise our standards for stability still farther, particularly if the table is to be used for making reflection transfers.

Though there is no particular reason nowadays why they should continue to be, large master holograms have been and still are usually made on plates rather than film; we shall therefore use the word 'plate' when discussing large holograms. There are in fact advantages in using film rather than plates: it is cheaper and easier to process, and it can be mounted on heavyweight glass or acrylic sheet, reducing both sympathetic vibrations (*microphoning*) and reflections from the base (*halation*). Microphoning can be a serious problem with large reflection holograms; with transmission holograms it is less important. However, *in-plane movement* caused by small changes in temperature is important with transmission holograms, and a temperature-controlled environment is therefore necessary.

Large holograms require heavy optical equipment, and when this is moved around on a table a surprising amount of distortion of the surface of the table occurs. Holographers who have a patron with a bottomless purse will be able to buy a research-grade steel table with pneumatic supports which sense any change in load and at once adjust the air pressure to compensate. Less fortunate workers may settle for a professionally-installed concrete table. The mix needs to change from ordinary concrete at the bottom to almost pure granite at the upper surface, and as the job must be completed in one more or less continuous pour, it is not a job for the amateur. Alternatively, 6 or 12 mm steel plate can be bonded in situ to ordinary concrete; this will usually be done by pouring the concrete onto the steel plate and turning the whole table over once it has set hard. The operation requires lifting tackle, so plenty of headroom is required. A table 3×0.5 m in size weighs about 10 tonnes, and for the supports you will need twelve inner tubes. Keeping these tubes inflated is a real problem, and one solution is to remove the valves and connect all the tubes to a central compressed-air cylinder. Another solution is to cut out the entire valve assembly from each tube, patch the hole, and fit the valves to the outsides of the tubes. Your local garage will be able to do this for you, though no doubt they will be curious as to the reason for this request. You may prefer to use the pneumatic supports designed especially for anti-vibration purposes, but they are often not designed with holography in mind, and you should take professional advice before installing such mountings.

If you have decided to work with a sand table, then provided it is large enough and rigid enough (a slate or concrete base is essential) there is no reason why you should not make large master holograms on it. Some professionals prefer to work with a sand table for master holograms and a hard-top table for transfers, and at least one worker has combined the two in a single large table.

Optical Equipment

The only new optical equipment you will need are a second spatial filter, several large plane front-surface mirrors and a collimator with a diameter not less than the diagonal of the largest plate you will be using. In most professional holographic laboratories a collimating mirror is preferred to a lens; however, some alternative layouts using a lens have been included. You will also need a plate-holder or transfer holder that is sufficiently large to take the biggest plate you are going to be working with. If you are intending to use an Ar^+ or Kr^+ laser you will need to be careful when buying your optical equipment. An unspread beam with a power of around 1 W will vaporize an aluminum coating, and before buying any reflector or beamsplitter for use with a source of power greater than 500 mW you must make sure it is suitable. Metallizing should be of titanium (aluminum is too volatile and gold too soft); dielectric mirrors and beamsplitters are better, as they do not absorb any energy, and therefore do not undergo any thermal movement.

Gravity Bases

This is a question you have no need to ask if you have decided on a sand table or a plain concrete surface. In the former case you will still use plastic piping for the lightweight components and timber platforms with spikes for the heavier ones. Wood is acoustically fairly dead, and you can use a hot-glue gun to fix the components to the platform. McNair (1) describes an ingenious holder for a collimating mirror, made from plastic piping components, as well as a support for a large plane mirror giving an overhead reference beam. For heavier work on a hard table, these designs can be duplicated in scaffolding (see below). If you have opted for a concrete table with a plain surface you will need to use such heavy supports; these are usually called 'gravity bases'. If you have a steel surface you have the choice between gravity bases and magnetic bases; usually a mixture of the two will be appropriate. Use magnetic bases for those small components that are most likely to be knocked accidentally in the dark. You will then be able to move them around easily and, when they are accurately in position, lock them firmly. Where a component is going to be moved only along a straight line, as with path adjusters ('optical sheepshanks')

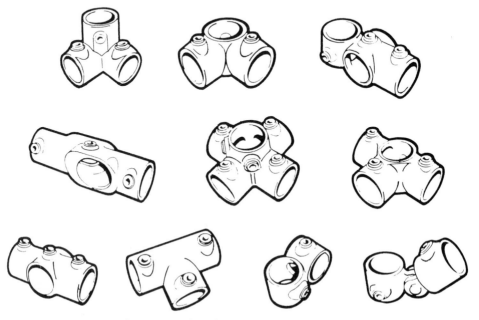

FIG 12.1 *Some examples of KeeKlamp joining pieces (see text).*

or the laser relay lens, a triangular bench is the most convenient support. Once in position, you can fix it with a shot of hot glue at the ends. For heavy components such as a collimating mirror, a versatile and cheap material is the type of scaffolding assembly known as KeeKlamp. This is a system of cross-pieces, T-pieces and L-pieces for joining standard scaffolding tube; the connecting pieces are assembled loose, then locked in position with a hexagonal key. The result is massive and rigid, and can be made heavier still by filling the tubing with sand. Many types of joint and clamp are available, and with a little ingenuity you can make adjustable stands for all your heavier optical equipment (Fig 12.1).

Size 8 tubes (approximate outside diameter 48 mm) are recommended for heavy-duty components and size 6 (34 mm) for lighter components. Kee Systems, the manufacturers, also make clamps for size 5 (26.9 mm) and size 4 (21.3 mm). Different sizes can be combined by filling the larger pipe with Isopon car body filler, pushing the smaller pipe inside, and, when the filler has set, sealing with hot glue or epoxy resin. Kee Systems will cut scaffolding pipe to any required length for a very small charge, and it is suggested that if you are going to make up support systems on an ad hoc basis, you should order a supply of pipes with lengths ranging from 150 mm to 600 mm in increments of 50 mm. (All measurements are metric, at

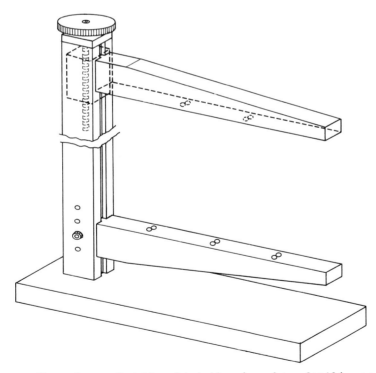

FIG 12.2 *Heavy-duty adjustable plate-holder for plates* 8×10 in *up to* 40×50 cm. *The locating pieces are 6 mm ball-bearings fixed in drilled depressions with epoxy resin or hot glue.*

least in the United Kingdom and Europe.) The address of the company, and of its outlet in the United States, are given in Appendix 12.

Plate and Transfer Holder

If you have access to high-quality metal-working facilities you can build a plate-holder similar to that shown in Fig 12.2. This design has been used repeatedly, and has been found to give good support. The five-point holder ensures that no sympathetic vibrations can occur except at frequencies so high that any displacements can be ignored. With the odd badly-cut plate it will revert to three-point support.

Transfer holders are adjustable parallel pairs of plate-holders, which, of course, you can also use for single plates, and for single-beam reflection holograms with the subject matter mounted in the rear plane. David Pizzanelli has produced an excellent transfer holder using KeeKlamp units, and this is described (with permission) in the Box.

A heavy-duty exposing frame using scaffolding clamps

This frame is one of a number of heavy-duty holders and stands designed and built by David Pizzanelli for See-3 Holograms. It is made from 48 mm (2 in) scaffolding using KeeKlamp components, and be used as a single plate-holder, a transfer holder, or a jig for single-beam reflection holograms. The construction is shown in Fig 12.3. The diagrams are self-explanatory. The plate is held in position by three pairs of 9 mm (3/8 in) ball-bearings affixed to the lower support and two pairs affixed to the upper support with a suitable fixative. In order to allow for plates that are cut slightly out of true, it is desirable to have a certain amount of 'give' in the support bearings. This can be achieved by using hot glue rather than epoxy resin (which would otherwise be preferable) to secure them. A better way is to use staggered pairs of ball-and-socket door catches sunk into the pipes. For 8×10 in and smaller plates three-point suspension is adequate. A leveling of adjustment for the frame is provided by a large ball-bearing affixed to the front member of the base; the front of the frame can be raised or lowered by simply turning the front member in its sockets. Because of the sideways force exerted on the ball-bearing it needs a secure seat, and this is provided by a stack of washers glued together, into which the ball-bearing fits snugly, the whole being secured by epoxy resin. By loosening the hexagonal-socket grubscrews in the connectors the front member can be rotated as necessary, then locked back in position.

When intended for use with a sand table, the vertical members are extended downwards and terminated with cross-bars which can be buried in the sand. The collars under the upper member of the frame are important, as they prevent the member from dropping down when the plate is removed. They are also used as a fulcrum for raising the member by hand when inserting the plate. When the plate is in position there should be a gap of a few millimeters.

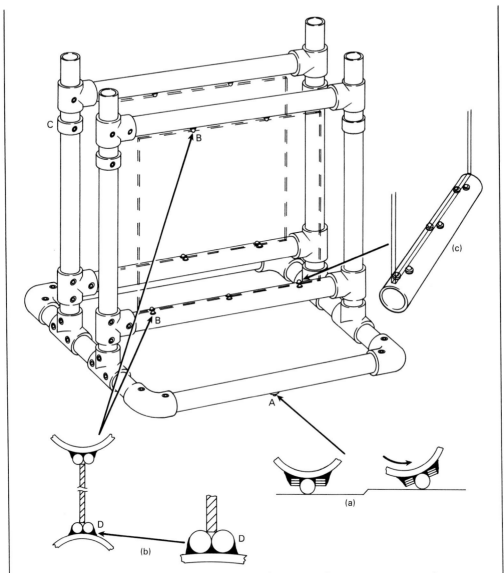

FIG 12.3 *A double plate-holder made from KeeKlamp components. A, a kinematic support made from a large ball-bearing cemented into a socket made from washers; B, a pair of ball-bearings supporting the plate: two above and three below for large frames, one above and two below for smaller frames; C a collar which supports the upper frame member when the plate (interrupted lines) is removed. Inset (a) shows how the kinematic support is rotated to level the frame; inset (b) shows how the ball-bearings support the plate. Inset (c) shows a recommended modification which allows five-point support for plates that are not cut true. Spring balls from ball-and-socket door catches are let into the pipe and secured with epoxy resin.*

For somewhat lighter structures you can get components in kits, made to assemble with clamps somewhat like those used in building chemistry set-ups. These are intended for professional photographic layouts, and are supplied by Climpex Ltd. Addresses are given in Appendix 12.

Collimators

Collimators are necessary because the master hologram has to be flipped in order to form a real image; to avoid distortion it is necessary to use a reconstruction beam that is the conjugate of the original reference beam. If you make the master hologram with a collimated beam the conjugate will also be a collimated beam, and you can use the same optics – and very nearly the same set-up – for the transfer.

Lens or Mirror?

In the realm of conventional optics a collimator is usually a lens, whereas in holography it is more often a mirror. However, the geometry of a mirror set-up means that there is a total beam path equal to almost twice the focal length of the mirror, whereas a lens can be close to the plate, minimizing the effect of optical imperfections. The lens need not be corrected for chromatic or *off-axis aberrations*, and *spherical aberration* can be kept small by using a plano-convex lens with the flat side facing the laser. The transmittance of such a lens is typically better than 95%. A mirror, on the other hand, has a reflectance of less than 90%. It needs to be used off-axis, which means that unless the offset is very small (which limits the scope of the set-up) there will be off-axis aberrations such as *coma*, which result in unevenness of illumination. The long distance between the hologram and the mirror means that any imperfections in the mirror also have plenty of space to form their own far-field diffraction patterns, which means that for good clear master holograms it is necessary to use a high-quality mirror. It might be argued that lenses of sufficient diameter are more difficult to come by than mirrors of comparable size, but there are simple ways of making perfectly satisfactory lenses of almost any desired size, as we shall see. Incidentally, for those whose ideas are ambitious enough to consider making a holographic lens or mirror, there are instructions for doing this in Chapter 15.

At present, most serious holographers seem to have settled for collimation by mirror, and this system does have the advantage that holograms larger than the mirror can be made if the perspective distortion caused by a small amount of divergence in the reference beam can be tolerated. Such a set-up using a lens involves either a very long table or a folding of the diverging beam with a plane mirror. So that you can make a choice, this chapter includes examples of practical layouts using both mirrors and lenses as collimators.

Full-aperture Transmission Master Holograms

This term 'full-aperture' refers to master holograms with dimensions that are at least as large in both directions as those of the final hologram (as distinct from masters for restricted-aperture transfers, which are dealt with on p. 174 ff). The geometry is broadly that of a conventional laser transmission hologram, though the object illumination can be as complicated as desired and the reference beam can come from any angle that is convenient. The only stipulation is that the reference beam should be collimated. For reasons that will become clear in Chapter 13, it is usual to have a reference beam angle of incidence of about 30–45° for reflection masters and about 60° for transmission masters. Using a mirror of about 320 mm (12.5 in) diameter and 1200 mm (48 in) focal length, covering an 8 × 10 in plate, a typical arrangement would be as in Fig 12.4. This arrangement has the subject matter set on its side and illuminated from 'above'.

You can use any similar arrangement that has the subject matter upright; the reference beam then becomes 'side' instead of 'overhead', and will demand a different beam geometry at the transfer stage. An alternative layout is shown in Fig 12.5. Again, the arrangement will serve for either overhead or side reference beams, depending on the orientation of the subject matter, and you can use the spare portion of the reference beam to produce backlighting.

There are, of course, endless variations, and every book published on practical holography has had its own, ranging from the excellent to the clearly unworkable.

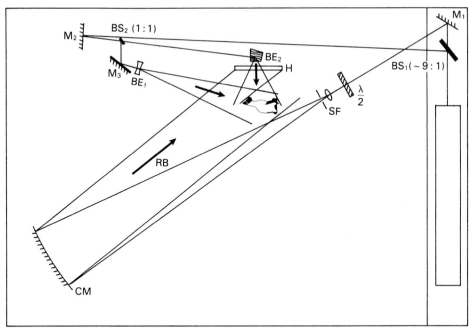

FIG 12.4 *Layout for a transmission master hologram. The symbols have their previous meanings; new symbols are* CM *(collimating mirror) and* λ/2 *(half-wave plate).* NOTE: BS_1 *must be a metallized beamsplitter (preferably variable).* BE_1 *produces direct light,* BE_2 *is a diffuse reflector, and is elevated. Carding-off is not shown, but the lower half of the reference beam provides rim-lighting and should not be carded off.*

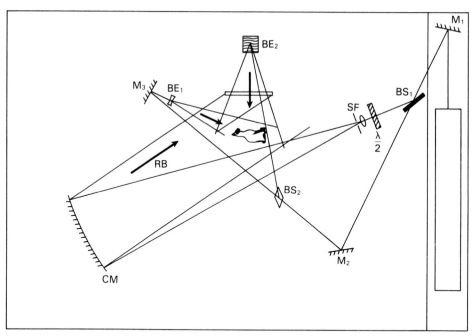

FIG 12.5 *Alternative layout for a transmission master hologram using plain glass beamsplitters, both at about 70° incidence. Note that in this layout the reference beam is the beam reflected from BS_1, whereas in* FIG 12.4 *it is the transmitted beam. With a metallized beamsplitter the reflected beam should always be the stronger, to reduce second surface reflections.*

In general, you will find that the lengthy path of the reference beam leaves you plenty of room to play with the object illumination, and perhaps this is the strongest argument for a mirror rather than a lens for collimating purposes. With variations of these two basic arrangements you will probably find that it is easier to work with the subject matter upright and a side reference beam. Unfortunately, this makes the transfer geometry awkward, as the final hologram will then require an overhead reference beam; it is a good deal more difficult to make transfers with the planes of the replay and reference beams perpendicular than it is with them parallel. A modification which will turn the reference beam into an overhead one, with a small increase in path length, is shown in Fig 12.6.

You can adapt any side reference beam to make an overhead beam by tilting the collimating mirror slightly upwards at a large relay mirror which re-directs the beam downwards at 45° or 60°. The mirrors are not particularly easy to set up as they are both skewed. Set them up with the beam unspread and make sure that the relay mirror is well clear of the hologram, otherwise you may be able to see it in the holographic image.

Set up the reference beam first, including the beamsplitter, roughly aligned. Once you have lined up all the components with the undiverged beam, insert the spatial filter. Make sure the collimating mirror is uniformly illuminated and that its beam is as nearly parallel as possible. A mirror such as that described above will be

TRANSMISSION MASTER HOLOGRAMS

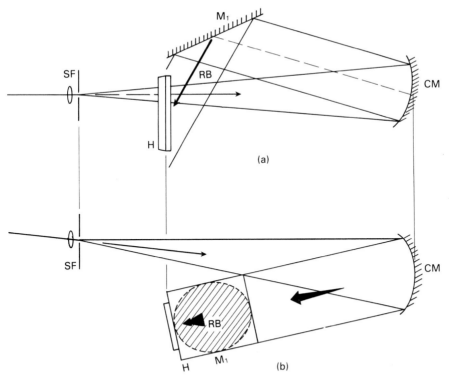

FIG 12.6 *Elevating a collimated beam. The collimating mirror CM is tilted upwards and slightly sideways, and the beam is reflected down at the required angle of incidence by the large plane mirror* M_1. *(a) and (b) represent side and plan views respectively.* M_1 *is shown at its minimum possible size and H at its maximum size for a circular collimating mirror.*

adequately filled by a beam expanded by a ×40 microscope objective. Examine the shadow of the mirror on the wall and make any necessary adjustments to the beam by means of one of the relay mirrors in the path so that the mirror is centrally placed in the beam; then go over to the patch of light reflected by the mirror onto the wall, and measure it. It should be the same size as the mirror, to within a few millimeters. If it is too large, move the spatial filter away from the mirror; if it is too small, move the spatial filter nearer to the mirror. When you are satisfied, insert the plate-holder at the appropriate angle and load a dummy plate of white card. Examine its shadow on the wall and ensure that it is positioned symmetrically in the ellipse of light.

Now set up the subject, then its illuminating beams, one at a time, checking distances to obtain the best match you can to the optical path of the reference beam from the beamsplitter to the plate-holder. With the layout of Fig 12.4 you will need a variable beamsplitter, or at least a metallized 9:1 beamsplitter; with the layout of Fig 12.5 you should manage with plain glass or a 3° wedge. With most subjects you will get the best lighting effect with directed light as key light and more diffuse light as fill-in light. Before checking the beam ratio, check the polarization of the object and reference beams as described on p. 85, and adjust the polarization of the reference beam as necessary by rotating the half-wave plate. Now check that the beam ratio lies between 3:1 and 10:1 (if you have a variable beamsplitter adjust it so that it is 8:1), and you are ready to make a test shot.

Processing a Transmission Master

Some workers prefer not to bleach-process their transmission master holograms, but simply to develop and fix them, on the grounds that any other process allows migration of the silver ions in the emulsion and can also cause a differential cross-linking in the gelatin, both of which effects increase noise. They also assert that bleach-processing causes the linearity of the emulsion's response to light to be seriously affected over the whole range of densities. There are two approaches to non-bleach processing, both of which give excellent results:

1 Non-physical development Here the developer contains a developing agent such as ascorbic acid and an alkali such as sodium or potassium hydroxide buffered with sodium carbonate, none of which compounds forms any complexes with silver ions. Fixation is in plain sodium or ammonium thiosulfate after a thorough rinse in two successive baths of de-ionized water. An alternative to fixing is stabilization with an acid stopbath. Formulae are given in Appendix 10.

2 Physical development Here the developer contains hydroquinone as the main developing agent, in combination with metol or phenidone, and comparatively large quantities of sodium sulfite. This combination, together with an alkali such as sodium carbonate or metaborate ('Kodalk'), has a strong tendency to form unstable complexes with silver ions. The net effect is that silver ions are removed from the unexposed regions of the emulsion and deposited as atoms on the developing specks of silver, enhancing contrast and retarding the formation of filaments. Little silver is lost in fixation (some workers do not fix at all), and the angle of the fringes to the emulsion surface does not change significantly.

Both systems have their supporters, and both systems work well; if the final density is around 0.7 (a medium grey) the signal-to-noise ratio will be very high and the image brilliantly defined. Indeed, the sight in a darkened lab of the real image with its extraordinary perspective hovering in the space between the two plateholders is an experience that in itself makes the work of setting up worthwhile. However, unbleached master holograms demand the use of a powerful laser for making the transfer hologram, and in view of the considerable improvements in signal-to-noise ratio obtainable with the more recently-developed techniques of bleach processing, most holographers now prefer to use one of the bleach processes recommended for transmission masters in Appendix 10.

Restricted-aperture Master Holograms

In both transmission and reflection holography, master holograms are often made with the vertical dimension restricted. In reflection holography the restriction is slight, its purpose being simply to make the transfer image brighter, but in transmission holography the horizontal aperture may be no more than a narrow slit; this eliminates the vertical parallax, replacing the vertical information by a holographically-generated diffraction grating. In both cases the final image is very much brighter than that generated by a full-aperture transfer hologram. The reason for this is that in the transfer process all the light that forms the 'object' for the final hologram has come from a restricted aperture, so when the hologram is flipped for viewing, all the light forming the image is concentrated back through the real image of the aperture, and the image is brightened in proportion to the narrowness of the aperture. Restricted-aperture transfer also provides a method of producing multiple images in both transmission and reflection holograms (see pp. 184–5).

Lens as collimator As argued earlier, there are good reasons for using a lens rather than a mirror for collimation purposes. It is possible to pick up ex-Government aerial reconnaissance lenses with diameter of about 150 mm and a focal length of about 1 m, and such lenses are excellent for 4 × 5 in holograms. You may find that the iris diaphragm is damaged, and if so, it is best removed altogether. The spatial filter has to be placed at the principal focus of the lens if it is to produce a collimated beam; as such lenses are usually of a telephoto configuration (ie the focal length is greater than the physical dimensions of the camera) this distance is often less than the marked focal length. Larger lenses are usually condenser lenses, sold as pairs of inward-facing plano-convex elements. Only one of these is needed. The largest size you are likely to find is about 305 mm (12 in) diameter, with a focal length of about 1 m. Larger plano-convex lenses are difficult to find, and expensive when you do find them. The best source for oddball optical items is the Edmund Scientific Corporation (see Appendix 12).

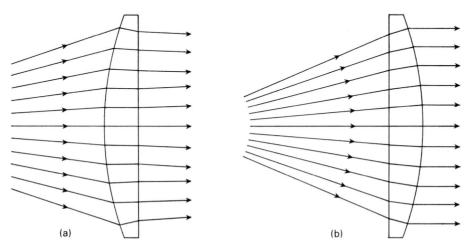

FIG 12.7 *When using a plano-convex lens for collimating, position the plane surface so that the divergent beam falls on it* (b) *Dividing the refraction equally between the surfaces minimizes spherical aberration. Configuration* (a) *will not give a uniform beam.*

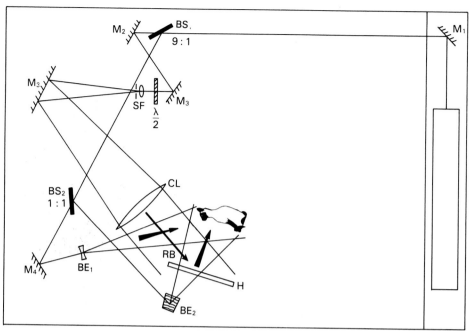

FIG 12.8 *A rearrangement of the geometry of* FIG 12.4 *to accommodate a collimating lens* CL *instead of a mirror. The object lighting remains the same, but there will probably be little overspill from the reference beam to give additional illumination. This geometry is suitable for metallized beamsplitters.*

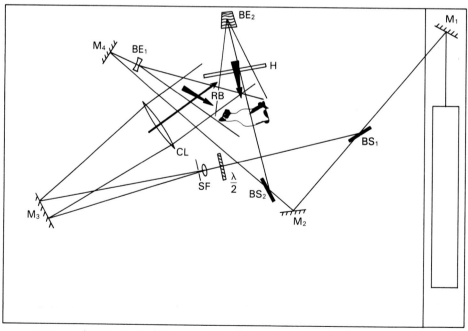

FIG 12.9 *Arrangement of the geometry of* FIG 12.5 *to accommodate a collimating lens* CL *instead of a mirror.* BS_1 *and* BS_2 *are plain glass.*

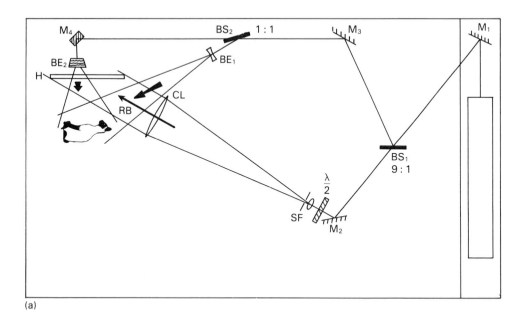

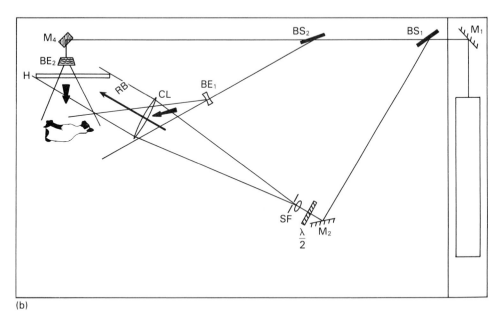

FIG 12.10 *With a longer, narrower table it is not necessary to fold the diverging beam from a collimating lens. These geometries are* (a) *for metallized beamsplitters,* (b) *for plain glass beamsplitters.*

Plano-convex lenses are preferable to double convex lenses as collimators, because if they are oriented correctly (the plane face towards the diverging beam) the deviation of the beam is divided almost equally between the two faces, and this is the condition for spherical aberration to be a minimum (Fig 12.7).

If you have a table with proportions of around 3:2, you can modify the mirror configurations of Figs 12.4 and 12.5 by using a combination of plane mirror and lens instead of the concave mirror. As the constraint of a small angle of incidence no longer holds, you have more flexibility of layout (Figs 12.8 and 12.9).

On the other hand, if your table is at least twice as long as it is wide, you will find it unnecessary to fold the optical paths, and thus there will be no need of the large front-surface mirror (Fig 12.10). On the debit side, you need a particularly stable table because of the long throw.

Slit Transfer Master Holograms

Slit master holograms can be used for special kinds of reflection-image effects (see pp. 184–5), but their usual purpose is for rainbow-hologram masters. Some workers make a full-aperture master hologram and decide by inspection the most suitable angle; others feel this approach to be wasteful and slipshod, as they believe the truly creative holographer should know exactly what is required long before switching on the laser. Whatever your personal feelings about this, there is no doubt about the financial advantages to be gained from the ability to make three or four perfectly good transmission masters from a single 8×10 in or 30×40 cm plate. You can make your tests on 35 mm film; such 'tests' are often good enough to be used as master holograms in their own right. However, if you are going to make restricted-aperture masters you will have to learn how to cut up glass plates in semi-darkness. You will find step-by-step instructions for this in the Box.

Cutting glass

People unfamiliar with the glazier's trade often associate glass-cutting with diamonds. Certainly, there are diamond glass-cutters, but they are expensive, difficult to use and easily damaged. Professional glaziers seldom use them; they use steel wheel cutters, which are cheap and reliable. You can buy a wheel cutter at any do-it-yourself counter in a multiple store, and as it has six wheels (when one is worn you just move on to the next) it will last for years.

If you are new to glass-cutting, try out your wheel cutter on some pieces of clean scrap glass. Place a piece of glass on several thicknesses of newspaper. Put a thick rule on it (a boxwood meter rule or a thick Perspex rule, not a carpenter's steel rule), and hold it pointing towards you firmly with the thumb and first two fingers of your left hand (right, if you are left-handed). Take the wheel cutter in your other hand, hold it upright with the flat side against the rule, and rest it on the far side of the glass with the wheel just over the edge and touching it. Pull the cutter towards you a little to start the scratch, then draw it steadily and quite quickly towards you across the glass. Don't press hard; just do it confidently, at a steady speed. You will hear a hiss from the glass as you draw the wheel across it. Put down the wheel cutter, pick up the glass, and place it with the scratch uppermost along the edge of the rule. Put the palm of your hands on the two sides of the scratch and press down gently. The glass will snap like a biscuit. If you need to use force there is probably a gap in the scratch, and when it finally gives way the break will curve away from that point. You will probably be able to break the other part away, and will be left with

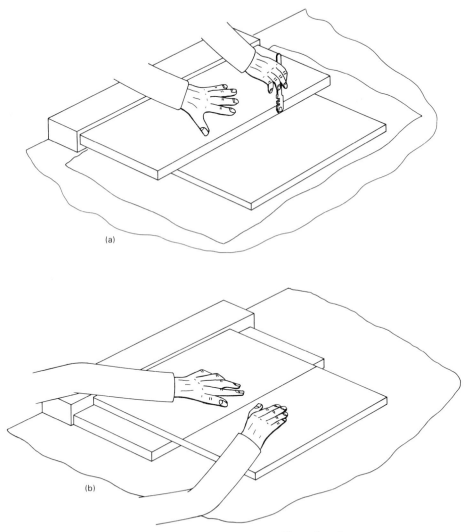

FIG 12.11 *Cutting glass plates in the dark. (a) Place the plate on newspaper, emulsion side down, and set it against the stop. Hold the template against the stop. Hold the cutter upright between your forefinger and thumb, flat side towards the template, and draw it steadily towards you. Then place the glass plate on top of the template and push down on both sides of the crack with the flat of the hand (b). The glass should snap cleanly and easily.*

a shallow cusp, which you will have to crack off bit by bit with the notch in the glass cutter, or with pliers, leaving a ragged edge and a sense of defeat. Keep trying; you will quickly acquire the knack of exerting precisely the right pressure, and you will soon hear that satisfying hiss every time.

To cut glass in the dark you need a template. There are many ways of making one, but the simplest template is probably made from two pieces of plywood or acrylic sheet that are the width and height you want to cut, but 1.5 mm narrower in both dimensions, to allow for the width of the cutter. If your workbench has a raised edge you can use this as a stop; if not, nail a batten to it. Cut the plate on newspaper or print blotting paper (Fig 12.11).

Before you cut up a plate, brush the paper to make sure there is no dirt or glass chippings on it. The template should be at least 3 mm thick, otherwise the cutter may ride up on it and lose contact with the glass. This can happen quite easily in the dark, and is the reason why steel rules are not a good idea. Professional glaziers use a meter rule with a strip of baize glued to the underside to prevent slipping.

If your master plate is only a few centimeters wide you have a good deal more freedom with the lighting, as there is more room. Keep the reference beam at an angle of incidence of about 60° (ie 30° to the plane of the plate); this will simplify

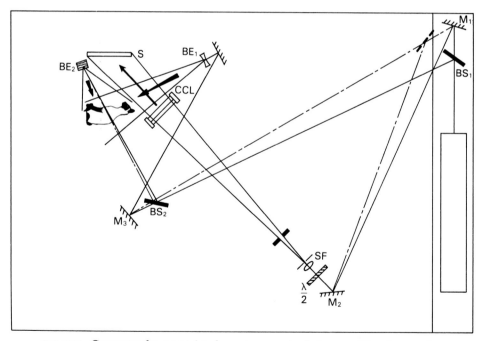

FIG 12.12 *Geometry for a restricted-aperture master hologram (S) using a cylindrical collimating lens (CCL). The beamsplitters are metallized; the configuration for a plain glass beamsplitter is indicated by broken lines. As this general principle holds for all geometries the remaining layouts are for metallized beamsplitters only.*

any calculations you may have to make if you want to launch into multicolor imaging (Chapter 18). One considerable simplification in slit layouts is that you no longer need a collimator that operates in two dimensions, and you can make a one-dimensional lens collimator very easily. If you already possess a collimating lens or mirror you can still use it, of course, carding off the unwanted light, but if you are interested from the start only in slit master holograms you can save a good deal of expense by making simple liquid-filled optics from acrylic sheet (see Chapter 15). The optical layouts are similar to those of Figs 12.8–12.10 except that this time the excess light is either carded off or used as extra rim-lighting for the subject matter, and only a narrow strip of the master hologram receives the reference beam.

Fig 12.12 shows a simple layout using a liquid-filled cylindrical lens. As suggested above, you can open out the carding-off on the object side to produce a rim-lighting effect.

Holographic Collimators

One of the most elegant ways of making your own collimator is to make it holographically. Holographically-made mirrors and lenses are called holographic optical elements (HOEs), and methods of making them are described in Chapter 15, pp. 228–33.

One of the more important characteristics of HOEs is that light falling perpendicularly on them is not in general transmitted (or reflected) along the same axis. The emergent beam angle depends on the geometry of the beams used when the HOE was made. This means, for a holographic mirror, that minimum aberrations will occur when the incident beam strikes the mirror at the same angle as one or other of the beams that were used in making it. Thus a mirror made with one beam at 90° incidence and the other at 45° incidence will always reflect light at about 45° to the incident beam, but the most effective reflection will occur when the mirror is set up so that the incident beam is normal to the emulsion surface (Fig 12.13). This angle is particularly convenient, as a holographic plate at an angle at 45° to the beam (in either direction) will be illuminated by a circular patch of light rather than an

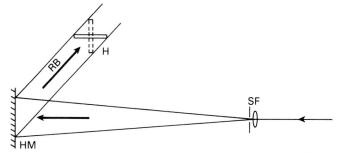

FIG 12.13 *If a holographic mirror (HM) is made using beams at 0° and 45° incidence (see pp. 230–2), a beam at normal incidence will be reflected and collimated at 45°. This will give a circular patch (not elliptical) on a holographic plate at either of the orientations at H.*

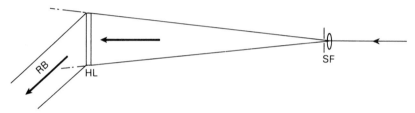

FIG 12.14 *A holographic lens* (HL) *made using beams with angles of incidence of* $0°$ *and* $45°$, *as with the set-up in* FIG 12.13 *both collimates the light and deflects it through* $45°$.

elliptical one. Similar principles apply to holographic lenses. If the two beams are at angles of incidence of $0°$ and $45°$ respectively, the emerging beam will be turned through $45°$ with respect to the incident beam (Fig 12.14).

With a holographic lens, a large proportion of the light remains undeviated, and you can use some of this light as extra object illumination, by placing a large mirror (not necessarily front-surface) in the beam.

As HOEs are so cheap, in terms of money if not of effort, it is a good idea to make several lenses and mirrors with different emergence angles.

Figs 12.15 and 12.16 show two schematic layouts for HOE collimators. For

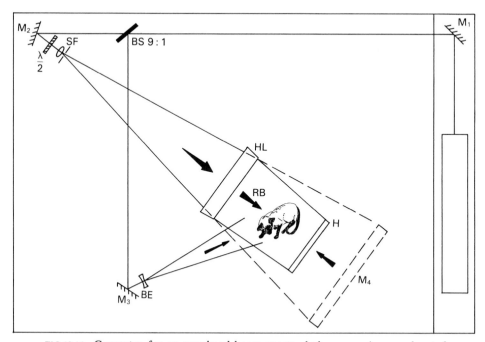

FIG 12.15 *Geometry for an overhead-beam master hologram using an elevated holographic lens* HL *which collimates light and directs it downwards.* M_4 *is an optional mirror elevated to catch the zero-order beam from* HL *and provide additional object illumination. It may itself be a* HOE. *For clarity only one main object illuminating beam is shown.*

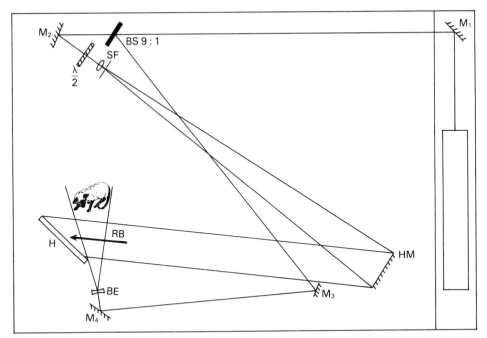

FIG 12.16 *Geometry for a side-beam master hologram using a holographic collimating mirror* HM.

simplicity only one object-illuminating beam is shown. To convert an overhead beam to a side beam or vice versa, all you have to do is to rotate the HOE through 90° on an axis perpendicular to its surface. You will also need to check on any rotation of the plane of polarization, and correct this as necessary with your half-wave plate.

Reference

1 McNair D (1983) *How To Make Holograms*, pp. 192–8. Tab Books Inc.

CHAPTER
13

Transfer Holograms

'Would you tell me, please, which way I ought to go from here?' said Alice.
'That depends on a good deal on where you want to get to,' said the Cat.
LEWIS CARROLL, *Alice in Wonderland*

For our purposes, the term 'transfer hologram' is taken to mean a hologram that is made using the real image from another hologram as the object, with the plates separated. Holograms made with the two plates close together or in contact are referred to as 'copies'; those made by using the real image formed by a lens are called 'focused-image holograms' and are treated separately (Chapter 14).

Basic Principles for Transfer Holograms

To make a satisfactory reflection image from a transmission master hologram, the master should be at least as wide as the transfer hologram and if possible wider, in order to give maximum parallax.* This is because in looking at a transfer image (which has necessarily been flipped) you are looking at it through the real image of the master hologram, ie through a rectangular hole. You can see this effect clearly if you stand several meters away from the hologram and move sideways. At a certain point on each side a black shadow will move across the image in the opposite direction, indicating that the edge of the real image of the master hologram is in front of both the plate and the holographic image. You can see the vertical boundaries, too, if you move up and down. Because of this restriction the amount of parallax in a transfer hologram is always less than that of the original virtual image. It varies with the separation of the plates; the more the separation, the less the parallax. Virtual-image transfers have the most parallax and real-image transfers the least, with hologram-plane images in between.

The general layout of the transfer area for the three types of image is shown in Fig 13.1.

---

* This is not applicable to the same extent when considering large master holograms (30 × 40 cm and larger) where there is adequate parallax even in a 90 × 90 cm transfer.

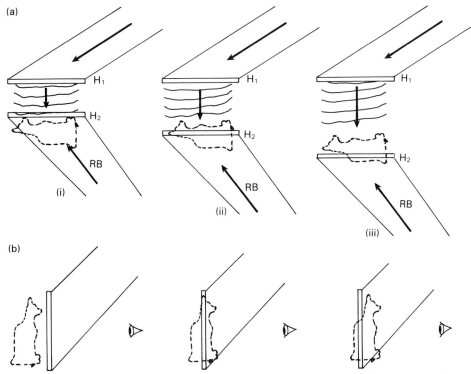

FIG 13.1 *Reflection transfer relationships: the separation of the master and transfer holograms for* (i) *virtual,* (ii) *hologram-plane,* (iii) *real image.* (a) *Shows the transfer geometries with the master flipped to give the pseudoscopic real image,* (b) *the final holograms flipped for viewing.* H_1 *is the master hologram,* H_2 *the transfer hologram.*

Aperture Restriction

You can make the final image noticeably brighter by masking down the master of a reflection transfer hologram so that it is narrower in a vertical direction (Fig 13.2). This restricts the vertical parallax, but this is not particularly important in most cases as the eye does not receive cues about depth from vertical parallax, and even the most extravert viewers do not usually jump up and down in front of a hologram. A restriction of the master hologram to half the height of the transfer hologram is sufficient to increase the *luminance* of the final image by a factor of two.

A general layout for reflection transfers is shown in Fig 13.3. If you want to modify this basic layout to use a collimating lens or mirror or HOE, you should simply follow the general rules laid down on pp. 228 ff, regarding a convex lens or mirror as the equivalent of the collimating mirror and the HOEs as non-axial lenses or mirrors. For example, using a HOE mirror, the reverse Z in the upper right-hand corner of Fig 13.3 would be replaced by a much flatter Z (Fig 13.4).

The beam intensity ratio for a transmission or reflection transfer hologram will

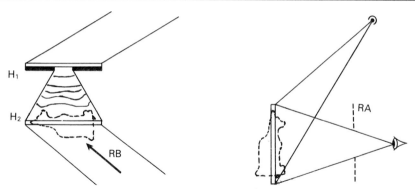

FIG 13.2 *Restricting the aperture of the master hologram in a 'vertical' plane during transfer gives a brighter reconstruction, as all the image-forming light passes through the real image of the aperture* (RA).

vary considerably across the H_2 plane, especially for a hologram-plane transfer. It should not go below 1.5:1 at the brightest spot, and it should not exceed about 4:1 in the neighborhood of the image. You should bracket your beam ratios to obtain the brightest result, in the same way as you bracket your exposures. Don't alter two variables at the same time; get the optimum exposure first, then the optimum beam ratio second. Log both, for future reference.

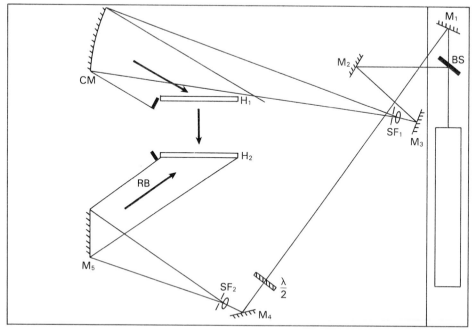

FIG 13.3 *Typical reflection-transfer geometry.* M_2–M_3 *is a path-equalizing zigzag or optical sheepshank. The beamsplitter* BS *should preferably be variable-ratio. The mirror* M_5 *can be a collimating HOE for very deep-image holograms.*

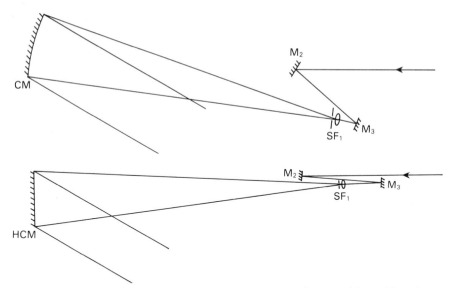

FIG 13.4 *Slight changes in geometry are necessary when working with* HOE *collimators.*

Manipulation of Cut-off Effects

The undesired cut-off that occurs when the viewer moves outside the real-image window of the master hologram has already been mentioned above. This effect can be made use of: just consider what it would be like if, instead of a black shadow, a new holographic image were to appear. This would have no effect on the original image, as it would be using a part of the emulsion at present not being used at all. The simplest version of this is just two master holograms side by side. This technique was used very effectively in 'Yalta' (Fig 13.5). This was probably the first political hologram. In order to have the parallax as large as possible, two holograms were made, one from each side of the model of the razor blade. The edge of the blade can be seen as a narrow black line imaging the butt-joined edge of the two holograms, well in front of the plane of the hologram.

A striking application of the technique is that of the appearing/disappearing image. As the viewer moves from left to right, the inner workings of a mechanical or electrical device appear suddenly, interpenetrating the original. This is simply a double exposure: the transfer hologram is exposed to a full-size master hologram of (say) an electric motor, and is then given a second exposure to a master hologram of the armature assembly, with the left half carded off. When the processed hologram is exhibited, the viewer sees the outside of the motor from the left, but on moving past the center to the right, the ghostly armature appears. Another apt and amusing example of a disappearing image can be seen in Plate 11.

If both master holograms are masked off with just a slight overlap, one image can switch into another. This can be done either vertically or horizontally.

 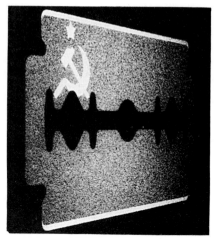

FIG 13.5 *Two views of 'Yalta' by AP Holographie. Two master holograms were used in order to obtain a wide angle of view.*

Multiplexing

By masking master holograms down to narrow slits it is possible to produce a reflection transfer hologram containing images from up to about nine masters. If the strips are horizontal, full horizontal parallax is preserved and the image changes as the viewpoint moves up or down. An early example of this technique was 'Portrait of Nicole' (Nicole Aebischer) (Plate 12), a composite of a number of pulse-laser portraits showing different fleeting expressions on Nicole's face.

Multiplexing has now become a standard technique in display holography, and many novelty holograms of this type have been made for sale or exhibition. In principle it is no more difficult to make a multiplexed reflection transfer hologram than a simple restricted-aperture transfer. The difference is that there are several restricted-aperture master holograms sitting side by side on the same plate, and they are all transferred at once. Although the composite master could be made up from pieces cut from separate master holograms mounted side by side and exposed together, it is usual, and somewhat simpler, to make the masters side by side on a single plate. The procedure is as follows:

1 Decide how many images are to appear in the final hologram, and whether they are to appear as the viewer moves horizontally, vertically, diagonally, etc. Remember that you can only put two or three images in for horizontal switching if there is to be any degree of horizontal parallax. In a vertical direction you can have as many as nine. Make a template the size of your master plate, and divide it into the required number of strips.
2 Construct a slit to the appropriate width, with a means of moving it accurately to uncover each of the strips in turn. Test this with your template.
3 Expose the master strips, changing the scenario as appropriate between exposures,

and moving the slit to its new position each time. Do not move the reference beam between exposures.

4 Process the master hologram, set it up to project the pseudoscopic real images into the transfer plane, and make the transfer hologram. When this is flipped for viewing, the images of the slits will be projected out into the viewing space, and the different images will appear in turn as the eyes are lined up with the different slit images.

If the slits are aligned accurately with no overlap, the scene will change abruptly as the viewpoint is changed. If the original slit positions overlap to a small extent, one view will dissolve into the next.

By sacrificing some aspects of horizontal parallax, the technique can produce a kind of jerky animation, as in 'The Logos' by Michael Wenyon and Susan Gamble, in which a black-hole image of hands makes a shadow of a dog which snaps and moves its ears as the viewer moves past (Plate 13). Perhaps the ultimate in this technique was achieved in 1982 by John Wood, who designed a device which exposed a narrow slit moving down the plate in synchrony with another device which opened and closed a pair of scissors. The result was an image of the scissors which snapped wildly as the viewpoint shifted vertically, alongside a perfectly sober teapot (Plate 14). The geometry of the system is correct from only one distance; from a more or a less distant viewpoint the scissors appear distorted. This is a phenomenon known as time-smear (see p. 252). Multiplexing is only a short step away from the holographic stereogram, which is discussed in Chapter 17.

Full-aperture Transmission Transfer Holograms

A full-aperture transfer hologram uses the whole area of the master hologram, and thus has vertical as well as horizontal parallax. However, this carries with it a number of restrictions. A hologram can be thought of as a kind of lens, and like lenses it suffers from chromatic aberration. Red light is deviated more than green light, and green light more than blue light (a glass lens deviates blue light more than red light, but the effect is similar). The focal length of a hologram is thus greater for green light than for red light and greater still for blue light, so that the green and blue images are larger and farther away from the plane of the hologram than the (true) red image. This is so for both virtual and real images. Only when the image is in the plane of the hologram will the images of all hues coincide (because in this case the focal length of the hologram is zero). Full-aperture transfers should therefore be made for shallow images close to the plane of the final hologram: 10 mm before and behind the plane is about the maximum. If the replay beam is filtered with an amber gel, a little more depth can be tolerated. If the fringes are made as slanting as possible (by using a reference beam at about $80°$ incidence) the Bragg condition will assist wavelength selection; if you try this, you will have to increase the beam intensity ratio, measured directly up the beams, to about $20:1$, because the light from the reference beam is much spread out on the surface of the emulsion and a lot of light is lost by surface reflection. A typical layout using a lens

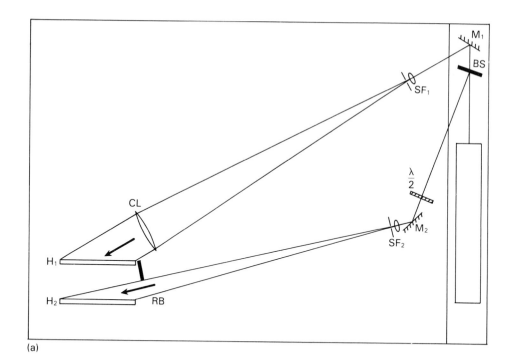
(a)

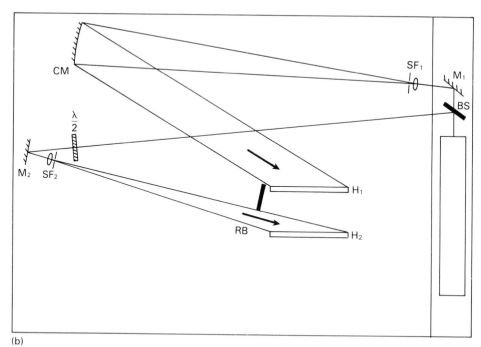
(b)

FIG 13.6 *Transfer geometries for full-aperture white-light transmission holograms,* (a) *using a collimating lens* CL *and* (b) *using a collimating mirror* CM. *The high-incidence reference beam* RB *gives a measure of the wavelength selection in reconstruction: the Bragg requirement helps to suppress color fringing of the image.*

collimator is shown in Fig 13.6(a) and one using a mirror collimator is shown in Fig 13.6(b).

Earlier authors have usually recommended that the reference beam for the transfer of a full-aperture hologram should be converging, presumably so that the diverging replay beam will be the exact conjugate. There does not seem to be any good reason for this; in fact, it is preferable to use collimated beams for making and reconstructing the master hologram, and a diverging beam for the transfer reference beam. This puts the real image of the aperture close to the viewing plane.

In making full-aperture transfer holograms, one of the difficulties is to avoid the directly-transmitted replay light (the zero-order beam) from falling on the final hologram. This is best dealt with at the master stage; you need to allow enough space between the object and the master plate, and a sufficiently large angle of incidence for the reference beam for the necessary clearances. If you have the master hologram reference beam at $60°$ incidence, the transfer plate can be at a distance from the master plate equal to just over half its 'vertical' dimension. However, if you need to have the plates closer together than this, you can card off the lower part of the beam; the small sacrifice in vertical parallax will not matter much, and the final image will be brighter.

Full-aperture transmission transfer holograms reconstruct very easily, in almost any light. This is a consequence of the geometry of the system. A transmission image which is wholly in the plane of the hologram will reconstruct with light of any wavelength and coming from any direction, provided the hologram is thin (ie the Bragg condition is unimportant).

Slit Transmission Transfer Holograms

The principles underlying slit transfer holograms are given on pp. 50–1, but a recapitulation is probably in order; at the same time, a slightly different approach may give a little more insight into what is going on.

If you flip a laser transmission hologram in its original plate-holder, and illuminate it with an unspread beam, you can put a piece of card on the side of the hologram away from the source and see the real image projected on the card. It is a flat image, not three-dimensional, and there is no clear plane of focus. It is as though the lens that is the hologram is stopped down to a pinhole. The image is simply a pinhole photograph. If you move the laser beam around the plate, you will see the image change as if the viewpoint had changed; this confirms that each point on the hologram codes the information from a particular viewpoint.

Now move the spot right across the horizontal axis of the plate. The image will rotate as you do this, giving you a full set of views in a horizontal aspect. There will be no change in the vertical aspect, of course. If you now expand the beam to fill the horizontal aspect the vertical aspect will still be a pinhole image. So a slit transfer hologram is a true hologram from a horizontal consideration, whereas from a vertical consideration it is a pinhole photograph. When you look at a slit transfer image in the usual way it may appear to be behind or in front of the holographic plate; but if you turn your head sideways the image jumps into the plane of the plate and becomes flat. The difference in distance between the horizontal and vertical aspects is a variety of astigmatism, and it raises problems in some types of white-light transmission holograms. These will be discussed as they occur.

When you view a white-light transmission hologram, you can see an image only when your eyes are in line with the real image of the slit. Because this image is formed in different positions for different wavelengths, you see the image at one height through a 'red' slit, from another height through a 'yellow' slit, and so on. In general, one wants to see as bright an image as possible and it is therefore usual in a one-color hologram to arrange that the image as seen from an average height is yellow or yellow-green, the range of wavelengths to which the human eye is most sensitive.

Why does a slit transfer hologram appear sharp all over, while a deep full-aperture hologram gives a blurred image? The answer is in the basic conditions for viewing a hologram, namely that it should be viewed by light of one wavelength only. What the Benton geometry does is to replace the vertical parallax with a device which disperses white light into a spectrum, so that from any viewpoint you see the image by light of only one wavelength. It may be a different wavelength for different parts of the image, but it is nevertheless only one wavelength for any part of the image that you are looking at.

Geometry of a basic slit transmission transfer hologram In order to get the maximum parallax in a holographic image you have to have the subject matter as close as possible to the hologram. Likewise, when you make a transfer hologram, you need the master and transfer hologram to be as close as possible, so that the cut-off from the edge of the real image of the master hologram occurs at a viewing incidence that is the largest possible. But if you are looking at an image through what amounts to a narrow letter-box, you also want your eye to be as close to the letter-box as you can get it, otherwise you will not be able to see the whole image from one viewpoint. This means that the real image of the letter-box (ie the slit) should be quite a long way out from the hologram while the image of the subject matter remains close to the plane of the hologram. This sounds a tall order, but in practice it is fairly easy to achieve.

We have seen that in order to reconstruct a geometrically-correct though pseudoscopic image we need to illuminate the flipped master hologram with the conjugate of the reference beam; in practice, for simplicity we make both beams collimated. As far as the final image is concerned, we want to exaggerate the distance of the real image of the slit while keeping the image of the subject matter largely undisturbed. It is possible to do this if we remember that although a hologram may not be quite the same thing as a lens, it does focus an image, and it does obey the *lens laws* (see Appendix 2). If we change the divergence of a beam illuminating a hologram, we shall change the position of the image according to the lens laws. If we flip a hologram that was made with a diverging reference beam and then illuminate it with a further diverging replay beam, anything in the image that is out of plane will be shifted along the axis by an amount that is related to the change in position of the point of illumination. If the image was close to the hologram plane this will not have much effect, but if the image was several centimeters away there will be a considerable shift in its position: it will be pushed out

PLATE 11 *As the viewer moves past the hologram the Cheshire cat vanishes – all but the grin! Hologram commissioned by Light Fantastic (see p. 183.)*

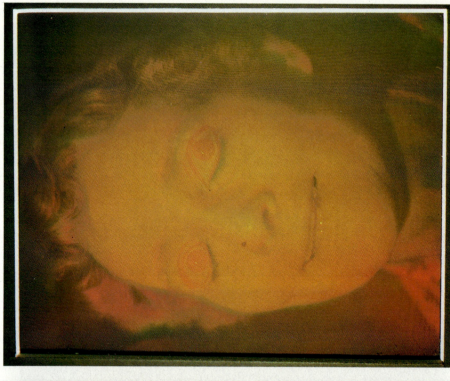

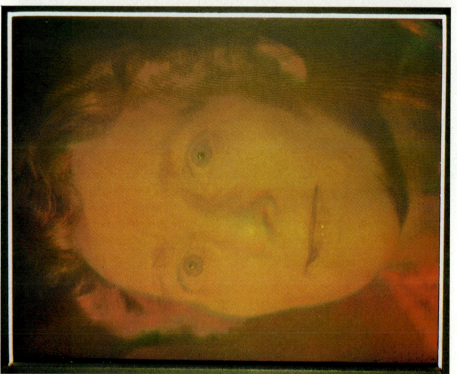

PLATE 12 *Portrait of Nicole*, by Nicole Aebischer (see p 184), is a slit transfer reflection hologram containing nine separate images with different expressions; these come into view successively as the viewpoint is moved up or down.

PLATE 13 *The Logos*, by Michael Wenyon and Susan Gamble (see p. 185). The hands form a black-hole image of a 'dog', which moves its mouth and ears as the viewer moves past the hologram. Photograph by Susan Gamble.

PLATE 14 *Scissors and Teapot*, by John Wood (see p. 185). The scissors open and close as the viewpoint is moved up and down.

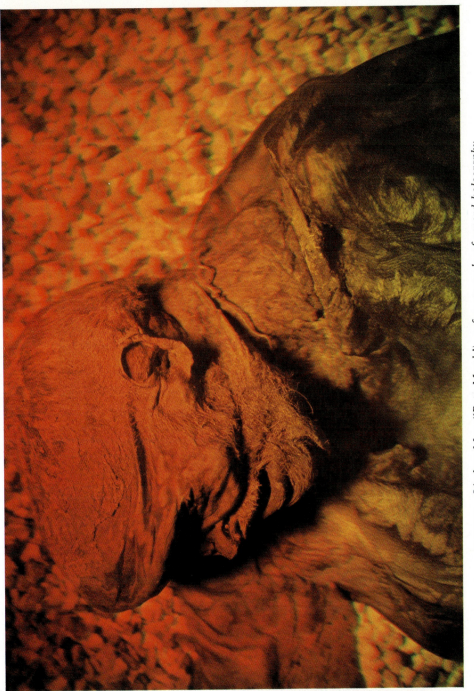

PLATE 15 *Lindow Man ('Pete Marsh'), a fine example of record holography. The pulse hologram was made by Richmond Holographic Studios to a commission by the British Museum for public exhibition, the original figure being too fragile to be used for this purpose. Photograph by Tim Hawkins.*

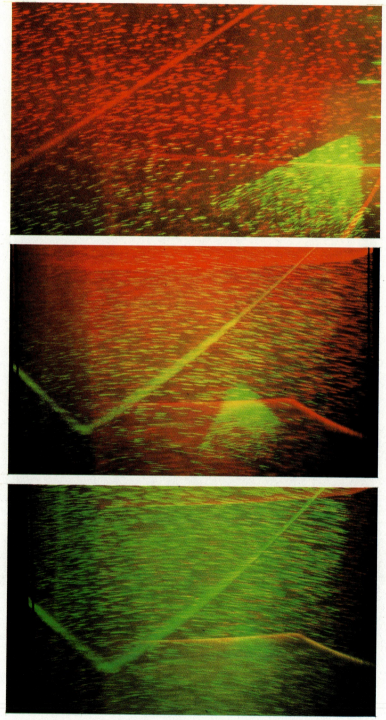

PLATE 16 *Photon Studies 10,* by Rudie Berkhout, *a pseudocolor white-light transmission transfer hologram (see p. 189) made using a number of slit master holograms positioned so as to produce interpenetrating images of great depth in*

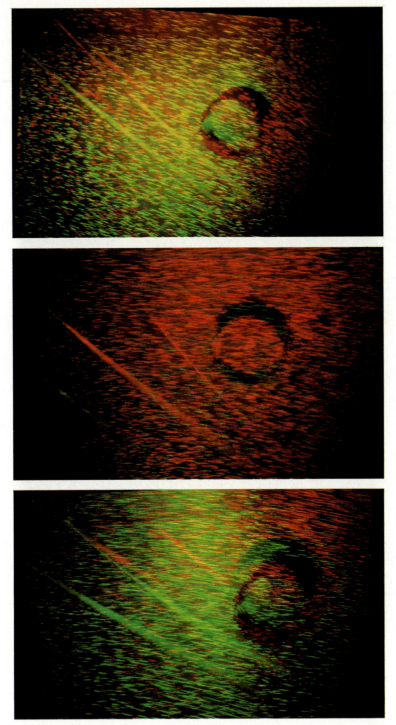

varying spectral hues. In order to give some idea of the scope of such images the hologram has been photographed from six different viewpoints, with both horizontal and vertical movements.

PLATE 17 *Portrait of the Artist as a Dead Man*, by Adrian Lines (see p. 239). A real-image Denisyuk hologram of a plaster cast made of his whole body by Lines, it is made up of 23 8 × 10 in. plates. Photograph by Adrian Lines.

from the hologram plane by anything up to a meter. This is exactly what we want. It is therefore usual to use a diverging beam for both reference and reconstruction beams (Fig 13.7).

Deep images will also be exaggerated in perspective, ie the parts of the image nearer to the eye will be formed closer than they would otherwise be, and they will also be enlarged to some extent. The effect will be a maximum for images that are wholly real. The effect has been exploited by artists such as Rudie Berkhout who work with abstract subjects (Plate 16). Exactly the same arguments apply to multiplex reflection holograms (see pp. 246–50). In representative work such distortions are unacceptable, and the transfer needs to be made with minimum (or no) divergence in the reference and replay beams.

The geometry of white-light transmission transfer holography has been dealt with fully by Benton (1); Steve McGrew (2) and Suzanne St Cyr (3) have prepared worksheets for practical situations. McGrew's geometrical method (edited) appears

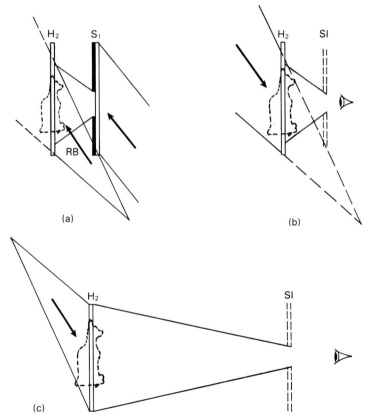

FIG 13.7 (a) *and* (b). *In order to reconstruct a geometrically-accurate image it is necessary to use the conjugate of the reference beam. However, if the image is in the plane of the hologram it will reconstruct in its correct position with a non-conjugate (diverging) beam, while the image of the slit* SI *will be projected farther away* (c).

in Appendix 7, with the author's permission. Using the configurations described in this book, the 'letter-box' through which the image is seen is at least 100 mm high, so that if it can be projected to about 450 mm (18 in) in front of the hologram, a complete image will be seen from as far away as 1 m for a hologram 300 mm high and from more than 2 m for a hologram 150 mm high. When the viewer is close to the real image of the slit the image of the object will appear to be all of one hue. A low viewpoint will give a blue image and a high viewpoint a red one. From farther away the image contains all the spectral hues, with blue at the bottom. Nearer, the hues are reversed, with blue at the top, and real images may be difficult to see because they are so close. In some cases they may actually be behind the viewer.

Because of the difference in deviation of the different wavelengths the slit images are formed at different distances from the hologram. Red light is deviated more than blue light, so that the red slit image is formed nearer to the plate and higher up than the blue slit image. The slit images form an angle known as the *tip angle*; for a beam incidence of 45° the tip angle is 35°, and for a beam incidence of 60° it is 41° (Fig 13.8). The tip angle becomes important when we try to bring the red, green and blue images together to form an achromatic image, or to form multicolor images with all colors in accurate register.

In a monumental paper (1) on the mathematics of white-light transmission holograms, Stephen Benton looks at some of the inherent distortions in the process. It is clear from this paper that McGrew's geometrical method (Appendix 7) is only approximate, especially in the case of the projection of a real image by a non-conjugate beam. Benton's paper comments on the difficulty and expense of obtaining high-quality collimators of large aperture, and shows that with a throw of ten times the larger plate dimension, perspective distortion is negligible and little is to be gained by collimation. He also shows, on the other hand, that a master image is displaced by about 50% (eg a subject at 220 mm is reconstructed at 310 mm) when

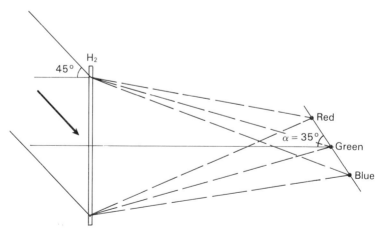

FIG 13.8 *For white-light reconstruction the images of the slit are at different heights and distances from the hologram H_2, lying on a line at the tip angle α. For a replay beam incidence of 45° $\alpha \cong 35°$ measured from the normal.* NOTE: *In this diagram the spacing has been exaggerated for clarity.*

a master hologram is both made and replayed (flipped) with a diverging beam source 1500 mm away. This is the kind of throw associated with a ×40 microscope objective. One problem thrown up by the use of non-conjugate beams is that the real image suffers from astigmatism, leading to unsharpness and distortion when the image is viewed at an angle. The final hologram, having suffered three successive generations of perspective distortion and astigmatism all in the same direction, is suitable only for admirers of the grotesque. The moral is that if you have a table of limited size and have to use a ×40 objective to fill your plate area, you need a collimator.

Slit width There are two ways of producing slit illumination. The first is to illuminate the whole master hologram, but to cut off all but a narrow slit by masking with opaque card. The second is to illuminate the master hologram with a narrow strip of light, using masking only to exclude stray light. The first method is not very satisfactory: apart from the waste of light, the sharp-edged aperture produces its own diffraction pattern which adds high spatial-frequency information about the slit to the hologram, and uses up valuable bandwidth. The second method is preferable, and is easy to produce using a one-dimensional beam expander. The actual width of the slit is important. As with so many of the variables in holography, it is a compromise. A narrow slit gives a sharper image but more speckle; a broad one makes the vertical aspect unsharp. The best width seems to be about 6–7 mm. This is because the speckle size is a function of the diameter of the smallest aperture in the image beam. It is therefore logical to choose a slit width equal to the diameter of the (dilated) pupil of the eye, ie 6–7 mm. (The author is indebted to Hariharan for this insight.) It is a good idea to mask down the master with opaque PVC tape, which you can stretch into a slight curve, so that the slit width is about 4 mm at the center, broadening to 8 mm at the outermost part. This will help to compensate for any fall-off in intensity of the illuminating beam towards the edges.

One-dimensional beam expanders As has been shown, it is important that the illumination of the master hologram should be collimated, otherwise the geometry of the real image will be incorrect. Whether you use a mirror or a lens collimator, the beam expander must be located in its principal focal plane, and its focal length must be such that the collimating optical element is illuminated from top to bottom as uniformly as possible (side to side in an overhead beam). To achieve this it is necessary to have a beam spread of between two and three times the diameter of the collimator.

One-dimensional beam expanders are simply cylindrical elements, either lenses or mirrors. Mirrors have the advantage of combining the function of relay mirror and beam expander in one component. The focal length of a cylindrical mirror is approximately equal to half its radius of curvature; thus a 25 mm diameter reflecting surface has a focal length of 6.25 mm, approximately the same as that of a ×40 microscope objective; it will cover a 320 mm (12.5 in) collimator with a focal length of 1.2 m (48 in). A steel roller bearing makes a good reflector, as it has a surface that is finished to better than a wavelength; however, its reflectance is not much better than 50%. Unterseher *et al* (4) recommend a metallized glass rod, but it may

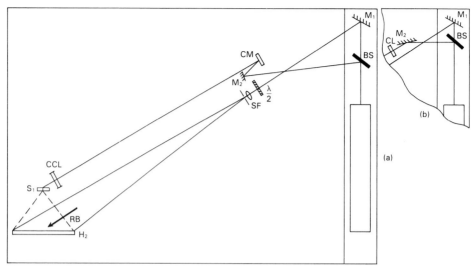

FIG 13.9 (a) *Slit transfer geometry for a rainbow hologram using a liquid-filled cylindrical collimating lens (CCL). (CM) is a cylindrical mirror (eg a roller-bearing), used as a one-dimensional beam expander. (b) Shows the arrangement where a cylindrical lens CL (eg a liquid-filled glass tube) is substituted for CM.*

be difficult to obtain one with the necessary optical quality. If such cylindrical reflectors are used more than a few degrees off-axis, the slit-illumination they provide is in general curved, which makes the satisfactory illumination of a pre-masked area difficult. By careful adjustment it is possible to get a straight slit, but it is not easy. A typical layout for a slit transfer using a one-dimensional reflector and a collimating lens is shown in Fig 13.9.

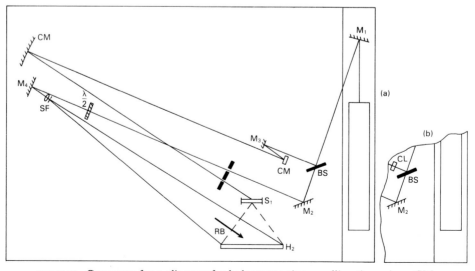

FIG 13.10 *Geometry for a slit transfer hologram using a collimating mirror CM. Symbols as in* FIG 13.9.

TRANSFER HOLOGRAMS

If you use a reference beam angle of about 60° for the master hologram and an angle of 54.5° for the transfer, a replay beam angle of 45° will give a yellow-green image when the final hologram is viewed normal to its surface. The beam ratio for the transfer should be about 5:1 at the center of the plate, with an upward tolerance to about 10:1. In order to avoid image distortion, the slit-illuminating beam should be collimated. You can use the same collimator as you used in making the master hologram (lens or mirror), or you can use the one-dimensional collimating lens described on pp. 220–3; this can also be used for making slit master holograms.

An alternative to a cylindrical mirror is a cylindrical lens. A piece of glass or acrylic rod will do, though a sample tube filled with glycerol is better. The focal length is approximately equal to the radius, so that a 12 mm diameter rod or tube has a focal length of about 6 mm, and will fill a 320 mm diameter collimator at a distance of about 1.3 m. If used in conjunction with a relay mirror, the configuration would be similar to that of Fig 13.9(b). An alternative arrangement using a mirror collimator is shown in Fig 13.10.

Master hologram illumination and the Bragg condition

When a transmission hologram is developed and fixed, some silver is lost from the emulsion, which therefore shrinks a little on drying as compared with its original thickness. This results in a change of the angle of the fringe slats which lie along the bisector of the object and reference beams. If the object and reference beams made equal angles with the emulsion this would not matter, but as the object beam is at roughly 0° incidence and the reference beam is at 60° there will be a discrepancy between the grating spacing (1.15 times the wavelength) and the spacing that satisfies the Bragg condition (equal to the wavelength) (Fig 13.11).

If there is a shrinkage of, say, 10%, the angle of the fringe slats will change from 30° to 32.7° and the Bragg condition will be satisfied for an angle of 65.4° instead of 60°. This will result in an image which is slightly smaller and nearer to the hologram than the original object. This change in angle is not always a disadvan-

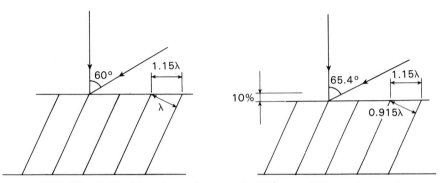

FIG 13.11 *When the object beam is normal to the emulsion surface and the reference beam is at 60° incidence, the separation of the Bragg planes is equal to the wavelength λ. If the emulsion shrinks as a result of processing, the optimum angle for reconstruction changes. If the wavelength is reduced the original angle of incidence is restored.*

tage, however; the Bragg condition is satisfied if the wavelength of the reconstruction beam is reduced by about 7.5%, which, as it happens, is almost exactly the difference in wavelength between a pulsed ruby laser and a HeNe laser. This simplifies the geometry when making transfers from pulse-laser master holograms. However, when a HeNe laser is used for both master and transfer holograms, it is desirable to keep this shrinkage to a minimum, and for this reason some workers prefer not to fix their master holograms. In setting up a master hologram for maximum image brightness any change in angle from the original reference beam angle, or change in wavelength, may result in some parts of the image being less bright than others, because it is not possible to find an angle such that the Bragg condition is fully satisfied for all points in the reconstructed image.

NOTE: For simplicity the effect of the refractive index of the emulsion has been ignored, as it is not relevant in this context; but see Appendix 5.

Beam angles for the final hologram If you are going to be able to predict the hue of the final hologram for a normal viewpoint, you should establish a constant angle for the master reference beam and stick to it. It is a good idea to use a 60° angle of incidence (30° to the plate). This angle has several advantages: it is close to the Brewster angle, so that internal reflections can be minimized; it allows the transfer plate to be fairly close to the master plate, and the sine of (60°/2) is exactly 0.5, which simplifies calculations. If you simply want the brightest image, go for yellow-green, for which the transfer reference beam is between 51° and 54° and the replay beam angle is 45°.

Focused-slit transmission transfer holograms As has previously been pointed

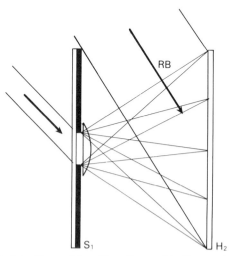

FIG 13.12 *If a cylindrical lens of focal length equal to the spacing between S_1 and H_2 is placed on S_1, the image will be focused rather than being a pinhole image, in the vertical sense. This will give better vertical resolution and a brighter, speckle-free image.*

out, a rainbow hologram is a hologram only in the horizontal dimension. In the vertical dimension it is a pinhole photograph, so that the vertical information is in the plane of the hologram; the horizontal information is, of course, in its correct position. Now, the problem with a pinhole photograph is that the bigger the pinhole the more blurred is the image. With a pinhole hologram, the bigger the pinhole the bigger is the effect of chromatic aberration, which results in a different type of blur. At the other extreme, the smaller the pinhole, the worse is the speckle. For the usual type of rainbow hologram the optimum slit width, as discussed on p. 191, is around 5–7 mm. However, Leith and Chen (5) have considered the possibilities of an image in the vertical plane that is focused, rather than being a pinhole image; this uses a 'slit' of up to 30 mm width, and gives a speckle-free image. This is achieved by covering the slit with a cylindrical plano-convex lens of focal length equal to the distance between the master and transfer plates. Such lenses can be bought from specialist dealers (see Appendix 12); alternatively, you can cut one from a Perspex cylinder (see p. 224) (Fig 13.12).

Achromatic Images

The fact that holographic images are in general monochromatic can be a serious disadvantage, and a number of methods have been tried in an attempt to produce an achromatic (uncolored) image. The following two methods have been among the more successful.

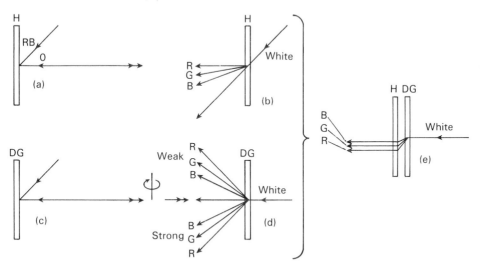

FIG 13.13 *Dispersion compensation. When a laser transmission hologram H (a) is replayed (b) by white light, the image is dispersed, red light being diffracted most and blue least. If a diffraction grating DG (c) is made using the same geometry and spun through 180°, if illuminated normally with white light (d) it produces a strong spectrum in the same direction as the hologram but with the hues reversed. If H and DG are placed close together and illuminated normally with white light (e) the two dispersions cancel out and the image is achromatic.*

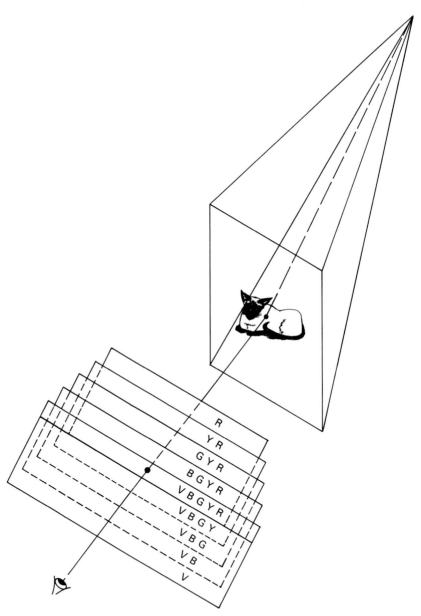

FIG 13.14 *Benton's achromatic system. The final hologram is combined with a HOE which projects not one but five slits superposed in the same plane and overlapping in such a way that the spectral hues at eye level add up to white.*

Dispersion compensation If an ordinary transmission hologram is illuminated by white light, each point in the image becomes spread into a spectrum. In lens optics this effect is neutralized by incorporating a negative lens of weak deviating power but high dispersive power into the system. This provides a dispersion that is equal and opposite to that of the converging elements, while not canceling the focusing power of the system. A similar system can be used for holograms; it is called dispersion compensation, and was previously mentioned on pp. 53/55. The hologram itself is made in the usual manner. Next, a dispersion-compensating plate is made using exactly the same geometry (Fig 13.13).

To provide the dispersion compensation the diffractor plate is set up *spun* through 180° so as to disperse the light in the opposite direction to the hologram. It can be set up a sufficient distance from the final hologram for the zero-order and higher-order beams to miss the hologram altogether. Alternatively, the plates can be placed close together, with a piece of proprietary cellular material designed for directional displays sandwiched between them. Such material is made by 3M (see Appendix 12). The method is described by Bazargan and Waller-Bridge (6).

Benton's achromatic system The most elegant system for producing achromatic transmission holograms has come from the inventor of the rainbow hologram. Stephen Benton (7). His method uses an intermediate HOE which produces overlapping spectra that add up to white light (Fig 13.14).

Benton made a HOE which would construct virtual images of a number of slits along the tip angle (see p. 53). He did this by constructing four actual slits which interfered mutually to form a hologram of the slits (Fig 13.15). This arrangement produces twelve strong virtual slits and a number of higher-order images. This 'diffractor plate' is now placed between the master hologram and the first transfer

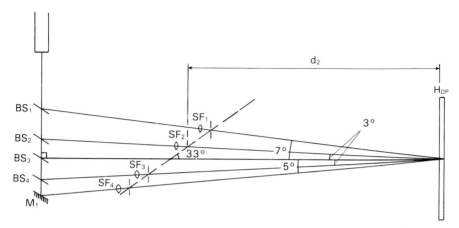

FIG 13.15 *The spatial filters are aligned along the tip angle. For clarity the beams are not shown expanded. d_2 is the distance at which the slits will all eventually be focused for normal incidence viewing. H_{DP} is the position of the holographic diffractor plate being exposed.*

hologram, which is set at the tip angle (Fig 13.16). Finally, the intermediate hologram H_2 is used (flipped) to produce a real image of the diffractor plate, whereby the images are re-registered and a third hologram is exposed (Fig 13.17). Illumination of the processed hologram gives a series of correctly-overlapping spectra (Fig 13.18). The difference in size of the overlapping spectra means that there

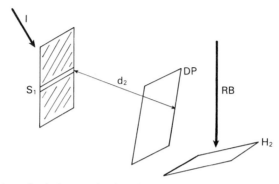

FIG 13.16 *A transfer hologram is placed at the tip angle, the diffractor plate DP intervening at an angle between the master hologram S_1 and the transfer hologram H_2. The illumination beam I is convergent and the reference beam RB is collimated.*

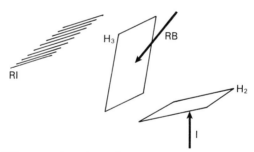

FIG 13.17 *The diffractor plate is replaced by the final hologram H_3. H_2 is illuminated by the conjugate beam I, producing an image of the diffractor plate in the plane of H_3 and a set of real images (RI) of the slits beyond H_3.*

FIG 13.18 *When H_3 is viewed (unflipped) by white light the projected spectra overlap to produce an achromatic image.*

are a few color borders, but these are at extreme angles of viewing and there is a considerable angle over which the image is colorless.

It is not at all easy to make the diffractor plate and Benton, in conversation with the author, said: 'If we ever broke the diffractor plate, I don't think we should ever make another.' This method is for the dedicated, and full details can be found in Benton's patent (7) and in his other relevant paper (8).

References

1. Benton S A (1982) The mathematical optics of white-light transmission holograms. *Proceedings of the International Symposium on Display Holography*, **1**, 5–14.
2. McGrew S (1982) A graphical method for calculating pseudocolor recording geometries. *Proceedings of the International Symposium on Display Holography*, **1**, 171–82.
3. St Cyr S (1984) A holographer's worksheet for the Benton math. *Holosphere*, **12**, 4–16. NOTE: there is a correction sheet for this paper.
4. Unterseher F, Hansen J and Schlesinger R (1982) *The Holography Handbook*, p. 266. Ross Books.
5. Leith E N and Chen H (1978) Deep-image rainbow holograms. *Optics Letters*, **2**, 82–4.
6. Bazargan K and Waller-Bridge M (1985) A practical portable system for white-light display of transmission holograms. *Proceedings of the SPIE* **523**, 24–5.
7. Benton S A (1985) US Patent No 4498729. Method and Apparatus for Making Achromatic Holograms.
8. Benton S A (1978) Achromatic images from white-light transmission holograms. *Journal of the Optical Society of America*, **68**, 1441.

CHAPTER
14
※

Focused-image Holograms

'One side will make you grow taller, and the other side will make you grow shorter.'
'One side of *what*? the other side of *what*?' thought Alice to herself.
            LEWIS CARROLL, *Alice in Wonderland*

Parallax in a Focused Image

In Chapter 1 it was suggested, not entirely facetiously, that inside every camera there is a three-dimensional image struggling to get out. It is the optical image formed by the camera lens: it is inverted but orthoscopic, and, as discussed on p. 49, it can be made the subject of a hologram. Such a 'focused-image hologram' can be made give a virtual, a real or a hologram-plane image. However, unless the camera lens has a very wide aperture ($f/1.5$ or larger) you will not be able to see this holographic image stereoscopically as there is insufficient parallax, and as a result the optical image of the object cannot be seen with both eyes at the same time (Fig 14.1).

The '$f/1.5$' mentioned above is called the *f-number* (abbreviated 'f/no'), or sometimes the 'f-stop'. It is numerically equal to the focal length of the lens divided by its effective diameter (ie the diameter of the *entrance pupil* of the lens), and the lower the f/no, the more parallax you get, for a given lens. A typical lens for a 35 mm single-lens reflex camera has a focal length of 50 mm and an aperture of $f/2$. If the lens is to form an image the same size as the object it must be placed two focal lengths away from the object, in this case 100 mm; the image will also be 100 mm away from the lens, on the other side of it. The effective aperture is now $f/4$, not $f/2$. The angle through which the image can be seen (the parallax angle) is equal to the angle subtended by the diameter of the lens aperture at the film plane. For a (nominal) $f/2$ lens at double extension as described above, this angle is approximately $14°$ and for an $f/1.5$ lens $21°$. The mathematical derivation is given in Appendix 2. Now $14°$ or, for that matter, $21°$, is not much parallax in terms of what we usually expect of a hologram. The real image used in making transfer holograms has much more parallax. The situation is analogous: a hologram, like a lens, has an f-number; this is equal to the object distance divided by the horizontal width of the hologram. In a first-generation hologram it is typically about $f/0.5$, giving a parallax angle of $90°$. If this image is used to make an image-plane transfer hologram, the latter will have about the same parallax, or a little less. If we want the same degree of parallax in a focused-image hologram we are going to need a very large lens.

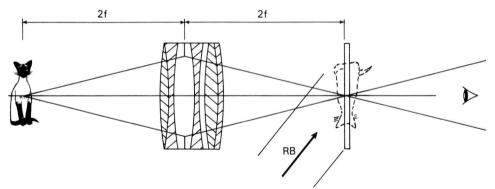

FIG 14.1 *The optical image formed by a lens is inverted but orthoscopic, and has full parallax. However, the depth of the image is correct only when object and image are at twice the focal length of the lens, and stereoscopic viewing is possible only if the lens has a very wide aperture (f/1.5 or less). By introducing a reference beam it is possible to make a focused-image hologram with the optical image as object.*

Perspective in a Focused Image

There is another problem, which was first noted in connection with stereoscopic photography (p. 4). The axial scale of the optical image is the square of the lateral scale. Only when the image is the same size as the object is true perspective preserved, and then only in the neighborhood of the focal plane. Thus as the scale gets smaller the optical image becomes flatter, until at about one-quarter scale there is scarcely any depth at all. If you use an $f/2$ or larger aperture lens from a 35 mm camera to produce the optical image, this effect is to some extent masked by the way the image swings in and out of the plane of the hologram because of the *curvature of field* of the lens. Such tiny images can be very attractive (Fig 14.2).

Lenses for Focused-image Holograms

Because of the limited parallax angle it is necessary to choose subject matter that is much smaller than the diameter of the lens, and which is imaged at near full-size. The lens should therefore be as large as possible, and fairly well corrected for spherical aberration. Plano-convex condenser lenses (which are mounted in pairs with the curved surfaces inwards) are quite well corrected, and have large apertures, around $f/1.2$. You may be lucky enough to find one of these in a junkshop; keep a look-out for windows with old photographic equipment in them. The biggest condenser lenses come from old horizontal enlargers, and may be up to 250 mm in diameter. They may be scratched, and it is worth paying to have the flat surfaces repolished (the curved surfaces will have been protected). You can also buy condenser lenses from mail-order suppliers such as Edmund Scientific, or from optical/electrical surplus shops such as Proops Bros of 52 Tottenham Court Road, London.

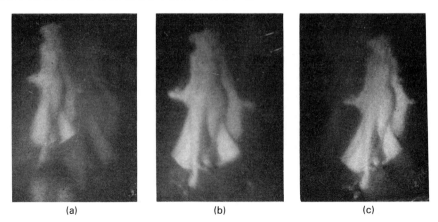

FIG 14.2 *Mitzi, a small focused-image hologram by Jeff Blyth, made using a wide-aperture* 35 mm *camera lens. Although the reduced size means that the image is almost flat, the curvature of field of the lens causes the image to swing in and out of plane as the viewpoint is varied; the effect is elusive and attractive. The double-image effect clearly visible in* (a) *results from the fact that a focused-image hologram contains a photographic image which is anchored in the plane of the film. It is a negative image, but appears positive when the bleached film is viewed by reflected light. The holographic image appears sharper to the eye than it does in the photographs.*

A second type of lens that is suitable is the double *Fresnel lens* that is used as a condenser lens in overhead projectors. These lenses are obtainable up to 300 mm (12 in) square, with a focal length of about 150 mm, giving an effective aperture when used at unity scale of $f/1$. In general they are well corrected for spherical aberration. They can be obtained from large educational suppliers, or from Edmund Scientific or Proops Bros. They are often constructed of two Fresnel lenses of different focal lengths, so that the image conjugates are at different distances; the side labeled 'this way up' is the longer conjugate, and if you are reducing the size of the image this is the side that should face the object.

A third type of lens is built from acrylic sheet and filled with liquid. Acrylic sheet is difficult to mold to an accurate convex shape, but it is very easy to bend into a cylinder. By having a vertical axis of curvature on one side and a horizontal axis on the other it can produce a satisfactory, if somewhat astigmatic, image for an effective aperture of about $f/3$. In this book such a lens is referred to as a *bicylindrical lens*. The construction lenses of this type is dealt with in Chapter 15.

Full-aperture Focused-image Holograms

These correspond to full-aperture transfer holograms and the conditions for a good image are similar, except that the image, being already orthoscopic, does not need

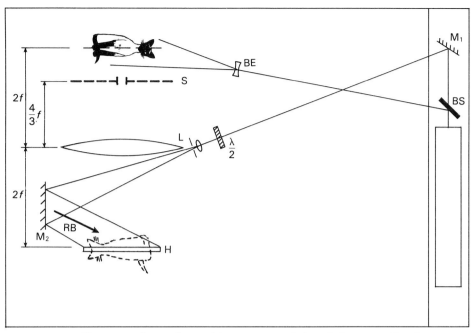

FIG 14.3 *Geometry for a focused-image transmission hologram. The large aperture lens* L *forms a full-size optical image in the plane of the hologram* H. *The insertion of a slit* S *between the object and the lens turns the hologram from a full-aperture hologram to a rainbow hologram. The hologram is viewed rotated but unflipped. Only one object illumination beam is shown. Other symbols have the usual meanings.*

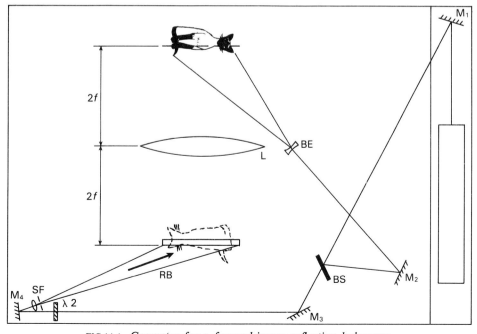

FIG 14.4 *Geometry for a focused-image reflection hologram.*

flipping (though it does need rotating). You should accordingly set up the reference beam to come from 'above' the optical image. Deep subjects are best reduced slightly in size, as this reduces the depth of the optical image. A reduction to two-thirds full size will result in a flattening factor of about 2.25; this is about the most you will get away with before the flattening becomes obtrusive. A divergent replay beam will restore some of the lost perspective; a typical layout for a focused-image transmission hologram is shown in Fig 14.3. Fig 14.4 shows the equivalent configuration for a reflection hologram.

In a focused-image hologram of the image-plane type the image information is strongly localized, and the beam ratio varies greatly across the hologram. For a transmission hologram the beam ratio within the optical image area should be about 4:1, and for a reflection hologram it should be about 1.5:1. However, some workers have reversed the ratio, with an object beam of higher intensity than the reference beam, and have obtained excellent results in the face of all the theory.

One-step Rainbow Holograms

As we saw in Chapter 13, the necessary criterion for producing deep white-light transmission holograms is an image of a 'letter-box' in the plane of viewing. This can be achieved for a focused-image hologram by placing a slit just beyond the principal focus of the imaging lens on the object side. At 1.5 times the focal length the 'letter-box' will reconstruct at 3 times the focal length in front of the lens, and at 1.25 times the focal length it will reconstruct at 5 times the focal length in front of the lens. If the reference beam diverges less than the replay beam the distance will be greater. The slit reduces the aperture of the lens and therefore reduces unsharpness due to aberrations in the 'vertical' plane, but as it is a comparatively long way from the lens there is some astigmatism and distortion in the optical image. The position of the slit is shown as a broken line in Fig 14.3. Suzanne St Cyr has investigated this layout extensively and has worked out a very full geometry for one-step pseudocolor holograms (see Chapter 18).

Benton *et al* (1) have adopted a somewhat different geometry in order to project the slit into the viewing space. Clearly, if the slit could be placed in the plane of the lens it would be more effective in reducing the lens aberrations. For this it is necessary to use a highly divergent reference beam, with an apparent source nearer to the plate than the lens itself. When the hologram is illuminated by a beam with little divergence, the slit forms a real image in the viewing space (Fig 14.5).

The geometrical requirements for this are straightforward. If you are using a lens of, say 150 mm focal length, the distance of the lens from the hologram will be 300 mm. If you spread the reference beam, using a supplementary lens, so that its effective point of origin is 150 mm from the hologram, you will be creating a holographic lens of focal length 300 mm. If the replay source is 1200 mm from the hologram, the slit will be imaged 400 mm from the hologram on the viewing side. An effective reference source at 200 mm from the hologram, with a replay light source at 1800 mm, will reconstruct the slit image at 900 mm from the hologram; this would be suitable for larger holograms. The method of calculating the distances

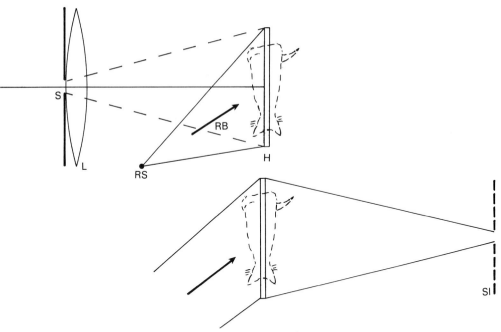

FIG 14.5 *The slit S and the reference source RS form a holographic lens in H. If RS is close enough to H, reconstruction with a near collimated beam forms a real image SI of the slit in the viewing space. The image of the object remains in the plane of H.*

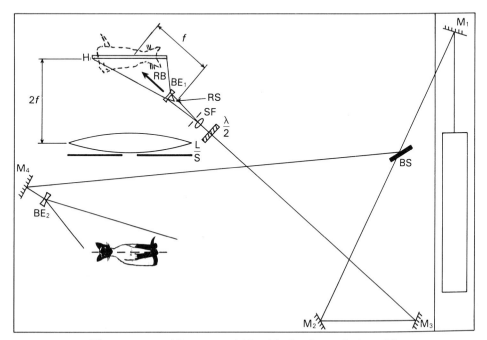

FIG 14.6 *The geometry of Benton et al (1) with the slit on the lens. The concave lens BE_1 provides additional expansion to the reference beam so that the effective distance f of the reference source RS from the hologram H is half the distance from the hologram to the lens (2f).*

is given in Appendix 2. A suggested layout for this type of hologram is given in Fig 14.6.

Focused-image Holograms using an Optical Mirror

Hariharan (2) has devised a method for obtaining wide parallax and a bright image without using expensive lenses. He used a concave mirror of wide aperture to produce images essentially in the same way as that shown in Fig 14.4 above (Fig 14.7).

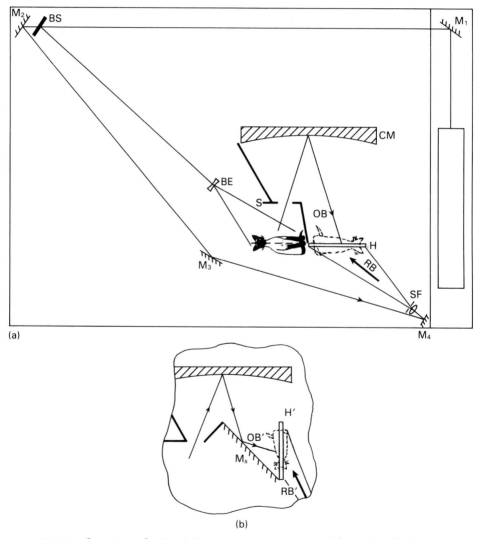

FIG 14.7 *One-step reflection hologram using a concave mirror (after Hariharan (2)). CM is a large-aperture concave mirror. The slit S is an image-brightening restrictor of 'vertical' parallax. (a) Produces a pseudoptic image, (b) shows how to produce an orthoptic image by means of a supplementary 45° mirror* M_5.

This method, while compact and aberration-free, has the disadvantage of producing an image which is either pseudoscopic or pseudoptic. In many cases this does not matter, eg for abstract images or symmetrical subjects. It can be rectified by the introduction of a 45° mirror into the system (Fig 14.7 (b)).

Hariharan's mirror had a diameter of 600 mm and a focal length of 275 mm, giving an angle of parallax of about 50°. The slit was positioned so as to produce a real image about 1 m from the mirror. It was a wide slit allowing about 10° vertical parallax, and its purpose was to give a brighter reconstruction and to reduce the information content of the hologram sufficiently to allow three-color images to be made.

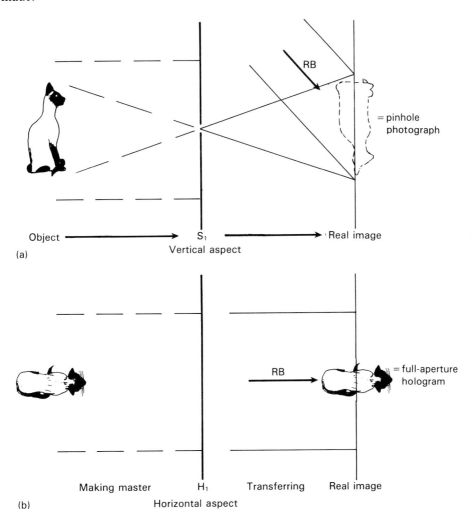

FIG 14.8 *A slit transfer image, viewed from the side* (a) *is equivalent to a pinhole photograph, as can be seen when making and transfer geometries are combined in a single diagram. Viewed from above* (b) *it is a full-aperture transfer hologram.*

Astigmatic One-step Focused-image Holograms

Once it had been appreciated that a Benton hologram contains two separate optical configurations, one in the horizontal plane and the other in the vertical plane (Fig 14.8), a number of workers got down to improving the slit geometry independently of the horizontal geometry. The first step was to broaden the slit to reduce speckle, and to compensate for the color blur by focusing the pinhole image into the transfer hologram plane. Because an image-plane hologram does not need a spatially-coherent reference beam, the vertical aspect of the hologram can use a diffuse reference beam, giving a speckle-free achromatic image, and Leith *et al* (3) employed a one-dimensional diffuser made by scoring a piece of acrylic sheet in a single direction with abrasive paper.

It was not long before it was realized that the intermediate hologram could be dispensed with. Chen, Yu and Tai variously worked on several methods (4, 5, 6), and in 1985 Chen (7) proposed a one-dimensional version of Benton's focused-image geometry combined with a conventional hologram in the horizontal plane (Fig 14.9). This produces a white-light transmission hologram which is almost achromatic. The addition of a cylindrical focusing lens with its center line vertical converts the system into a focused-image hologram.

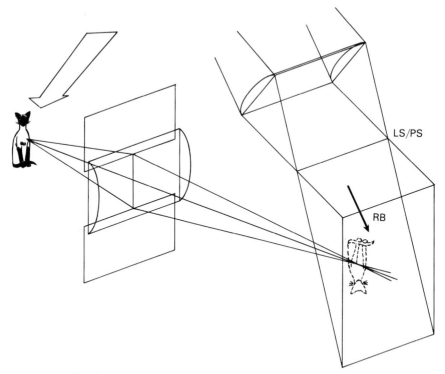

FIG 14.9 *Chen's hybrid geometry. In the horizontal plane the hologram is a focused-image hologram (Benton's configuration); in the vertical plane, a conventional hologram. The reference beam is a point source which is close to the hologram in the vertical plane but at infinity in the horizontal plane.*

Fourier-transform Holograms

There is nothing aesthetically appealing about a Fourier-transform hologram, and so perhaps the technique of making one does not really belong in a section on display holograms, although it has a number of uses in data storage and processing (see Chapter 24). On the other hand, it is a focused-image hologram of a kind, and it does have some curiosity value. If you are involved with the teaching of modern optics it is very useful to have one available for demonstration purposes, and they are not to be bought off the shelf. The full theory of optical Fourier transforms falls outside the terms of reference of this book, though an intuitive description of its predictions is given in Appendix 3. When light is reflected from or passes through an object, the behavior of the emergent light can be described in terms of diffraction theory. Briefly, the diffracted wavefront can be thought of as the sum of a large number of plane wavefronts of various amplitudes and with various phase relationships, traveling in different directions. Those representing the highest spatial frequencies (roughly speaking, the finest detail), are diffracted through the largest angles. Now, if the object is in the front focal plane of a lens, all these plane wavefronts will be brought to a focus in the rear focal plane of the lens (Fig 14.10).

The diffraction pattern can be recorded by a photographic emulsion, but the phase information is lost. However, by adding a reference beam this information can be recorded and retrieved (Fig 14.11). In the retrieval process two images are recovered. They are at the principal focus of the second lens, and one of them is inverted and pseudoscopic. In the absence of the second lens the two images are formed at infinity. The layout for a Fourier-transform hologram is shown in Fig 14.12.

An identical image can be obtained rather more easily by using a so-called lensless configuration; this was first described by George Stroke (8). This bears some resemblance to Benton's configuration for focused-image holograms in that it

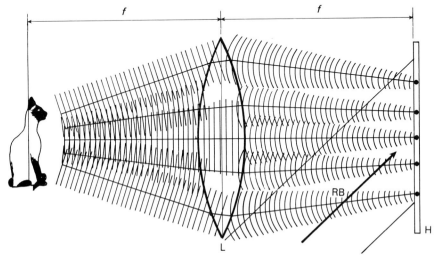

FIG 14.10 *Recording a Fourier-transform hologram. The hologram* H *is in the* FT *plane.*

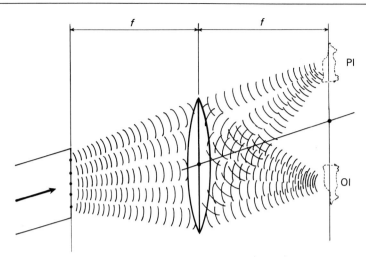

FIG 14.11 *Replaying a Fourier-transform hologram.* OI *is an orthoscopic image;* PI *is a pseudoscopic image.* NOTE: FIGS 14.10 *and* 14.11 *are* not to scale.

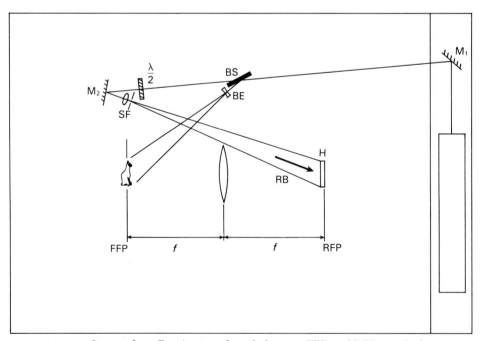

FIG 14.12 *Layout for a Fourier-transform hologram.* FFP *and* RFP *are the front and rear focal planes of the lens* L. *Strictly speaking, the object should be flat and in the* FFP *and the reference beam* RB *should be collimated.* RB *should be at as small an angle of incidence as possible.*

FOCUSED-IMAGE HOLOGRAMS

exploits the lens-like qualities of the hologram itself. Stroke showed that the interference pattern produced was the same when the lens was placed in the reference beam rather than in the object beam. In order to produce the two images sufficiently close together to be seen at the same time when viewing the hologram, the reference source must be close to the object. In addition, it must be in the same plane. A typical layout is shown in Fig 14.13.

A wide-aperture ($f/2$) lens of 50 mm focal length from a 35 mm camera is the most suitable lens. If you replay this hologram using a diverging beam it will form two virtual images in the plane of the original object, one inverted and orthoscopic, the other erect and pseudoscopic. If you replay the hologram with a collimated beam the images will be at infinity. Strictly speaking, the orthoscopic image is virtual and at $+\infty$, and the pseudoscopic image is real and at $-\infty$; this is the correct way to view the images. If you replay the hologram using a converging beam both images will be real.

Fourier-transform holograms are far-field holograms. In order to see the images you have to look directly into the spread laser beam, which appears as a bright spot between the two images (Fig 14.14). (This is not dangerous with a HeNe laser.) As they are transmission holograms you need a monochromatic source (a sodium lamp masked down will do), but it is possible to make white-light copies by the standard transfer methods.

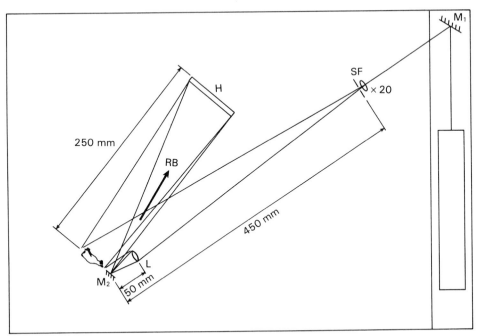

FIG 14.13 *Layout for a 'lensless' Fourier-transform hologram (after Stroke (8)). The spatial filter SF should be approximately ×20, and the lens L about 50 mm focal length, focused on the mirror M_2, which must be close to the object and in the same plane. When the hologram is illuminated by collimated monochromatic light at normal incidence the two images are formed at infinity.*

FIG 14.14 *A Fourier-transform hologram made using the layout of* FIG 14.13. *When the hologram is replayed using collimated light the two images appear at infinity. The difference in size of the two images in this illustration is an artifact of the photography. Visually they are the same size.*

There are several interesting points about Fourier-transform holograms. The first is that you do not need holographic film (Stroke *et al*, (9)). This is partly because of the small angle between the reference and object beams, and partly because of the nature of the coding of the information (successively finer detail is coded farther and farther out from the optical center of the hologram). The second is that the image does not move if the hologram is moved in its plane. This property has led to some interesting experiments in the use of continuously-moving film for holographic movies (Komar (10)). The third is that because of the nature of the coding of the information, the resolution of the hologram is dependent not on the resolution of the emulsion, but on the size of the hologram. Positional information is coded all over the hologram area, and it is this that depends on the resolving power. This is a situation opposite to that of a focused-image hologram, where positional information is coded locally, and detail resolution depends on the resolving power of the emulsion. Fresnel holograms (traditional holograms) are intermediate, and rainbow holograms, as we have seen, are a hybrid of Fresnel and focused-image holograms.

References

1 Benton S A, Mingace H S Jr and Walter W R (1979) One-step white-light transmission holography. *Proceedings of the SPIE*, **212**, 2–7.

2 Hariharan P (1980) Improved techniques for multicolour reflection holograms. *Journal of Optics (Paris)*, **11**, 53–5.
3 Leith E N, Chen H and Roth J (1978) White-light hologram technique. *Applied Optics*, **712**, 1957–63.
4 Chen H and Yu F T S (1978) One-step rainbow holograms. *Optics Letters*, **2**, 85–7.
5 Chen H (1979) Astigmatic one-step rainbow hologram process *Applied Optics*, **18**, 3728–30.
6 Yu F T S, Tai A M and Chen H (1980) One-step rainbow holography: Recent Developments and Applications. *Optical Engineering*, **19**, 666–78.
7 Chen H (1985) Paper presented at *International Symposium on Display Holography 2*, but not submitted for publication.
8 Stroke G W (1965) Lensless Fourier-transform method for optical holography. *Applied Physics Letters*, **6**, 201–3.
9 Stroke G W, Brumm D and Funkhouse A (1965) Three-dimensional holography with lensless Fourier-transform holograms and coarse Polaroid P/N film. *Journal of the Optical Society of America*, **55**, 1377–8.
10 Komar V G (1977) Progress on the holographic movie process in the USSR. *Proceedings of the SPIE*, **120**, 127–44.

CHAPTER
15

Home-made Optical Elements

All this time the Guard was looking at her, first through a telescope, then through a microscope, and then through an opera-glass.
LEWIS CARROLL, *Through the Looking Glass*

In holography for creative or display purposes, as distinct from holography for scientific or technological purposes, it is seldom necessary to buy components of the highest optical quality. As the diameter of the unspread laser beam is so small, it uses only a small part of any reflecting or transmitting surface, and components that are by no means of top quality can be used, with a little care, to produce perfectly good holograms. In any case, many of the optical devices you may need are quite hard to come by in their best Sunday garb, yet a perfectly satisfactory substitute may well be lying in a cupboard full of photographic – or non-photographic – junk. In this chapter optical components are discussed under three main headings. The first is the range of reflectors, diffusers and other light spreaders that are used in the illumination of the subject matter; also, on occasion, for the reference beam. The second is the range of liquid-filled optical components, usually made from acrylic sheet ('Perspex' in the United Kingdom, 'Plexiglas' in the United States), and often possessing excellent optical properties. The third is the exciting range of holographic optical elements (HOEs) that you can make yourself, using the properties of a hologram to produce lenses, mirrors and beamsplitters.

Subject Illumination

Next to the subject matter itself, its illumination is the most important item, as anyone who has indulged in table-top photography will aver. In holography there are several basic kinds of illumination, depending on the apparent size of the source. There is point-source illumination, which corresponds to a photographic spotlight; partly-diffuse illumination from light that has been passed through (or been reflected from) a diffusing surface: fully-diffuse illumination, where an already-expanded beam is diffused; and a modified beam which produces dappled or otherwise non-uniform illumination and superimposes its own pattern on the subject matter.

Point-source illumination   To get true point-source illumination you need a spatial filter in the object illumination beam, just as you do in the reference beam. If you are doing scientific work or making HOEs this is mandatory, as both beams are required to be coherent in every respect. However, for creative or display purposes the subject illumination does not have to be spatially coherent. In fact, the less the spatial coherence the less obtrusive will be the laser speckle.

The simplest method of spreading out the light is by a concave lens of focal length about -6 mm. This spreads the light in a ratio of about 1 in 6, roughly the same as a $\times 40$ microscope objective. A second concave lens a few centimeters nearer the subject will diverge the light further and allow a shorter throw for the illuminating beam. You can use a convex lens and obtain a very similar effect (the beam diverges after converging to a focus); but, in the absence of a spatial filter, a concave lens seems to give a cleaner beam than a convex one. It is certainly preferable with high-power laser sources, and is obligatory for pulse-laser work. Point-source illumination gives very sharp shadow edges, and is comparable with the light given by a photographic spotlight (Fig 15.1).

If you use this kind of lighting only, you can expect a good deal of laser speckle and harsh shadows. You can do little about the speckle, but you can alleviate the harsh shadows by using stray light to illuminate them. Be sure that the optical path differences are within the coherence length of the laser beam, otherwise the fill-in lighting will not register at all; and ensure that none of the stray light falls on the emulsion. The general modeling effect is similar to that which you get when you light a photographic subject with a solitary spotlight; the result can be very dramatic.

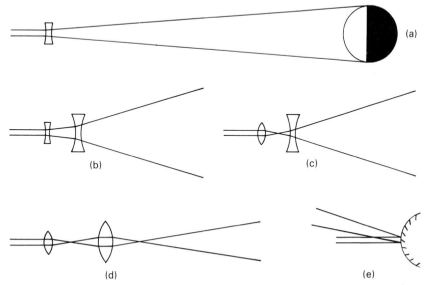

FIG 15.1 *Quasi-point-source lighting for object illumination. (a) The simplest point-source is a concave lens of focal length approx -6 mm, producing very sharp-edged shadows. (b) (c) and (d) Show the use of two lenses, positive or negative, to give a wider divergence. (e) Shows a reflection beam expander, which can be a front-surface convex lens or a trinket or Christmas decoration.*

Partly-diffuse illumination There are two ways of producing partly-diffuse illumination. The first is by passing the undiverged beam through ground glass; the best diffusion is obtained from acid-etched glass, which you can get from photographic specialist dealers. However, you can get equally good diffusion for nothing by using a piece of the glass envelope of an old pearl light-bulb. For a smaller amount of diffusion, a well-thumbed piece of cellulose tape stuck onto plain glass works well. The second method of diffusion is by reflection. For this all you need is a small piece of solid material that has been sprayed with aluminum paint. This causes little disturbance to the polarization of the light beam. If you use a can lid you can employ the irregularities of the surface to spread out the beam so as to cover the subject matter in exactly the way you want (Fig 15.2). The shadow of this beam will be soft-edged, and the speckle effect will be less obtrusive than with point-source illumination. However, the shadows will still be dense and the lighting contrasty and dramatic.

Fully-diffuse illumination This is produced by diverging the laser beam before passing it through a ground glass, or reflecting it off a diffusing surface. It produces a pattern of light and shade which depends on the angle subtended by the light source at the object; in other words, it depends on how large the light source appears when seen from the object. There is an area of full illumination, where the entire light source is seen by the object. Beyond this there is an area of decreasing illumination. Finally, from the point from which the light source is not visible at all, there is a region of total shadow (Fig 15.3).

The professional photographer's 'brolly-light' is an example of this type of source as used in photography; but although it serves the photographic entrepreneur well enough for such purposes as portraits of students on graduation day, it makes for very uninteresting lighting, and there are still black shadows which really need

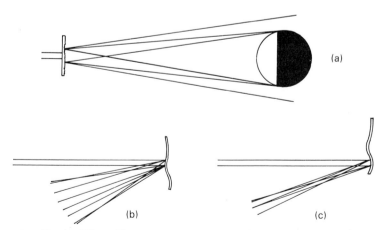

FIG 15.2 *Partly-diffuse illumination.* (a) *Fine-ground or etched glass gives a slightly soft edge to shadows.* (b) *and* (c) *An aluminum-sprayed can lid will give a slightly diffuse reflected beam, the spread of which is determined by the curvature of the surface.*

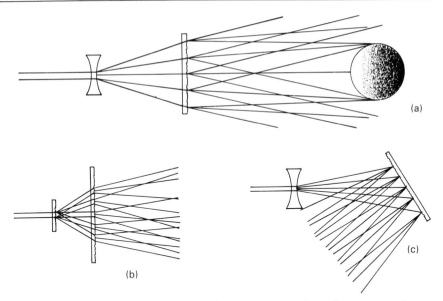

FIG 15.3 *Fully-diffuse illumination. The beam passes through two expanders, the second of which must be a diffuser.* (a) *Shadows are soft-edged;* (b) *and* (c), *alternative configurations.*

a second light to control them. The holographer, like the expert photographer in a properly-equipped studio, can make use of concentrated illumination to obtain modeling and texture, and diffuse illumination to lighten shadows, particularly in holographic portraiture, using the sort of imagination that is evident in the best studio photography. The special requirements of the lighting for holographic portraiture are discussed in Chapter 16.

Special lighting effects One of the more important lighting effects is textured lighting. You can obtain this by a number of methods. The simplest is to expand the beam using a simple lens or an aluminum-sprayed reflector then to put a textured plate into the expanded beam. This may be a simple array of lines (to suggest a venetian blind) or a criss-cross lattice, or letters, symbols or squiggles. You can paint these on a sheet of glass, using a spirit-based fiber-tip pen of the type used for overhead-projector drawings. The contrast will vary with the color of the pen: blue gives a high contrast, green somewhat less; black gives the highest. You can also obtain striking effects using hammered, reeded and fly's-eye glass or acrylic sheet.

One of the most dramatic effects is obtained by using an ordinary ball-bearing as an expanding reflector. A 25 mm ball gives about the same spread as a ×40 microscope objective. Smaller balls give correspondingly greater spreads. Interference between wavefronts reflected from adjacent crystalline domains on the surface cause the illumination area on the subject to be deeply mottled, with an appearance somewhere between greatly-enlarged laser speckle and veined marble (see Fig 11.8, p. 156). The size of the mottling depends on the distance of the ball from the object, and you can spread it out one-dimensionally (ie long and thin) by placing a concave cylindrical lens in the partly-expanded beam.

You may occasionally come across a ball-bearing which does not produce a mottled effect. Keep it for producing uniform lighting of the subject matter. By the way, don't ever handle a steel ball with your bare fingers; if you do, it will begin to corrode within minutes, and may be visibly pitted within 24 hours. If you accidentally handle a ball-bearing clean it at once with tissue. Store ball-bearings wrapped in tissue or cottonwool in a closed box, preferably with a sachet of desiccant (silica gel).

Another method of putting texture into a beam that has been partly expanded by a lens is to reflect it off a piece of aluminum kitchen foil that has been crumpled, then straightened out. The diameter of the incident beam determines the fineness of structure of the pattern; so a single well-crumpled piece of foil will provide large or small patterns as you wish. These methods are summarized in Fig 15.4.

You can also use some of these devices in the reference beam, in particular the painted glass and the ball-bearing. The textured result then appears in the plane of the hologram rather than in the plane of the image.

To produce darting gleams of light in an abstract image you can combine these techniques. By using lightly-crumpled aluminum foil as 'subject', reflecting the expanded beam directly towards the plate, and interposing a sheet of hammered or irregular glass, you will obtain an image formed of dancing lights; by differential swelling and multi-exposure techniques (see Chapter 18) you can extend your images into a variety of hues.

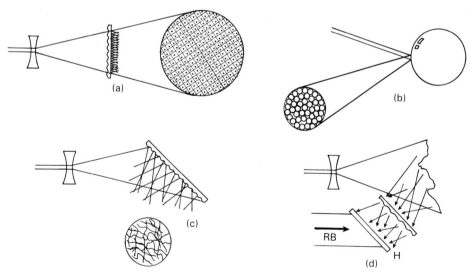

FIG 15.4 *Special lighting effects.* (a) *Expanding the beam through embossed-patterned glass or acrylic produces regular mottles.* (b) *Reflection off a ball-bearing gives an enlarged reticular pattern* (c) *Reflection off metal foil that has been crumpled and then smoothed out and mounted on a rigid surface gives random patterns.* (d) *Irregular glass plate interposed between a holographic plate and illuminated crumpled foil yields a hologram of dancing random light patterns.*

One-dimensional optics There are some devices which will spread a light beam in one direction only. These may be used for the purpose of providing an image reconstruction beam for transferring rainbow holograms from a transmission master hologram. Used on their own with the kind of table layout shown in Fig 13.9 they will provide a beam about 3 mm wide at the master. The width of the beam depends on the length of the optical path from the laser output port to the master plate; the spread is usually about 0.8 mm per meter. There are two basic types of one-dimensional beam expander, a cylindrical mirror and a cylindrical lens; these are described on pp. 191–2. As explained there, the best expander is probably a sample tube filled with glycerol. Use a rubber cork, leaving a small bubble for expansion, and wire the cork in position. Wash off any leaked glycerol with water, and when the tube is thoroughly dry seal the cork with epoxy resin or dip it in Tensol no 12 acrylic cement. The bubble will act as a spirit level, helping you to get the tube horizontal. Liquid paraffin, preferably medicinal grade, is an alternative to glycerol.

Ordinary glass rod is often suitable as a cylindrical lens, but as it is not of optical quality you may have to try a number of positions in order to find one that gives a really uniform spread of light. Check the beam by running an exposure meter up and down it. There should be a gradual and symmetrical rise and fall of the light intensity.

As cylindrical lenses have a positive focus about one-quarter of their diameter from the outer surface, it is possible to spatially filter the beam. There do not appear to be any one-dimensional spatial filters on the market as such, but firms such as Graticules Ltd make narrow rectangular apertures that could be used for this purpose. You can experiment more cheaply using a couple of razor blades. Place them

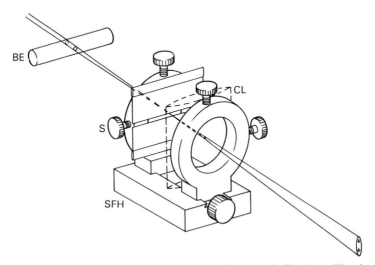

FIG 15.5 *One-dimensional beam expander with spatial filtering. The beam expander BE is a glass rod or liquid-filled sample tube. The slit S can be made from two razor blades affixed to a spatial filter holder SFH. The beam can be spread horizontally by a weak cylindrical lens CL (interrupted lines) attached to the other flange of the holder.*

on a pinhole holder from a spatial filter, using Blu-tack, with the blade edges touching the surface of the flange, and move them carefully together until the edges just touch one another. Now, by very gentle pressure, move them apart until you can just see a uniform slit of light between them. Mount the spatial filter, with microscope objective and pinhole removed, so that the slit is exactly parallel with the cylindrical axis of the lens, and adjust by successive approximations until the light of light is bright and uniform (Fig 15.5). As there is no need to use the x-adjustment (for a vertical slit) you do not need to look down the beam; indeed, it is inadvisable to do so, as it has a much higher intensity than when expanded in two dimensions. All you need to do is to hold a piece of card in front of the slit, and adjust the y-control until the streak of light on the card is brightest. Then use the y- and z-controls alternately to find the position of exact focus. This is the point at which the intensity of the central streak is at a maximum and that of the diffracted side-lobes is at a minimum. If you are using a focused-slit transfer as described on pp. 194–5, you will also need to spread the beam in a perpendicular direction to fill the slit. You can do this with a further cylindrical lens set at right angles to the main one-dimensional beam expander. A focal length of -25 mm will spread the beam to cover a slit up to 75 mm (3 in) wide at a distance of rather more than a meter. The unused microscope-objective holder of the spatial filter mount can be used to support this lens (CL in Fig 15.5).

Liquid-filled Lenses

As far as the world of orthodox optics is concerned, liquid-filled lenses as used in holography are an oddity, not because they are filled with liquid (there have been liquid-filled camera lenses), but because they are basically one-dimensional. This is because they are made from acrylic sheet, which is difficult to mold into spherical shapes without a special oven and a good deal of skill, but which bends into a perfect cylindrical arc under compression from its ends. As we have seen, there are many lenses that are appropriate to holography but which do not need to be spherical, at least in the conventional sense.

One-dimensional collimators The first of these is a lens which can be used for collimating a slit beam. As the expansion is one-dimensional, we need only a one-dimensional collimator. So that the beam can be shifted (or expanded) laterally to make multicolor rainbow holograms without disturbing the lens, this should be made 100 mm wide. The simplest type of lens, and optically the most effective, is the plano-convex type. For a master hologram 10 in long (ie one cut from an 8×10 in plate) it should be 300 mm long, with a focal length equal to its distance from the spatial filter when set up. As the focal length of a plano-convex lens of refractive index 1.5 is equal to twice its radius of curvature, a radius of curvature of approximately 600 mm is required for the type of one-dimensional optics described earlier. This is achieved by making the lens in the form of an open box, using acrylic sheet 3 mm (1/8 in) thick, and inserting in it a piece of 1.5 mm (1/16 in) sheet cut to width but a few millimeters oversize in length. It will naturally take up a circular arc shape (Fig 15.6).

To make the box you will require the following pieces of 3 mm acrylic sheet:

        25 × 106 mm   2 pieces
        25 × 300 mm   2 pieces
        100 × 300 mm   1 piece

and for the curved surface

        100 × 303 mm   1 piece

You can obtain the acrylic sheet from a shopfitter, who will also cut it to size for you. You will need a can of cement suitable for acrylic sheet; Tensol no 12 is suitable. You can also make your own cement by dissolving clear scrap acrylic pieces in dichloromethane or chloroform. To hold the box together while the cement is drying you will need a carpenter's vise and a cramp.

Building a one-dimensional collimator First, mark the center point of one of the long sidepieces, by drawing diagonals on the protective paper, and bore a hole at this point using a 9 mm ($\frac{3}{8}$ in) drill. Remove the protective paper and any swarf. Remove the protective paper from the other sidepiece and the endpieces, and tear off strips about 10 mm wide from the edges of the sheet which is going to form the plano side of the lens; leave the rest of the paper for the time being, to protect the optical surfaces.

Clamp the sheet and the two long sides between two sufficiently-long pieces of 25 mm square timber in the vise. Now use the carpenter's cramp to hold the two end pieces in position. Make sure all the edges are 1 mm (but no more) proud of the surface of the flat sheet and that all the sidepieces are square to it. Do not over-tighten. The easiest way to get the 1 mm rebate is to place a piece of cardboard of the right thickness and a little smaller than the face on the vise shafts, then rest the edges of the sidepieces on the shafts (Fig 15.7).

Now apply the cement to the inside edges that are to be joined, using a small watercolor brush. If the cement is thick and syrupy it may not penetrate the space between the pieces (try it first on two scrap pieces), in which case you will have

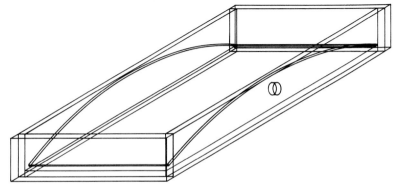

FIG 15.6 *Cylindrical acrylic lens. The frame and flat surface are 3 mm, the curved surface 1.5 mm thick.*

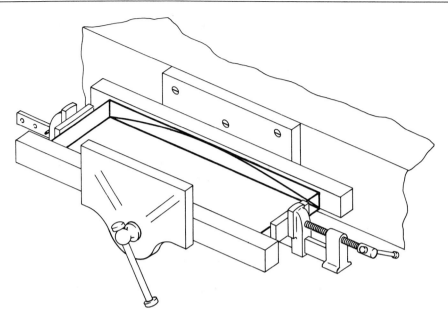

FIG 15.7 *Holding the completed box together with vise and cramp.*

to dilute the first coat with dichloromethane; alternatively, you can run neat dichloromethane or chloroform down the cracks before applying the cement. Apply several coats of cement, allowing half an hour between coats, to ensure that the joints are liquid-tight. When the final coat is dry remove the box, turn it over and apply several coats in a similar manner to the outside of the joins. Set the box aside overnight.

The following day remove the protective paper from the inside of the box and from one side of the sheet that is to form the curved surface. Tear off the marginal strips from the other side. Make sure the inner surfaces are completely clean (you may have to use a drop or two of proprietary window-cleaning fluid) and insert the sheet into the box with the paper side outwards. It should take up a curve that is a perfect arc of a circle. If it does not, test it for freedom of movement by pushing it gently, and if the sheet sticks at any point (indicating a high spot), remove this with fine glasspaper, carefully removing any dust before re-insertion. Once you are satisfied, apply cement at the edges as you did with the flat sheet. When the cement is thoroughly dry, remove the rest of the protective paper and clean the outside surfaces as necessary. Be very careful when handling the surfaces, as the material is soft and is easily abraded.

Now fill the lens through the hole with glycerol or liquid paraffin at room temperature, leaving a small bubble to allow for expansion. Plug the hole with a synthetic rubber bung, and the lens is ready for use. If you treat it with the same care as you reserve for your front-surface mirrors it will last indefinitely.

What to do in case of leaks It is a great pity that the two liquids most suitable optically for filling lenses, namely glycerol and liquid paraffin, should have such a penchant for finding the tiniest leak. If one does occur it will probably be at a corner; it is usually easy to trace. If you do find one, empty out some of the fluid, stand the lens so that the leak is uppermost and thoroughly clean off any fluid that has leaked out, using tissue dampened with water for glycerol or with methylated spirit for liquid paraffin. Allow the area to dry out thoroughly, and apply three more coats of cement. When the cement is completely dry, mark the position of the leak on the outer frame with a wax pencil, just in case it goes again, and refill the lens. Keep the marked corner at the top when you are using the lens.

Other sizes and focal lengths If you intend working with master holograms of different sizes you may find it worthwhile to have several one-dimensional collimators of different sizes and focal lengths. If so, you can calculate the required dimensions of the various pieces using the formula given in the Box.

Calculations for designing a liquid-filled lens

To find the length of the curved side, you need to be able to find the length of the arc of a circle relative to that of the straight line (chord) which cuts it off, in terms of the radius of curvature. Let us call the length of the arc a, the length of the chord c and the required focal length f; then the length of the arc is given by the formula

$$a = f \sin^{-1}\left(\frac{c}{f}\right)$$

The construction is shown in Fig A2.13 (p. 372).

NOTE: you must set your calculator to 'radians' before starting the calculation. On some calculators the term $\sin^{-1}$ is shown as 'arcsin', and with some you obtain this function by pressing the 'inv' button followed by the 'sin' button.

The width of the sidepieces and endpieces needed to give the required rebate is given by one of the following formulae:

$$\text{Width (mm)} = \frac{f}{2}\left(1 - \cos \sin^{-1}\left[\frac{c}{f}\right]\right) + 6.5 \text{ for a plano-convex lens}$$

$$\text{or } f\left(1 - \cos \sin^{-1}\left[\frac{c}{f}\right]\right) + 5 \text{ for a double convex lens}$$

The volume of fluid v required for a plano-convex lens of depth x is given by the formula

$$v(\text{litres}) = \frac{xf^2}{4}\left(\sin^{-1}\left[\frac{c}{f}\right] - \frac{1}{2}\left[\frac{c}{f}\right]^2\right) \times 10^6$$

For a double convex lens of edge a (mm) the volume is given by the formula

$$v(\text{litres}) = \frac{af^2}{2}\left(\sin^{-1}\left[\frac{c}{f}\right] - \frac{1}{2}\left[\frac{c}{f}\right]^2\right) \times 10^6$$

All these formulae are derived in Appendix 2.

Focusing lens for broad-slit transfer holograms To make one of these you need to perform the same calculations as those given in the Box above for a plano-convex lens. If you make your lens from an acrylic cylinder as suggested on p. 195, all you need to know is the distance between the slit and the transfer hologram: this is the required diameter for the cylinder. The thickness of the lens is not important provided the finished lens is at least as wide as the slit. The slice should be cut off with a circular saw. Don't attempt this yourself; get the shopfitter to do it for you. He will have a circular saw with the right kind of teeth for cutting acrylic material without leaving marks, and will be able to polish it for you too. Otherwise you will have to make passes of the lens over successively finer grades of glasspaper, finishing with 'Crocus' paper and, finally, metal polish. You can make the lens more cheaply by taking a piece of 3 mm acrylic sheet, laying it on a former and placing it in an oven at $150°C$ ($300°F$) (gas mark 4). Watch it until you see it slowly collapse onto the former, then switch off the oven and allow it to cool with the door shut. You can then plane the lower (concave) surface down with a planing machine until it is flat, finishing it off as described above.

Two-dimensional collimating lenses The lenses described so far have been cylindrical. For full-aperture transfer transmission and reflection holograms it is necessary to collimate the beam in two dimensions, for this purpose, the simplest lens to make is a bicylindrical lens. If you imagine a cylindrical lens 300 mm square (say), collimating light in a vertical direction, and immediately behind it another identical lens rotated through $90°$ so that it collimates the light in a horizontal plane, you have what amounts to an ordinary convex lens with spherical surfaces (Fig 15.8).

The lens itself is made along the same lines as the one described on p. 220. In principle it consists of two cylindrical plano-convex lenses back to back, though there is, of course, no need for the 'plano' surface. Assuming a focal length (as before) of 120 mm, and dimensions of 300×300 mm, you will require the following pieces of acrylic sheet:

$45 \times 306 \times 3$ mm   2 pieces
$45 \times 300 \times 3$ mm   2 pieces
$300 \times 303 \times 1.5$ mm   2 pieces

As there is no 'plano' surface you will have to build the sides of the box like a honeycomb frame, holding it together with stout rubber bands. To make sure it is square, draw a 300 mm square on a piece of paper and stand the frame on this while it dries. Better still, cut an accurate square from stiff cardboard and use it as an inside template. The first thing to do is to mark a deep line all the way along each sidepiece 1.5 mm from the midline, using a scriber or other sharp point. These lines will go on the inside to act as guidelines for the curved sheets. Drill the 9 mm hole as before, with its center 4.5 mm from the scribed line so that it just touches it (on the side nearer the center). When the frame is assembled, these lines all go on the inside to act as guidelines, one pair above the midline and one pair below it; if you have made the scratches deep enough the curved surfaces will lock into them and

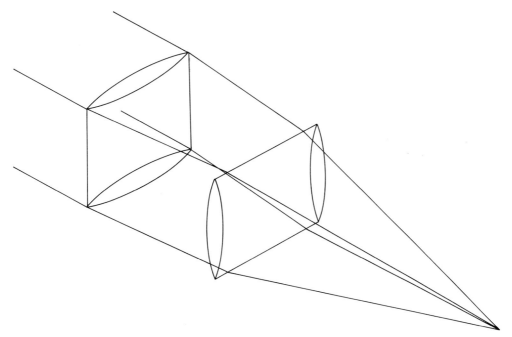

FIG 15.8 *The effect of two orthogonal cylindrical lenses is the same as that of a single spherical convex lens. In practice the lenses should be close together.*

ensure a good seal. Paint the cement onto the edges before you insert the curved portions, and scrape out any excess at the corners before it dries out completely.

If your rubber bands are stout enough you may not need to wait overnight, though if you have the time it is better to do so. Peel off all the protective paper from what will be the inside of the curved surfaces and clean them as described on pp. 221–2, but remove only a strip approximately 20 mm wide from the edges of the surfaces that are to be outside. Insert the first curved piece so that the ends fit into the upper of the grooves you have scribed, with the convex side uppermost. Test for high spots and remove them as necessary. Once you are satisfied that the curvature is uniform, seal the edges on both sides. It is a good idea to have a dummy run for high spots before using the cement, for both curved pieces. When you come to put in the second curved sheet you will not be able to seal the inside, so paint around the approximate curved line on the inside of the frame before you insert the acrylic sheet. Finish off the lens by running cement all round the edges, but don't remove the rubber bands until 24 hours after you have applied the last coat (Fig 15.9).

Crazy lenses By using a higher degree of curvature on the curved sheets you can shorten the focal length of your lens to a minimum of about 400 mm, and use the lens for making focused-image holograms. The curved edges will be longer and the edge pieces wider (Fig 15.10).

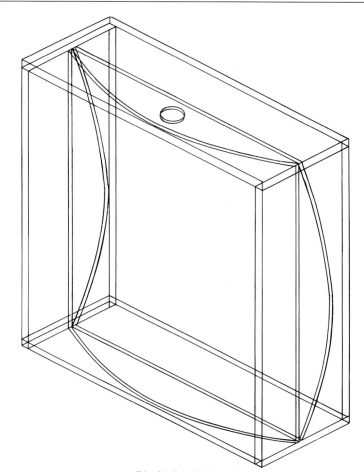

FIG 15.9 *Bicylindrical liquid-filled lens.*

For a lens of 400 mm focal length you will require the following pieces of acrylic sheet:

    140 × 306 × 3 mm 2 pieces
    140 × 300 × 3 mm 2 pieces
    300 × 340 × 1.5 mm 2 pieces

Because of the high curvature of the surfaces the pieces may require pre-bending for up to 24 hours before insertion into the box. It is a good idea to lightly scribe a circular arc making the correct position of the outer surface on the insides of the pieces making up the frame so that you can check the uniformity of the curve while the cement is drying. If you are dedicated enough, you can adjust the curvature to approximate a parabola, which is closer to the ideal curvature than a circle, but you will have to make a pair of templates to hold the material steady while the cement is drying. If you are even more dedicated you can draw out a ray diagram to produce a very exact curve using the laws of refraction, as suggested by McCormack (1).

FIG 15.10 *The author with some of his liquid-filled lenses.*

Such lenses hold a considerable amount of liquid, around 7 litres, so expensive fluids such as glycerol and medical-grade liquid paraffin are out. Fortunately, you can get technical-grade liquid-paraffin, usually labeled 'For oil baths', and known in the United States as 'mineral oil'. It is amber in hue, but is fully transparent to HeNe laser light; most important, it is cheap.

The reason the lenses have been dubbed 'crazy' is that they suffer from vertical–horizontal astigmatism, because one surface is so much nearer to the object than the other. If the center-line of the lens is at approximately the same distance from the subject matter and its image, ie twice its focal length, the horizontal and vertical aspects of the subject matter will be brought to a focus in the same plane; however, the scale will be different, so that the image may be stretched out horizontally or vertically. You can avoid this distortion by reducing the curvature of the surface that is away from the subject matter. In the case of the lens described above, for example, you can use a piece of acrylic material 328 mm instead of 340 mm in length for the side away from the subject. As usual there is a snag: although the image is no longer stretched, the astigmatism is worse than ever, as the focus for the rear component is 150 mm nearer to the eye than the focus for the front component. However, provided the viewer is a meter or so away from the image the discrepancy does not show. After all, this sort of discrepancy is present in all rainbow holograms. Benton (2) gives a full account of all these difficulties with astigmatic optics.

All crazy lenses are fundamentally simple lenses, even though they may have aspherical (or, more correctly, acylindrical) geometry; and they all show some of the aberrations of simple lenses. With wide-aperture lenses of any kind, the image seems to twist and vanish as if in flames as the viewpoint approaches the limiting angle. With this technique there are possibilities which as yet have scarcely begun to be exploited pictorially. Holographers with access to acrylic moulding facilities and to shaping ovens can dream up their own crazy-lens shapes without being confined to one-dimensional curves.

Holographic Optical Elements (HOEs)

Principles of HOEs The most satisfying way of making your own optical elements is to make them holographically. A transmission hologram behaves in many ways like a lens, and a reflection hologram like a mirror. When a hologram is illuminated by collimated light the image is formed at its principal focus, a distance from the hologram equal to its focal length. The usual holographic image

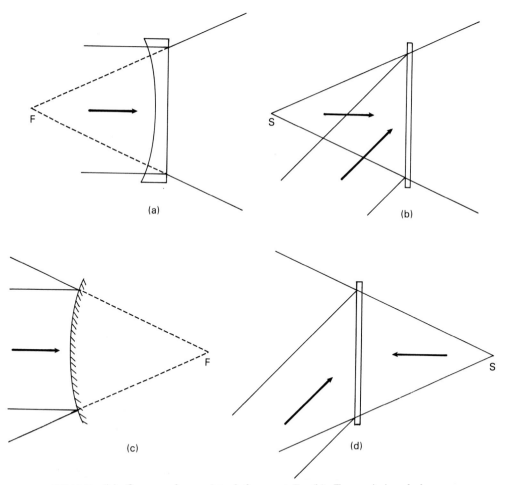

FIG 15.11 (a) *Concave lens, virtual focus at* F. (b) *Transmission hologram equivalent.* S *is the second source, which becomes the virtual focus on replay.* (c) *Convex mirror and virtual focus* F. (d) *Equivalent reflection hologram.*

consists of a very large number of points, and for each point the hologram has its own independent focal length. Let us, then, simplify the situation: consider a hologram of a single point source of light, made with a collimated reference beam. On replay, the virtual image appears where the original point-object was. But what happens to the transmitted (or reflected) beam? In effect the hologram is acting in exactly the same way as a concave lens or, in the case of a reflection hologram (Fig 15.11), a convex mirror. There are, however, two important differences. The first is that with a hologram the image is not in line with the incident beam. The second is that, unlike a lens or mirror, a hologram can be reversed (flipped) in the beam; it will then have a positive focal length, and will form a real image of the point source (Fig 15.12).

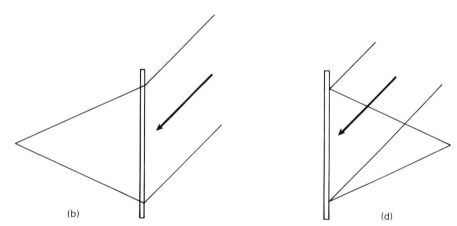

FIG 15.12 *The holograms of* FIG 15.11. (b) *and* (d) *illuminated by conjugate beams behave respectively as a convex lens and a concave mirror, of the same (but positive) focal lengths.*

A hologram of a point source made in this way shows all the characteristics of a convex lens or concave mirror. Most important, it obeys the lens law which relates the distance of the object and image from the lens or mirror and the scale of the image to that of the object (see Appendix 2). It is possible to produce a holographic lens or mirror simply by placing a holographic plate in the interference pattern produced by two diverging laser beams (see Box).

Holographic mirrors and lenses: calculation of focal length

A holographic mirror or lens is a *Gabor zone plate*, which focuses images by constructive interference. Although a zone plate may operate in a different manner from a conventional optics system, it obeys the laws of geometrical optics, in particular the basic lens formula

$$\frac{1}{u} + \frac{1}{v} = \frac{1}{f}$$

where u and v are the distances of the object and image respectively from the zone plate and f is its focal length. Clearly, if $u = \infty$ (a collimated beam), $v = f$, and the simplest method of making a holographic lens of a given focal length would be to have a point source (ie a spatial filter) at a distance f from the holographic plate, and a collimated reference beam. However, this is not necessary. A hologram made using two point sources at finite distances still has a focal length which can be calculated to have any desired value, depending on the respective distances of the two sources from the plate. If these distances are d_1 and d_2, the focal length f is given by the formula

$$\frac{1}{f} = \frac{1}{d_1} - \frac{1}{d_2}$$

provided d_1 and d_2 are fairly large compared with the diameter of the plate. The minus sign occurs because the reference beam distance, d_2, is reversed in direction on replaying the hologram.

If you want a zone plate with a focal length of 1000 mm you simply have to arrange the values of d_1 and d_2 such that $f = 1000$. One way of doing this (there are many others) is to have $d_1 = 600$ mm and $d_2 = 1500$ mm, and this can readily be achieved with two spatial filters, one ×20 and one ×60, set at appropriate distances. However, it is easy to remember that whenever $d_2 = 2d_1$, the focal length will be equal to d_2 – hence if the distances are, say, 500 and 1000 mm respectively, the focal length will be 1000 mm. This can conveniently be achieved if you have two microscope objectives one of which has twice the magnification of the other. You can use the method with either transmission or reflection holograms. The latter have a higher diffraction efficiency, but the processing techniques are restricted to those which give a reconstruction at the original wavelength. If both beams are collimated, the focal length is infinite, and the result is a plane diffraction grating or a plane mirror.

Holographic diffraction gratings If you have not made a HOE of any kind before, it is a good idea to start off with a holographic diffraction grating. This is the holographic equivalent of a prism, as it turns the beam through an angle. When collimated white light falls on a diffraction grating made with a HeNe laser, red light will be diffracted at the original angle, but shorter wavelengths will be deviated progressively less, so that the light is spread out into a spectrum. All you need to make a diffraction grating are two beams of approximately the same intensity, one perpendicular to the plate and the other at 30–45° incidence (Fig 15.3).

Provided the spatial filters are equidistant from the plate and not less than about 1 m from it, the grating space will be uniform. Develop and bleach the plate as for a transmission hologram. The grating can be used as a beamsplitter, the ratio of the beams being between 4:1 and 16:1 depending on your holographic technique.

If you make a grating with an angle between the beams of less than 45°, particularly if you are making a symmetrical rather than an asymmetrical or 'blazed' grating, the non-linearity in the processing will lead to higher-order diffraction. This means that when you put a laser beam through the diffraction grating you will get

HOME-MADE OPTICAL ELEMENTS

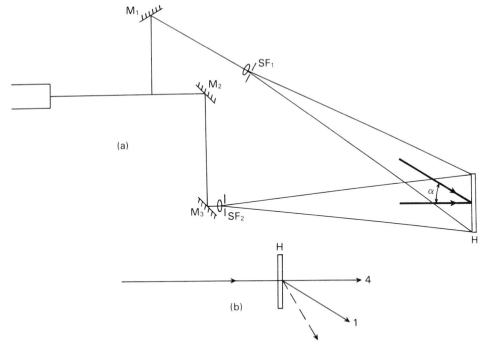

FIG 15.13 *Holographic diffraction grating.* (a) *Shows the table geometry; the beam ratio should be* 1 : 1. *The spatial filters* SF_1 *and* SF_2 *should be equidistant from the hologram* H. (b) *Shows the grating used as a beamsplitter, with a minimum ratio of about* 4 : 1. *If the angle between the beams in* (a) *is less than* 45° *there will be a weak second-order beam.*

a row of spots each side of the central spot, and if you use white light you will get a row of spectra. The only way to avoid this is to develop and fix the plate without bleaching; even then there is only a narrow range of density (usually somewhere between 0.4 and 0.7) that will give only one spot each side of the zero-order (directly-transmitted) beam. This does not happen for angles greater than 45° as the conditions for higher-order spectra are not satisfied.

Holographic lenses A holographic lens is only one step away from a holographic grating. It requires two beams of differing divergences to fall onto the holographic plate. The calculations required are explained in the Box above. The table layout is shown in Fig 15.14. The beams should be of approximately the same intensity. If there is any difference, that of the longer beam should be the higher. Bleach-process the HOE as for a transmission hologram.

Much of the incident light will still pass through the collimator undeviated, and you may be able to use this for extra object illumination. You can custom-design a holographic collimator for a specific table arrangement, as the deviation of the transmitted beam is governed by the angle between the two beams used to make the HOE.

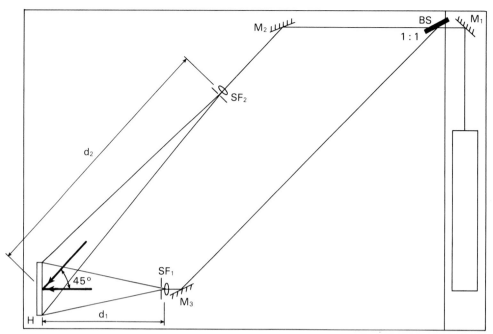

FIG 15.14 *Geometry for a holographic lens. The beam ratio should be close to 1 : 1. In general $d_2 = 2d_1$, giving a focal length of d_2.*

Holographic mirror beamsplitter Holographic mirrors are generally more useful than holographic lenses, as their diffraction efficiency can be much higher. A beamsplitter of the conventional type is no more than an inefficient mirror, and the best mirror beamsplitters are dielectric-coated. Holographic mirrors operate on the same principle as dielectric mirrors; you can make one by making a Denisyuk hologram of an ordinary front-surface mirror. The easiest way to do this is to use the single-beam frame described on pp. 96–9. You should set the relay mirror at 30°, and the 'object' mirror at a further 15°. The light passing through the holographic plate will have an angle of incidence of 30°, and the returning beam will strike it at an angle of incidence of 60° (Fig 15.15).

When the processed hologram is set up as a beamsplitter, with the beam at an angle of incidence of 30°, the transmitted beam will pass through undeviated, and the reflected beam will be at right angles to it. The reason for producing the asymmetry is to get the specularly-reflected beam well out of the way. If you use reflection master processing (see Appendix 10) the angles will be correct and the efficiency high. A small rotation of the hologram out of plane will alter the transmission ratio in favor of the transmitted beam. If you use an off-center beam to make the hologram, so that the exposure varies across it, you can vary the beam ratio by sliding the hologram in its own plane. As with the transmission HOE, you can make custom-designed beamsplitters to operate at any angle you wish to suit a given table geometry.

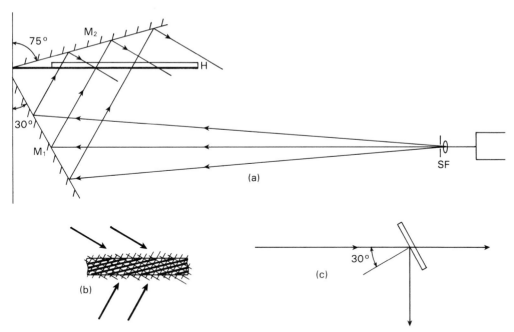

FIG 15.15 *Making a holographic mirror beamsplitter with a single-beam frame. (a) Shows the geometry: the mirror M_1 is at 30° to the vertical and the 'object' mirror M_2 is at 75° (ie 15° to the hologram). (b) Shows the formation of the Bragg planes at 15° to the plane of the emulsion. (c) Shows the two orthogonal beams formed when the HOE is used at 30° incidence.*

You may calibrate a variable beamsplitter in terms of intensity ratio and optimum angle of incidence (which may differ from that of the original geometry, depending on the processing), using a photographic exposure meter or a photometer. Use a fine pen of the type sold for overhead projectors to mark the positions of the various ratios on the emulsion side, near the edge. Don't forget to record the original angles of incidence of the two beams as well.

Holographic reflection collimator The simplest method for making a holographic reflection collimator is simply to borrow a collimating mirror and make a Denisyuk hologram of it. Tilt the relay mirror to about 43° so that the incident beam is at nearly normal incidence, and tilt the collimating mirror about 3°. You can make a holographic reflection collimator without a mirror if you use a method similar to that described above for a collimating lens. Such a mirror has the advantage over a conventional mirror that it need not be confined to small angles of incidence: it can be designed to reflect at 45° or more. For a mirror of 1000 mm focal length reflecting at an angle of 45° the optical layout is as shown in Fig 15.16.

You will notice that one of the distances is exactly twice the other. This holds for all focal lengths. Thus for a focal length of 1200 mm the two distances should be 600 and 1200 mm respectively; for a focal length of 1500 mm, 750 and 1500 mm, and so on. This is explained in the Box on pp. 229–30.

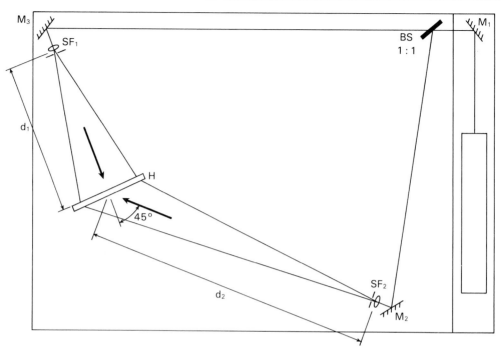

FIG 15.16 *Geometry for making a holographic collimating mirror. The beam ratio should be approximately 1 : 1. If $d_2 = 2d_1$, the focal length of the HOE will be d_2.*

At the correct angle of incidence this HOE will produce a collimated beam 45° to the incident beam when flipped. You can produce further HOEs to produce beams at steeper and shallower angles as required. You can sometimes pick up the unwanted transmitted beam to give extra object illumination, provided you can keep the optical path difference within the coherence length of the laser.

References

1 McCormack S (1982) Liquid-filled cylindrical lenses. *International Symposium on Display Holography*, **1**, 149–62.
2 Benton S A (1982) The mathematical optics of white-light transmission holograms. *International Symposium on Display Holography*, **1**, 5–15.

CHAPTER
16
✳

Portraiture and Pulse-laser Holography

> He was black in the face, and they scarcely could trace
> The least likeness to what he had been.
> While so great was his fright that was waistcoat turned white –
> A wonderful thing to be seen!
>
> LEWIS CARROLL, *The Hunting of the Snark*

Portraiture has always had a fascination for artists: some, like Gainsborough and Reynolds, painted little else but portraits. Others, such as Rembrandt and Van Gogh, obsessively painted portraits of themselves. Sitters had to be patient, often being required to remain motionless for hours at a time. With the advent of photography times came down from hours to minutes, then to seconds and less. Then, around 1950, came the flashtube, and exposure times dropped dramatically to a millisecond or so. Quite suddenly it became unimportant for the sitter to remain even moderately still. In a way this simplified the task of the portrait photographer, who was now able to concentrate on catching the instant when the sitter had just the right expression; though the photographer now had to learn from the beginning how to control this new type of lighting. At the same time the quality of lenses and films was improving to the point where it was becoming possible to use 6×6 cm or even 35 mm formats for portraiture, and to make a dozen exposures in rapid succession, thus increasing the chances of a really successful shot. The early flashtubes contained no modeling lamps to enable the photographer to judge the lighting effect in advance, and the photographic portraitist had simply to rely on experience to get the modeling and lighting balance right.

Much the same problem faces the holographic portraitist today. Fortunately, there are excellent books on photographic portraiture which go into detail about all the possible types of lighting, their effect, and their application in a huge variety of situations involving the making of portraits. This chapter therefore deals only with the more basic types of lighting for portraiture.

The Need for a Pulse Laser

Photographic flashes for studio portraiture have a flash duration of the order of 1 ms (10^{-3} s), which is adequate for producing satisfactorily sharp portraiture even of a sitter who is talking and gesticulating animatedly. However, durations even as short as these are not satisfactory for holographic portraiture. Even the blood moving under the skin produces enough movement to spoil the image. At the time of writing, the only laser suitable for portraiture is the Q-switched ruby pulse laser, (see pp. 30–2). Such lasers have pulse durations of the order of 25 ns (25×10^{-9} s). During this time even the bullet leaving the muzzle of a rifle travels only 7 μm (7×10^{-6} m), about ten times the wavelength of laser light. Plainly, there is no need with a pulse-laser source to have the kind of stability associated with CW holography, though it must not be supposed that there is as much freedom as there is in a modern photographic portrait studio.

Safety Considerations

There are two aspects of safety to be considered: that of the sitter, and that of the equipment. Although the total energy contained in the pulse is low, its very short duration means that during the actual pulse, power is being dissipated at up to 40 MW, enough to light a medium-size city, in a beam only a few millimeters in diameter. Even a partly-expanded beam can be hazardous to the eyes; in order to be safe it must be diffused so that the diameter of the beam on the diffuser is not less than 50 mm for a 1 J pulse or 150 mm for a 10 J pulse. The path of the undiverged beam should be protected so that there is no chance of any part of the body intercepting it, and there must be a safety device ensuring that under no circumstances can the laser be fired accidentally. A set of safety recommendations is included in Appendix 1.

The dangers of lasers, even pulse lasers, are often exaggerated, and it must be said that even a slackly-run holographic studio is a less hazardous place to work in than the average home handyman's workshop, as far as the risk of eye injury is concerned. Nevertheless, there are certain safety precautions that are mandatory when working with lasers, just as there are when working with home power tools; this is particularly so when a second person who may not be familiar with lasers is present. The most dangerous part of a pulse laser is not in fact the beam but the power supply, a huge bank of electrical capacitors with virtually zero internal resistance. Anyone making contact with the high-voltage output would have roughly the same chance of surviving it as of surviving being struck by lightning. Fortunately, the makers of pulse lasers take particular care over this, so it is the light beam that chiefly concerns us; the operator, the sitter and the laser itself all need protection from its effects.

The main hazard to the laser equipment comes from the possibility of retroreflection. A beam reflected back along its own path will almost certainly destroy the ruby rod of at least one of the amplifiers. This does not happen only with carelessly-placed reflecting surfaces; it can happen if an aluminized mirror or beamsplitter is used in the undiverged or partly-diverged beam. The metal evaporates instantaneously, and retroreflection can occur from the metal vapor. It is also important not to have any reflecting surfaces, particularly those of diverging lenses, with the curvature of their surfaces positioned so that they can retroreflect. Take particular care over this when setting up. Put curved surfaces slightly off-axis, and plane surfaces at a small angle of incidence. A ruby rod with both ends blown off is not a pretty sight, either for you or your bank manager.

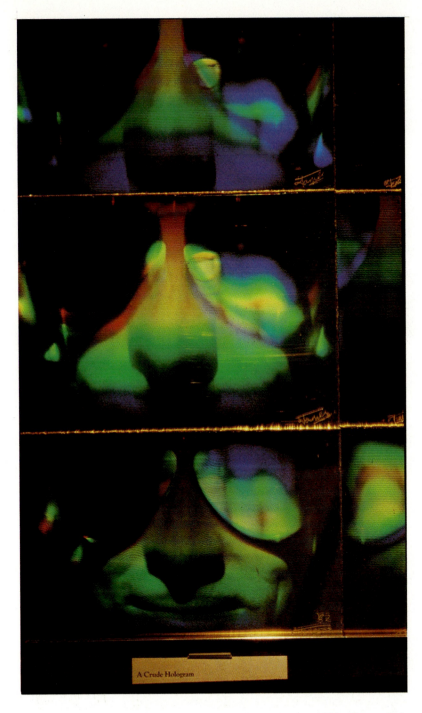

PLATE 18 *A Crude Hologram, by James Copp (see p. 239). Transmission hologram montage with dispersion; the face is that of a waxwork model of Copp's head.*

(b)

(a)

PLATE 19 (a) The type of make-up needed for ruby-laser portraiture (see pp. 239–40). (b) 'Emma', a holographic portrait made by a professional studio. (c) shows a master transmission hologram of a Richmond portrait. (d) is an achromatic (neutral-color) reflection transfer hologram (see caption to PLATE 26). Holograms by Richmond Holographic Studios. Photographs by Tim Hawkins.

(d)

(c)

PLATE 20 *Wildlife studies (see p. 244). The lion cubs and snake are reflection holograms by Richmond Holographic Studios. The vulture is a pulse-laser master hologram and the barn owl is a white-light transmission transfer hologram, both by Rob Munday, made at the Royal College of Art holography facility by arrangement. All photographs by Tim Hawkins.*

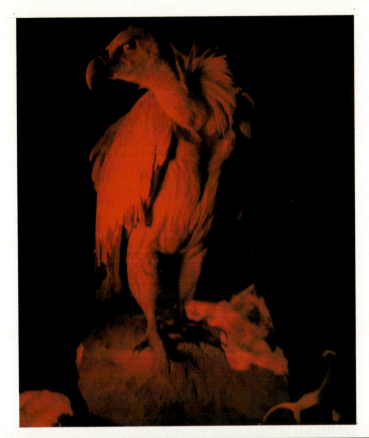

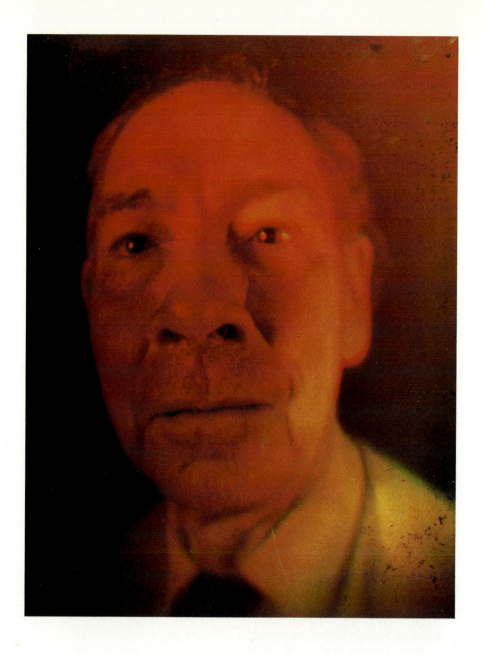

PLATE 21 *A test shot of the author made during the setting up of the Royal College of Art's pulse-holography facility. Notice the distortion, similar to wide-angle distortion in photography. This was caused by using a reconstruction beam that was diverging when the master hologram had been made with a collimated beam. The low side-lighting, which was all that was available at the time, is also less than flattering. (See p. 240.)*

PLATE 22 *Two views of a Cross hologram. The original film was shot by Nigel Abraham, using montage techniques. Note the distortions inherent in the image. (See p. 253.)*

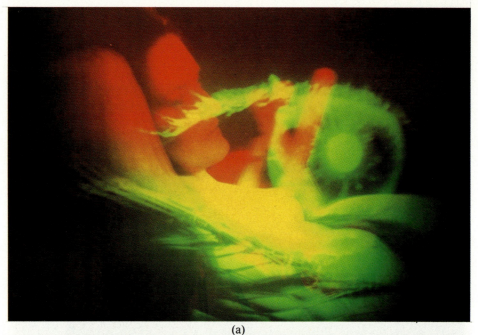
(a)

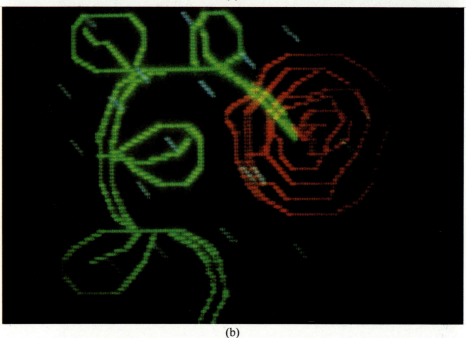
(b)

PLATE 23 (a) *The sculptor Alexander, working with Hariharan (see p. 253), has produced a number of very large (1 × 1.5 m and more) holographic stereograms exploiting montage techniques and using such artifacts as time-smear to produce creative effects. 'Danielle's Dream' is a triple montage of a nude woman in a hammock, a sea-creature and a huge eye. (b) 'Digital Rose' by Bill Molteni. This is from a color transparency of Molteni's hologram produced on a home computer as described on pp. 255–7.*

Optical Requirements

All the optics up to and including the beamsplitter, and the diverging and collimating optics for the reference beam, must be of the highest optical quality, with a surface accuracy of one-tenth of a wavelength or better, as it is not possible to use spatial filtering. A spatial filter is in fact installed within the laser immediately after the oscillator element, but thereafter the beam is too powerful for spatial filtering; if converging optics are placed in the beam the air will become ionized in the focal area and will be opaque to the beam. Diverging optics must therefore be used throughout, with dielectric coatings. Front-surface mirrors must also be of the dielectric type, and should be at the correct angle of incidence to give 100% reflection. *Dove prisms* are an alternative to dielectric mirrors. Variable beamsplitters should be of the polarizing-cube type (see Box); fixed beamsplitters can be dielectric-coated.

Polarizing-cube beamsplitters

In this type of beamsplitter a cube is made up of two right-angle prisms separated by a multiple dielectric layer. Unpolarized light incident normally on one face of the

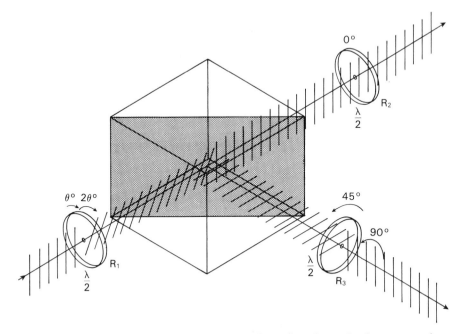

FIG 16.1 *Polarising-cube beamsplitter. Light reflected at the hypotenuse is s-polarized, transmitted light is p-polarized. Half-wave plates R_2 and R_3 are used to bring polarizations of the emergent beams into the same plane. The half-wave plate R_1 rotates the plane of polarization of the incident beam, in order to vary the intensity ratio of the two emergent beams.*

cube emerges as two beams, one from the opposite face of the cube and one from an adjacent face. The two beams are equal in intensity and are linearly polarized orthogonally to one another, the transmitted beam being *p*-polarized with respect to the hypotenuse surface. However, if the beam entering the cube is already linearly polarized, the ratio of transmitted to reflected light intensity will be determined by the direction of polarization of the incident light, the range of ratios being from 98% to 2%.

This beam ratio can be controlled by a half-wave plate situated at the entry port; rotation of this by $\theta°$ results in rotation of the plane of polarization by $2\theta°$. As the emergent beams are orthogonally polarized, further half-wave plates are required at the output ports in order to line up the direction of polarization of the two emerging beams (Fig 16.1).

Energy Requirements

The amount of energy required depends on the size of the holographic plate and its sensitivity, as well as the distance of the object from the plate. For a minimal portrait (head only, on a 24 × 30 cm plate at a distance of about 250 mm from the plate) something under 0.2 J will be sufficient (Bjelkhagen (1)); for a full head and shoulders at 600 mm from a 50 × 60 cm plate, about 3 J will be needed; this is on finegrain emulsion such as Agfa-Gevaert 8E75HD or Ilford SP673. The largest plates (1 × 1 m or 1 × 1.5 m), for three-quarter length or full-length seated portraits, require up to 10 J and may also demand a higher-speed emulsion such as Agfa-Gevaert 10E75HD or Kodak SO-125. The beam intensity ratio at the beamsplitter should be about 9:1, the larger figure making up the subject illumination (as more than 95% of the light incident on the subject will not actually reach the hologram). 10 joules may seem a lot of energy, but it is only about the same as would be used for a photographic studio color photograph*, and it must be remembered that holographic emulsions are about 200 times less sensitive than professional color photographic emulsions – about the same sensitivity as photographic printing paper. This is why it is necessary to illuminate the subject from as close as possible.

---

* Photographic flashes are measured in terms of *electrical* energy dissipated; their luminous efficiency is of the order of 10–40%, depending on the gas employed and the tube loading.

Portraiture with a CW laser?

It is not impossible to make holographic portraits (of a kind) using a CW laser. The most dramatic genre is the three-dimensional silhouette. Granted that a true hologram is out of the question – it is difficult to remain totally stationary for as much as a millisecond, let alone several seconds), it is still possible to produce an 'anti-hologram' or black-hole image (see p. 124). To make this type of hologram the background should be uniformly illuminated. The illumination of the subject matter is immaterial. The only important thing is that the subject should be as nearly still as possible, as would be required for a photographic portrait, otherwise the image will be blurred. Near-profiles seem to give the most striking effects with this technique.

Another method is to use a life mask. This can be a cast of the face; it can also be a hollow mould, giving a Denisyuk real image when flipped. Adrian Lines was probably the first person to achieve a hologram of a full-length nude by this method, using his own body (Plate 17).

Workers such as Hans Bjelkhagen have used both pulse-laser technique and plaster casts; others include Karl Reuterswaard and James Copp (Plate 18).

It is also possible to make holograms of stereoscopic pairs of photographs using a masked double-exposure technique with focused images. The geometry is described in the Newport Corporation's 'Projects in Holography' (2), but the viewing conditions are exacting and the results not very exciting. Much better results can be obtained by making straight transfer holograms of stereoscopic pairs using a left–right slit-masking technique (p. 183), so that from an average viewing distance each eye sees each appropriate image only.

Special Problems with Holographic Portraiture

There are three problems specifically associated with holographic portraiture, and until recently these have combined to make the results barely acceptable in aesthetic terms. The first was that the pupils of the subject's eyes were fully dilated, giving a curious staring appearance. The cause of this was the need to spend some time in near-total darkness while the plate was positioned. This can be overcome by illuminating the sitter with green light to which the emulsion is almost totally insensitive, but to which the human eye is very sensitive. The most effective way of achieving this is to use a 35 mm slide projector with a 50 mm square 514 nm narrowband filter in its slide carrier, directed at the sitter's face. This, while being totally inactinic, also helps the portraitist to judge the appropriate moment at which to fire the laser. Incidentally, this type of filter is the only really satisfactory safelight filter for pseudocolor reflection work, where you need to repeatedly examine the emulsion close to the safelight. Another quick and cheap solution is to flash a torch into the subject's eyes immediately before exposure. However, as might be expected, this is not always conducive to the adoption of the desired facial expression.

The second problem is one which, in a somewhat different form, also faced early cinema and television technicians: the color sensitivity of the detector did not match that of the eye, and skin tones were falsified. For example, in early cinema, the use of red-blind orthochromatic film resulted in blotchy complexions and black lips. The solution at the time was to use a greenish-blue make-up. Although this looked ghastly in real life, it achieved a natural balance in the photographic image. The difficulty with pulse holograms is that the wavelength of ruby-laser light, 694 nm, is close to infrared, which, as anyone who has sat close to an electric fire knows, penetrates the outer skin. This makes men's faces look unshaven and the complexions of both sexes waxen. As with the cinema, the solution is to use opaque make-up (Plate 19). There are a number of proprietary formulae, used nowadays mainly for television work. The important points in applying such make-up are (a) it should nowhere be applied so thickly as to disguise the texture of the skin, and (b) there must be no gaps in it whatever. (Holographers contemplating full-length figure studies should take note!) Unlike photography, in holography there is no way of retouching blemishes in the image. There are, however, indications that this problem can be minimized without recourse to all-over make-up, by the use of the recently improved processing techniques (see Appendix 6).

The third problem is the possible change in perspective on transfer to a final image-plane hologram. Any change in wavelength or geometry will result in a change in perspective. It is to some extent possible to compensate for an error in one direction by a similar error in the opposite direction. Thus a change in wavelength from 694 nm to 647 or 633 nm for shooting the transfer hologram, which would result in exaggerated perspective, may be balanced by illuminating the master hologram with a slightly converging beam rather than a collimated one. As a general rule it is best to position the transfer plateholder so that the eyes of the sitter are in the plane of the final hologram (Plate 21).

Lighting for Portraiture

The first rule is a safety rule, that the reference beam, or its reflection from the plate, must not be allowed to enter the sitter's eyes. This is because it is an undiffused beam. Fortunately, this is unlikely to happen with an overhead or side reference beam at an angle of incidence of between 45° and 60°, as the reflected beam will be well out of the image area.

In photographic portraiture, the simplest effective lighting is a single key light positioned about 45° away from the direction the sitter is facing, and about 45° above the sitter; in a three-quarter view it should be on the side away from the camera. The rule of thumb is to have the shadow of the sitter's nose just touching the corner of the mouth; this highlights the features and emphasizes the line of the cheekbone. In a full-face shot, the side from which the light should come depends on which side of the face needs more emphasis. Most people have somewhat asymmetrical faces, often with the right side more developed than the left. The purpose of the second light, the fill-in light, is to control the lighting contrast by illuminating the shadows.

It is usually a more diffuse source than the key light; it is placed close to the axis of the camera so that it does not cast a visible second shadow, on the side of the sitter away from the key light. A third light (when used) is a small spotlight behind and above the sitter. It is known as a *kicker*, and its purpose is to highlight the hair, and sometimes to outline the side of the face and cheekbone (Fig 16.2).

In holographic portraiture, although the key light is no longer the sole means of producing the impression of depth as it is in a two-dimensional image – hence the term *modeling light* often used for the key light – its positioning is still very important in bringing out the best in the sitter's features, not to mention suppressing those whose emphasis is less desirable. A small change in the position of the key light can make a considerable difference: lowering it to about 25–30° can emphasize 'character' rather than looks; moving it farther round to the side can make a broad

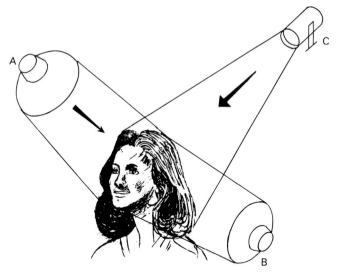

FIG 16.2 *Basic lighting for portraiture. A is the key light, slightly diffuse, producing shadows at nose and cheekbone and highlighting features, usually angled approximately 45° forward of the sitter and 45° above. B is the fill-in light, diffuse, with its axis close to the axis of view, controlling lighting contrast. C is a kicker or small spotlight with a concentrated beam, throwing catchlights on the hair and possibly the rim of the face.*

face appear narrower; a low position gives a theatrical effect and a high position a mysterious one. Not all effects are desirable, of course; a key light on the same side as a hair parting can exaggerate the appearance of an already receding hairline, as can an ill-positioned kicker. In general, the rules for key lighting are the same as in portrait photography, and a good way to gain experience is to practice lighting with a friend as the model, using a lamp on a stand in an otherwise darkened room.

One of the limitations in holographic portraiture is that the fill-in light cannot be placed frontally because the holographic plate is directly in front of the sitter. The simplest way of providing fill-in lighting is to place a large card sprayed with aluminum or gold paint on the side opposite to the key light, picking up the part of the key-light beam that bypasses the sitter. For this purpose it will be necessary to move the key light slightly off-center, so that the fill-in light is sufficiently strong. You can also pick up the light from the reference beam using white card positioned round the plateholder, to provide additional diffuse illumination. You do not have to match the path lengths too carefully, as the coherence length of a pulse-laser is more than a meter. A fairly unsophisticated lighting set-up, based on Bjelkhagen's paper (1) is shown in Fig 16.3. This would be appropriate for a 24 × 30 cm plate. The set-up can be approximated initially by using a beam of collimated white light, and corrected using a CW alignment system which is part of some pulse-laser systems; it may be difficult to use this from the start as the presence of diffusers seriously weakens the beam.

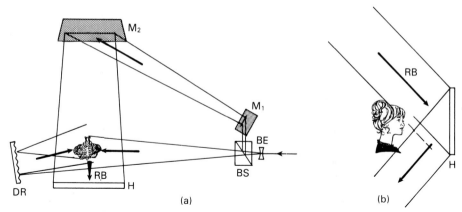

FIG 16.3 *Typical lighting for head-only portrait (after Bjelkhagen (1)). (a) Key lighting comes directly from the beamsplitter* BS; *fill-in lighting is reflected from the diffuse reflector* DR. *The reference beam* RB *is directed via the relay mirror* M_1 *and the overhead mirror* M_2. *(b) A portion of the reference beam acts as kicker. Notice that undiffused light is reflected away from the sitter's face.*

For more ambitious work you need further beamsplitters, to give lighting where needed for hands, etc (Plate 19). At this point you are on your own. It is the author's hope and belief that within the next few years there will be courses and books on holographic portraiture that rival those that are now available on photographic portraiture, describing systems incorporating mechanical linkages for moving the optical components, which will give the same flexibility with lighting that is currently available in any photographic studio; and we may soon see holographic portraiture becoming generally accepted, with a studio in every city. But whatever transpires, there will always be some simple rules to be getting on with, in the interests of both safety and the image:

1 Never position an undiffused beam where it can enter the subject's eyes.
2 Never allow the light from any beam other than the reference and object beams to reach the plate.
3 Always arrange the diffusers so that as much as possible of the light falls on the sitter.
4 Control the illuminance by adjusting the beamsplitter ratios, not by changing the distance of the illuminating sources as you would with a photographic portrait.

Exposure and Processing

Exposure is just as in photographic flash portraiture; the most important thing is to choose exactly the right moment for making the exposure. With professional models this is no problem, but with untrained sitters the onus is very much on you. Look at the sitter through the plate, if possible; otherwise, from a point immediately above it. If conditions allow, fire the laser several times with the plate out of the way. Try to retain what you see in your mind's eye while you check it mentally (this takes concentration and practice), and make any necessary small adjustments for the sitter's position. Also check the absolute value of the various beams using a photometer with a polarizing filter (you will have to use a suitably-calibrated energy-density meter for this), as a fill-in beam that is depolarized can look much stronger than it will be on the final hologram. When you make the actual exposure, concentrate totally on the sitter's face. If you are making a formal portrait, give a countdown, aloud, to make sure the sitter does not blink at the instant of exposure; this is a serious hazard in group portraits. If you have any doubts at all, don't be afraid of re-exposing; the plate is by far the cheapest item in the set-up.

Processing is similar to the processing of CW master holograms. If you are making a transmission master, there is no particular difficulty, but if you are making a reflection master, there is a danger that the tanning effect of a pyro developer may hold the gelatin in a swollen state so that the image reconstructs in the infrared. It is probably best to use a non-tanning developer and a rehalogenating bleach (see Appendix 10); research is still going on into better methods. There is still a debate as to whether transmission master holograms should be bleached or even fixed; this may perhaps never be settled to everyone's satisfaction. It probably depends on the precise method of working: beam ratio, exposure, angle of reference beam, and so on. Try out each method, and settle on the one which gives *you* the best results.

Making the Transfer

The Kr^+ laser is the best bet for making transfers; not only has it the right kind of power to keep exposures short for faster production, but its most powerful emission line is at 647 nm, which is reasonably close to the emission line of the ruby laser (694 nm). The emission line of the HeNe laser (633 nm) is at the other end of the red, and does not work nearly so well. The best way of avoiding the wavelength mismatch is at the master stage. If the master plate is exposed to a humidifying bath for 20 min–1 h before exposure, it will reconstruct at a wavelength that is close to the Kr^+ and HeNe wavelengths. Alternatively, Ilford materials (see p. 69) may be used, Richard Rallison (3) has made transfers of portraits using dichromated gelatin, and has achieved good flesh colors.

Double-pulse Holography

In industrial research, double-pulse holography is used to obtain information on the vibrational behavior of rapidly moving objects (see Chapter 23). In 1976 a double-pulse hologram was made for the Open University course *Images and Information* of Keith Hodgkinson of the Course Team playing the viola (Fig 16.4). It was noticeable that fringes could be seen not only on the instrument soundboard (as expected) but also on Keith's tie, shirt and face. Since then a number of artists, notably Margaret Benyon (4), have experimented with double-pulse portraiture, deliberately moving the head or grimacing in order to produce a particular pattern (Fig 16.5). The spacing of the pulses is a few microseconds, the optimum interval being determined by experiment. (Most pulse lasers nowadays can be obtained with double and triple pulse facilities.)

FIG 16.4 *An early double-pulse hologram made for the Open University by Loughborough University of Technology. The dark glasses were necessitated by the undiffused beam. Photograph courtesy of the Open University.*

Other Uses for Pulse-laser Holography

With a pulse laser you can make any kind of hologram without any stabilization beyond what you would need in a photographic studio. You could even lay out your studio in a similar way to a photographic studio, with a plate-holder on the tripod instead of a camera. Pulse lasers are very expensive, and safety requirements mean that the setting up of such a studio is a matter for the expert; it therefore seems likely that artists wanting to do pulse-laser work would hire facilities by the day, rather than set them up themselves. The idea of a pulse-laser studio is a fairly new one, and at the time of writing there are probably no more than half a dozen of them; but modern beam-steering techniques, including those that are still being evolved, such as monomode fiber-optics (see Appendix 8), make it likely that the holographic studio of the future will be as readily controllable as the photographic studio of today.

In the meantime, what are the possibilities of pulse-laser holography apart from portraiture? Its main use would seem to be the holography of any subject matter that cannot be expected to remain absolutely motionless for several seconds. Portraiture need not be confined to humans; first-class holograms of wildlife already exist (Plate 20).

Flower arrangements and other not-quite-stable still-life studies lend themselves particularly well to pulse-laser holography. There is now much interest in the

FIG 16.5 *Counting the Beats, a double-pulse self-portrait of Margaret Benyon and John Webster with technical assistance from Chris Mead; the translation fringes indicate that John is nodding and Margaret is shaking her head.*

application of high-speed photographic techniques such as triggered exposures to holography. Infrared and acoustic triggering can be used to fire flashes to record birds in flight, baseball bats striking balls, spectacular card shuffles, juggling, water splashes, breaking glass; all these are as amenable to holography as to photography. Double-pulse recording can be extended to all kinds of objects in motion; indeed, many double-pulse holograms made specifically for scientific purposes have already brought us images that have a beauty quite incidental to their original purpose. There is unlimited scope along these lines. It is a great pity that lasers that can bring us this enhanced world, and their associated optics, are so expensive. At least, unlike CW lasers they do not wear out, so the number available to holographers can only increase.

References

1 Bjelkhagen H (1983) Holographic portraits: transmission master making and reflection copying technique. *International Symposium on Display Holography*, **1**, 49–54.
2 Newport Corporation (1981) *Projects in Holography*.
3 Rallison R D (1985) Pulse portraits, the Holochrome process. *Proceedings of the SPIE*, **523**, 2–6.
4 Benyon M and Webster J (1986) Pulsed holography as an art. *Leonardo*, **19**, 185–91.

CHAPTER
17

Holographic Stereograms

> She noticed a curious appearance in the air. It puzzled her very much at first, but in a minute or two she made it out to be a grin. [...] In another minute the whole head appeared.
> LEWIS CARROLL, *Alice in Wonderland*

Holographic stereograms are hybrids of photography and holography that retain a little of the quality of both but add something of their own. They are a kind of parallax panoramagram (see p. 4), and they have their own set of virtues and vices. There are two kinds of stereogram; they are quite different in optical principle, and whereas one of these has virtually reached the end of its development, the other holds a great deal of promise. Both are records of ciné sequences, and can to a limited extent record movement. For this reason they have sometimes been misleadingly called holographic movies.

Many important inventions have come about as a kind of technological serendipity; the inventor was looking for something else. Holography itself is no exception: it originated as an idea for improving the resolution of electron microscopes (which it did not succeed in doing). The rainbow hologram was the consequence of an attempt to produce a hologram with its information content sufficiently reduced to permit transmission by TV signal (which it did not succeed in doing). Even the discovery of the three-dimensional nature of the holographic image came as a surprise. The concept of the holographic stereogram, however, did not arise by chance. It arose from frustration with certain of the limitations of existing types of hologram. Having seen what could be done with holograms in the way of realism, the questions arose: Why can't we make holograms of big objects, reduced in size like photographs? Why can't we make holograms out in the open? The temptation to try to use the holographic format to embrace what had been done in the field of stereoscopic photography, especially the parallax panoramagram, was strong. One could obviously make a hologram of a photographic stereogram, but that would clearly be a pointless exercise; the images should be encodable by holographic means rather than optical means. To see how this problem was tackled, it is necessary to discuss the principle of multiplexing.

The Multiplexing Principle

As with so many terms used in holographic technology, the term 'multiplexing' was first used in connection with communications theory. It refers to the simultaneous sending of more than one signal down a single channel (such as a TV cable). This can be done in one of two ways: by splitting up the signals on a time-sharing basis, or by sending the signals continuously but distinguishing them by using different frequencies. The photographic panoramagram is like the first, as the images are dissected into interlaced lines which are separated by a retroreflecting lenticular screen over the print. The (multiplexed) holographic stereogram is like the second, as all the image information is encoded into the entire area of the hologram, and the individual images are sorted out by the direction of the fringes through the emulsion, which direct the light from each image in a specific direction. This is the type of holographic stereogram pioneered by Dominic DeBitteto (1). A second type of holographic stereogram developed by Lloyd Cross (2) has the image information coded locally in narrow strip holograms, and so is more like a signal in which each message is sent separately, one at a time, but compressed, not multiplexed. Both types of hologram have been the subject of considerable research.

Making a Multiplexed Hologram

A hologram of a stereo pair You have already met the multiplexing principle in Chapter 13, where it was used to place two three-dimensional images into the same hologram. Let us look at the principle again, this time with a view to making autostereoscopic images from photographs, starting with a simple stereo pair, made from two photographs of a scene taken from positions separated by about 60–70 mm (ie the average interocular distance). Make a master transmission hologram from each of these, using a single plate masked off so that half of the plate is exposed to each photograph, at a distance of about 250 mm. Having processed the master, make a transfer hologram from it, either reflection or full-aperture transmission, with the real image of the two photographs in the transfer plane. When this is flipped for viewing, the real images of the master apertures will be side by side, standing out in front of the hologram, and if you stand with your two eyes approximately in the plane of these images of the apertures you will see the two images autostereoscopically. When you are making the masters you must make sure that the second photograph goes into exactly the same place as the first. The principle is shown in Fig 17.1.

Many-slit multiplexing From this point it is only a short step to making a hologram in which nine or more images are multiplexed. You need to start with nine photographs taken from nine equally-spaced points along a straight line centered opposite your subject matter, so that they cover an angle of approximately $60°$. Mark these points on a horizontal surface on which you can place your camera, and make your nine exposures from each in turn. Don't turn the camera to point at the subject; keep it pointing straight ahead. In order to get the subject matter into the end pictures you will need to be using a wide-angle lens – a shift lens, if you have one, is even better. You will then need enlargements made; trim these so that the subject is in the center of each, and they are all the same size.

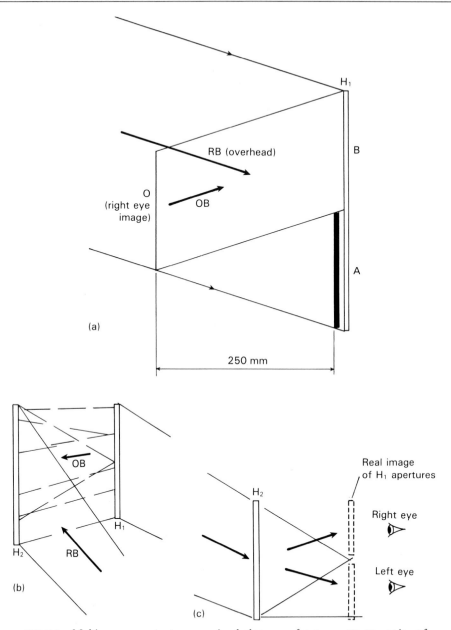

FIG 17.1 *Making an autostereoscopic hologram from a stereo pair of photographs. (a) A master hologram is made of each photograph in turn, covering half the plate for each exposure (A for the right-eye image, B for the left-eye image). For the second exposure the mask is moved to B and the left-eye image substituted. (b) Transfer made in plane for real image of object. H_2 contains both images multiplexed. (c) Reconstruction. When flipped for viewing the transfer hologram directs the light from each image through the appropriate real image of the master hologram aperture to give stereoscopic viewing.*

This time your masking slit should be just over one-ninth of the width of the plate. If you are using an 8 × 10 in plate, make the slit 30 mm across. Mark out the nine positions, allowing just over 1 mm overlap between positions. Now make your exposures, remembering to move the slit and change the photograph each time. Transfer the image in exactly the same way as you did for the two-exposure hologram. This time you will have an image with much more parallax, as the stereo effect occurs between *all* pairs of adjacent images. Walking past such a hologram you can observe parallax (albeit of a somewhat jerky kind) over 60° (Fig 17.2).

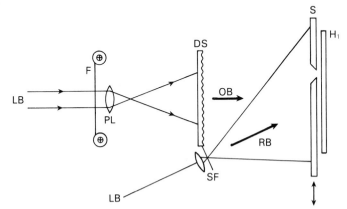

FIG 17.2 *Holographic stereogram (after DeBitetto (1)). LB = laser beam, F = ciné film, PL = projector lens, DS = diffuser screen, SF = spatial filter, S = slit (movable), H_1 = master hologram, OB = object beam, RB = reference beam. Slit is moved one step each time frame is changed in projector.*

This is the system that was developed by McCrickerd and George (3) in 1968, and, rather more completely, by DeBitetto (1) in 1969; DeBitetto went one stage further, and used ciné photographs taken from equally-spaced positions along a horizontal line to produce a much larger number of images. These were projected onto a ground glass screen using laser light, for making the master images (Fig 17.3). These early images were viewed by laser, but they can equally well be made into reflection or white-light transmission (open-aperture) transfer holograms.

Cross Holograms

Principle of the Cross hologram A somewhat different approach was adopted by Lloyd Cross (2). A Cross hologram does not use transfer techniques; instead it consists of a series of narrow vertical strip holograms made by focusing the ciné film image onto the holographic film by means of anamorphic optics. In designing the holographic stereogram Cross adopted what may seem a somewhat cavalier attitude to problems of astigmatic distortion, and his geometry is something of an optical horror story. However, these holograms have been marketed extensively and with a fair measure of success by Cross's company (the Multiplex Corporation), and a number of studios have been set up for shooting ciné originals for Cross holograms.

There are arrangements for making them from your own length of ciné film: you simply shoot it according to the requirements and sent it to one of the authorized companies (two of which are listed in Appendix 12). There is, of course, no reason why you should not build your own printing table provided you keep within the terms of patent law: there is plenty of information available (there are a great many papers on Cross holograms), but as all the optics have to be tuned by hand, you will have a hard slog ahead. The following account gives an idea of the geometry, and refers you to some of the more important papers on the subject.

Making a Cross hologram To produce the original film that will result in a Cross hologram you need to make a movie film on 16 mm or 35 mm stock while rotating either the camera axis or the subject matter through $\frac{1}{3}$ of a degree per frame. At 24 pps (pictures per second) this represents a total time of 15 s (360 frames) for a 120° hologram or 45 s (1080 frames) for a 360° hologram. The whole sequence needs to be carefully prepared. At the simplest level, all you need to do is to put the subject on a turntable and rotate it at the appropriate speed while you film it with a stationary camera. It is best to work at a fairly large distance using a long-focus lens, if space will allow. You may utilize the full height of the frame, but it is advisable to fill no more than one-third of the horizontal width of the frame. Alternatively, you can move the camera rather than the subject: in this way, using single-frame shots, you can produce 360° views of large objects such as buildings. This is, of course, a time-consuming exercise: positioning and aligning the camera 1080 times is not something you can do during your lunch break, and you need settled weather, too. Nevertheless, the principle is simple enough. At a slightly more sophisticated level you can allow some movement of the object during the exposure. For a 120° hologram this movement should be of a kind that would not normally last more than about 5 s, the time it takes to walk past the hologram. The movement must be carried out with exaggerated slowness, as it takes the camera 15 s to shoot a 120° rotation. Alternatively, you can run the camera at three times the normal speed, ie 72 pps, and the movement can then be made at the normal speed. At an increased level of sophistication, you can use fades, dissolves, montages, time-lapse, slow-motion and all the gimmicks of cinematography, including animation, though these effects can interfere with stereoscopy and make viewing uncomfortable unless they are done with care and restraint. Another possibility, which much more closely resembles a holographically-displayed movie, is simply to take a length of ciné film and make it viewable stereoscopically, with a very limited amount of parallax (and a good deal of time-smear), in a rolling stereogram. At the highest level of technical complexity, it is possible to construct the whole image by means of a computer, producing autostereoscopic images that have never existed, indeed could not exist. These images can be designed to compensate for the perspective distortions inherent in the format. Sharon McCormack (16) has covered this ground in detail.

Printing a Cross hologram A Cross hologram is a set of astigmatic focused-image holograms set side by side, with some overlap. It is also a slit hologram, both vertically and horizontally; in the vertical aspect it is a rainbow hologram. This is somewhat bewildering combination is achieved by an anamorphic series of lenses. The original Cross optics are shown in Fig 4.18(a) (p. 57). A condenser lens is positioned in the plane of the projected image, followed by a large-aperture cylindrical lens, usually liquid-filled. More recently, both the projection and collimating spherical lenses have been replaced by pairs of cylindrical lenses, and this has helped to offset the severe astigatic distortion seen in the early Cross holograms (Fig 17.3).

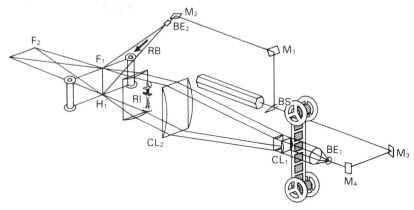

FIG 17.3 *Optical principle of a Cross hologram. The anamorphic cylindrical pair CL_1 forms a real image of the ciné film frame at RI. The astigmatic pair CL_2 forms a vertically-focused image hologram H at F_1, and a slit image at F_2 (the viewing slit). The reference beam RB is derived via the beamsplitter BS, the mirrors M_1 and M_2 and the one-dimensional beam expander BE_2. M_3 and M_4 are relay mirrors and BE_1 is a beam expander/collimator.*

In this system, a beamsplitter near the laser output port divides off the reference and object beams. The reference beam is diverged to a vertical line, which is accurately superposed on the object beam from above. The object beam is expanded, collimated and passed through the film frame, after which the image is focused astigmatically in two orthogonal planes by cylindrical lenses. Vertical image rays are focused in the plane of the holographic film or a little behind it, and horizontal rays are focused to a horizontal line about 1 m behind the film. When the film is spun through $180°$ for viewing, this horizontal line forms the viewing 'slit'. The hologram thus satisfies the conditions for a rainbow hologram, and when illuminated by the conjugate beam the image is reconstructed in the position it originally occupied, ie roughly in the plane of the astigmatic pair of lenses. In making the successive recordings the holographic film is advanced by the appropriate distance between exposures of successive frames. To avoid image-jumping, each frame is actually exposed twice, so that each movement of the holographic film corresponds to a rotation of $\frac{1}{6}$ of a degree, about 0.4 mm for a display diameter of 400 mm (16 in) (Fig 17.4).

It is important that the second lens be well corrected, as otherwise the focused line will be too wide, and the diffraction efficiency will be affected by the overlap of too many holograms. In addition, any aberrations will result in ray bundles near the focus which produce grossly different angular components, causing secondary interference patterns between successive holograms. It has been the usual practice to construct the lenses out of acrylic sheet with the space filled with liquid paraffin as described on pp. 220–8, the precise curvature being established by ray tracing.

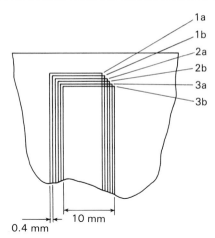

FIG 17.4 *In a Cross hologram the holographic exposures are approximately 10 mm, wide, and overlap by 0.4 mm. Each frame is printed twice, and the angular overlap is 1/6° (10 mm of arc). In this illustration the exposures have been displaced vertically, for clarity.*

In practice the building of such a lens is by no means easy. A full set of instructions has been prepared by Sharon McCormack (4).

Improving the image If you have examined a Cross hologram carefully you will no doubt be aware of its optical shortcomings regardless of its entertainment value (which is often considerable); apart from the gaunt appearance of the models and a marked lack of sharpness in the vertical aspect, faces are a sickly green and the whole image is overlaid with vertical black lines. In addition there can be a lot of distortion towards the edge of the image. This is due to what is called time-smear: this is an effect caused by the fact that when you look at the outside edges of the image area, you are looking at an image made earlier or later than the one in the middle, and therefore taken from a different viewpoint. This distortion is compounded by any movement the model may be making during the photography. Such distortions have led to the disparagement of cylindrical stereograms, and it seems a pity that they were launched on the market with so many important optical problems still unsolved. A number of workers and, notably Fusek and Huff (5 and 6), have devoted a great deal of research time to reducing the various distortions so as to enable the image space to be enlarged, for dispersion to be compensated so as to produce a more nearly achromatic image, and for simplifying the optics by replacing the awkward and unreliable liquid-filled optics with holographic optical elements. There are two reasons for continuing to take the Cross format seriously: first, it is readily capable of providing 360° of parallax, and thus lends itself to medical applications such as X-ray tomographic images, scientific applications such as electron micrographs in the round and, on the artistic side, holograms of whole buildings. Secondly, the format can be used for ciné tricks such as dissolves and montages, and, in particular, to animations which can be pre-distorted to give a better rendering of shape. The more important relevant research is listed at the end of the chapter (5, 6, 7, 8, 9). Benton (9), in a survey of holographic stereograms made in 1982, lists over thirty further papers on holographic stereograms.

In spite of these improvements, however, the Cross process seems to be somewhat of a dead end, among other things because it is not practicable to lift the image out from behind the hologram without employing an intermediate hologram stage. Many workers, most notably Benton and his team, have gone back to DeBitteto's principles instead, and developed these.

Stephen Benton *et al* (10, 11) have tackled the problem in a different way. To begin with, they move the ciné-camera along a straight rail rather than an arc of a circle (or rotating the subject). The camera lens is also moved according to the laws of perspective; the geometry will be familiar to photographers who use a shift lens for architectural photography. This allows the final hologram to be flat rather than curved. They have removed the large condenser lens from the Cross format; the viewing slit is now coincident with the position of the original projection optics. The reference beam is collimated, and viewing is by its conjugate, without spinning the film. Using this type of geometry it is possible to bring the image into the plane of the hologram. Rather than building a large liquid-filled lens, the team use a large-aperture cylindrical Fresnel lens (such lenses are available from mail-order science suppliers such as Edmund Scientific). Fig 17.5 shows the system and a typical table layout using this geometry; note that in the actual layout the optics are laid on their side so that an 'overhead' reference beam can be set up on a flat table.

Benton has also managed to eliminate the problem of green faces. He describes (11) a method developed from his achromatic white-light transmission system (pp. 53–4 and 196–9) to produce stereograms with an uncolored image. Bill Molteni (12) shows how to produce pseudocolor stereograms using a similar method. A paper by Jaffey and Dutta (7) summarizes work on perspective correction and offers a digital implementation of Benton's method. Okada *et al* (8) have also worked on computer-processed stereographic images. Much of this work overlaps into the area of applied holography, and is further discussed in Chapter 24.

In the early days of Cross holograms there were a number of variations of the blow-a kiss-and-wink routine, some of them decidedly less chaste than the original. Several holographic portrait studios were also set up to make stereographic portraits as a business. Some British holographers such as Nigel Abraham made film footage for Cross holograms involving double-exposure and montage techniques (Plate 22). The use of the medium as an art form interested a number of American artists, notably Harriet Casdin-Silver, who has used its distortions creatively in several well-known pieces. More recently, the sculptor Alexander, working in Australia with Hariharan, has produced a number of very large multicolor stereograms which exploit the inbuilt aberrations of the system and cinematic tricks (Plate 23). Other workers, such as Alexis Krasilovsky, have produced images including drawn animation.

Computer-drawn Images for Holographic Stereograms

From hand-drawn animation it is only a short step to computer-drawn animation. This is an exciting area, as it is well within the capabilities of the average home-

254 PRACTICAL DISPLAY HOLOGRAPHY

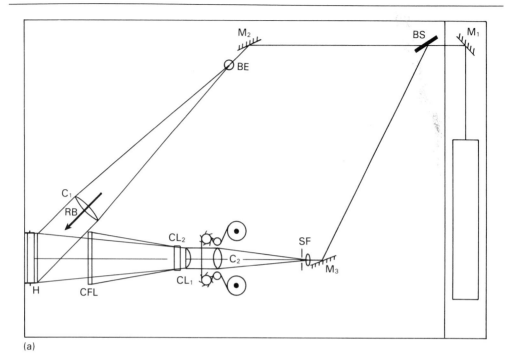

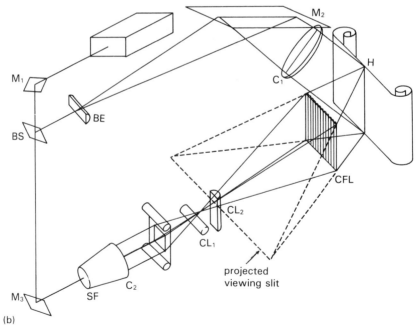

FIG 17.5 (a) *Flat table geometry for holographic stereograms. The reference beam* RB *is derived via the beamsplitter* BS *and the one-dimensional beam expander* BE, *and collimated by the lens* C_1. *The object beam is collimated by*

computer buff. Molteni (13) has written an instructional paper on the subject, the essential features of which are summarized below (with permission). Computers have been used in animation for a number of years; programs for animation that include in their menu both 'in-betweening', which connects two different drawings in a linear way, by a succession of intermediates, and 'rotating' which corrects the two drawings of different views of a subject by simulating rotation from one viewpoint to another, are readily available. If these intermediate drawings are displayed on a VDU and photographed on ciné film using single-frame techniques, the result can be made into a holographic stereogram. More sophisticated computer programs can be written to include instructions to modify the drawings so as to take care of perspective distortion.

Holographic stereograms designed by home computer The following is a précis of a paper (13) given by Bill Molteni at the 1985 International Symposium on Display Holography at Lake Forest.

The basic hardware is available from Tandy (Radio Shack in the United States) and consists of a Radio Shack color computer, joystick and Koala drawing pad. The software consists of OS9, a Unix-like operating system, and Basic09, a structured Pascal-like programming language. Owners of other computers, such as the various versions of the BBC microcomputer, can use similar software. Basic09 has the advantage over other Basics of being clear and extremely fast. The drawing pad and joystick are used to input the x/y and z-coordinates respectively to the 3-D input device, within a $64 \times 64 \times 64$ volume. As this is only half the required drawing area, the x/y input needs to be multiplied by two. It is not necessary to specify more than the endpoints of line segments, and so only these are saved. Thus when it is necessary to move the image only the ends of the lines need to be operated on: the inbuilt graphics routines do the rest.

Designing a stereo pair The first operation is to draw a side-by-side stereoscopic pair of two-dimensional images. There are a number of ways of doing this: pairs of photographs taken from two different angles and simplified in drawings; projections from architects', engineering, etc, drawings; abstract compositions where the displacement of lines is estimated from the artist's own three-dimensional mental picture. A word of warning: beginners tend to be over-ambitious in the amount of depth they try to portray. Molteni has worked out a simple guideline for avoiding eyestrain: the difference in eye convergence for the nearest and farthest points

---

the lens C_2 and passes through the film. The cylindrical pair of lenses CL_1 and CL_2 focus the 'horizontals' onto a 'vertically'- oriented cylindrical Fresnel lens CFL, and the 'verticals' onto the film. CFL is itself focused on the holographic film. The final image is in the plane of the hologram H and is open-aperture and more or less achromatic. The focal lengths of CL_1 and CL_2 are both 25 mm; that of CFL is 150 mm. Notice that the first component of the condensing lens system has been removed; when illuminated by a diverging replay beam the viewing 'slit' is projected to a plane corresponding to the rear surface of CL_2. (b) This is a perspective view of the optical system, depicted 'right way up' (after Teitel (10)).

should not exceed 1.25°, which means that the difference in distance between the nearest and farthest points should not exceed $\frac{1}{46}$ of the viewing distance. Using a viewing distance of 500 mm (20 in) and 100 × 100 mm (4 × 4 in) display areas the comfortable depth is about 80 mm (3.3 in). The maximum positional disparity is approximately 5 mm (0.2 in). If you are making a completely abstract drawing this is the only criterion you need to bother about.

Displaying   The system has three main display modes: a menu of commands; a plan and front elevation of the object; and a stereoscopic pair (Fig 17.6). Molteni suggests 'free viewing' of the (crossed) stereo pair from a comfortable distance with crossed eyes, a skill which becomes comparatively easy with a little practice and is generally believed to be completely harmless (the converse routine of learning to diverge the eyes for viewing of an orthoscopically-displayed pair causes eyestrain and is not recommended).

Drawing in three dimensions   It is this pseudoscopic pair that is used for drawing. The vertical and horizontal (x/y) positions of a flashing 3-D cursor are controlled by the drawing pad, while the depth (z) position is controlled by the joystick in the other hand. When the correct location in the volume is found, the joystick button is pressed and the location is stored. The process is repeated, a new location is stored and a line is drawn connecting the two points. Curves can also be followed by the flashing cursor. The program is complete when the entire outline has been traced.

It is possible to work in three colors with this system. The line color of the image is assigned an RGB code before drawing. The code can be changed as the drawing progresses, and the image can then be displayed in color separation.

A program called Runner generates and displays 100 two-dimensional perspectives for each color separation, in sequence. The 300 displays are photographed onto 35 mm film, the ciné camera being also controlled by the program. The processed film is then used to construct a holographic stereogram by the means described earlier in this chapter. (See Plate 24.)

Resolution   The area of screen generating the image for photography does not have a very high resolution, much lower than that of a television screen. It contains only 128 × 96 pixels (picture elements). The maximum displacement under the 1.25° rule

FIG 17.6   *The two graphic display modes of the Runner program:* (a) *is the front and plan view two-dimensional drawing mode and* (b) *is the stereoscopic drawing and viewing mode (after Molteni (12)).*

is 6 pixels. A shift of a point of less than 0.1° (6′ of arc) will not cause the point to move between views, whereas a shift of 0.11° (6.6′ of arc) will cause it to move a whole pixel. This leads to *aliasing*, or staircasing, of diagonal lines and curves, which varies from one perspective to another, resulting in an unpleasant sensation of scintillation.

The use of more powerful computers and programs can eliminate staircasing and other annoyances, and great strides have already been made. The reader who is well-versed in home computing is advised to keep up to date with the literature. This brief description has been presented as an example of what can be done with nothing more than a simple home computer and a few peripherals.

Alcove Holograms

The most recent development along these lines is the *alcove hologram*, which is currently being researched by the Spatial Imagery Group at the Massachusetts Institute of Technology Media Lab under Stephen Benton. This work is described by Michael Teitel (14), whose thesis is chiefly concerned with the ray-tracing program for eliminating perspective distortion, but which also contains an excellent account of the optics of the original Cross hologram, Benton's modification for flat rainbow stereograms, and the laser transmission alcove hologram. (At the time of writing a white-light-viewable version with multicolor imagery is under development.)

An alcove hologram, as its name suggests, is a hollow half-cylinder. The real image is projected to its center. The difference between this and other real-image holograms is that the alcove hologram gives an image with nearly 180° horizontal parallax, and the image space can extend back through the hologram to infinity. One of the first successful alcove holograms produced by the team was of a projected car design seen against the Boston skyline (the project is funded by General Motors). Of course, there is no particular difficulty in producing a real-image stereogram, given Benton's improved geometry. The trouble is that if you curve the stereogram to form a concave arc the resulting image shows severe perspective distortion (the team's first effort was dubbed 'Banana-Cam'. The problem has been tackled for photographs of real objects by Huff and Fusek (6), and this technique has been evolved by the Benton team into an image-processing technique which they call 'slice and dice', whereby each of the views is sliced into vertical strips, and images intended for printing are assembled from different strips for each image, with the magnification changed where necessary. The result is that the image is pre-distorted to counteract the distortion inherent in the projection system. This is clearly very difficult to do with actual photographs, but it is not particularly taxing for a computer working on drawings and given the right algorithm. This part of the work has been done by Michael Teitel (14). The most complete account of the whole process is given by Benton (15). Once the computer program is corrected (Teitel (14) gives an example from the program) the real image can be displayed without distortion. The main modification to the Benton flat rainbow stereogram geometry is the replacement of the cylindrical Fresnel lens by a holographically-generated high-efficiency diffusion screen. In addition the reference beam is converging rather than collimated, and the projection lens is of the normal spherical type, focused on the

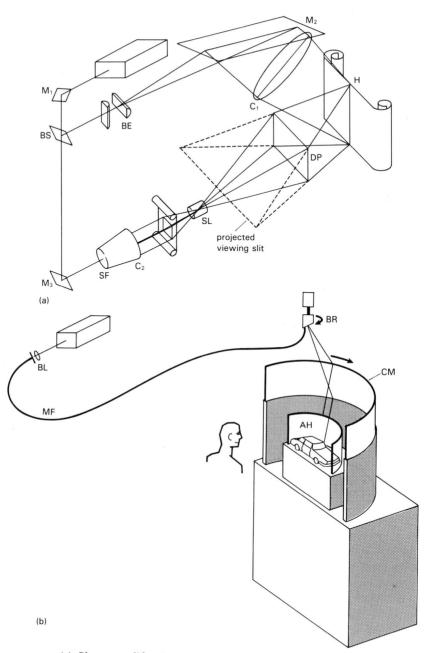

FIG 17.7 (a) *Shows modifications to the optical arrangement of* FIG 17.5 *for an alcove hologram:* C_1 *is now a converging element;* SL *is a conventional spherical projection lens;* DP *is a diffuser plate in the plane of the optical image.* (b) *Shows the laser transmission viewing mode. The beam launcher* BL *inserts the beam into a monomode fiber* MF. *The beam rotater* BR *spreads the beam into a rapidly-scanning line via the circular mirror* CM, *which generates an apparently continuous image through persistence of vision.* AH *is the alcove hologram itself, against a dark background (after Teitel (10))*.

diffusion screen. The viewing 'slit' is in the plane of the lens (Fig 17.7). When the hologram is illuminated with the conjugate beam, a real image appears in the plane of (image of) the diffusion screen. For laser viewing, a beam is spread out into a vertical line which is rotated rapidly by a mirror scanner so that persistence of vision gives an apparently continuous image. As mentioned earlier, it is expected that a white-light reflection version will be available before long. Pre-swelling techniques (see pp. 264-8) will then permit full-color images.

This new technique is an example of the way in which holography can advance in areas where it might have been thought that the last word had already been said. Alcove holograms have immense potential in design and development studies, as well as their more obvious uses in commerce and advertising.

References

1 DeBitetto D J (1969) Holographic panoramic stereograms synthesized from white-light recordings. *Applied Optics*, **8**, 1740-1.
2 Cross, L (1977) Multiplex holography. SPIE meeting (unpublished). See p. 60, ref 14.
3 McCrickerd J T and George N (1968) Holographic stereogram from sequential component photographs. *Applied Physics Letters*, **12**, 10-12.
4 McCormack S (1982) Liquid-filled Cylindrical Lenses. *International Symposium on Display Holography*, **1**, 149-62.
5 Fusek R L and Huff L (1980) Use of a holographic lens for producing cylindrical holographic stereograms. *Proceedings of the SPIE*, **215**, 32-8.
6 Huff L and Fusek R L (1981) Optical techniques for increasing image width in cylindrical holographic stereograms. *Optical Engineering*, **20**, 241-5.
7 Jaffey S M and Dutta K (1982) Digital perspective correction for cylindrical holographic stereograms. *Proceedings of the SPIE*, **367**, 130-40.
8 Okada K, Honda T and Tsujiuchi J (1983) A method of distortion compensation of Multiplex holograms. *Optics Communications*, **48**, 167-70.
9 Benton S A (1982) Survey of holographic stereograms. *Proceedings of the SPIE*, **367**, 15-19.
10 Teitel M A and Benton S A (1986) *Anamorphic imaging for synthetic holograms*. MIT Library.
11 Benton S A (1983) Photographic holography. *Proceedings of the SPIE*, **391**, 2-9.
12 Molteni W J Jr (1982) Black and white holographic stereograms. *Proceedings of the International Symposium on Display Holography*, **1**, 15-26.
13 Molteni W J Jr (1985) Computer-aided drawing of holographic stereograms. *Proceedings of the International Symposium of Display Holography*, **2**, 223-30.
14 Teitel M A (1986) *Anamorphic raytracing for synthetic alcove holographic stereograms*, MSc thesis for MIT.
15 Benton S A (1987) 'Alcove' holograms for computer-aided design. *Proceedings of the SPIE*, **761**, 1-9.
16 McCormack S (1986) Special effects techniques for integral holograms. *Proceedings of the SPIE*, **615**, 24-30.

CHAPTER
18

✹

Holograms in Color

> A large rose-tree stood near the entrance of the garden: the roses growing on it were white, but there were three gardeners at it, busily painting them red.
> LEWIS CARROLL, *Alice in Wonderland*

Natural or Simulated Color?

It took more than fifty years from the inception of photography before the first photograph in full natural color was made. The theory of color reproduction had been known for nearly a hundred years, but photography in monochrome was seen as such a marvelous thing that it occurred to no one that there was anything inherently unrealistic in it. Today, creative photography in monochrome is used only for images which, exceptionally, carry a message that can be adequately conveyed only in stark black and white. More than nine-tenths of today's photography is in full natural color. Indeed, it has become difficult to buy a black-and-white film from an ordinary retailer. The question is, will holography eventually go the same way, with monochrome the exception rather than the rule?

The answer is not entirely clear. Although it is theoretically possible to achieve a greater range of color in a hologram than in a photograph or a television picture (see Appendix 6), it is difficult to produce accurate colors in a hologram, especially when the subject matter contains color of high *saturation*. The reason for this is that an object of highly-saturated color reflects only a narrow band of wavelengths, and the wavelengths of the illuminating lasers may not fall within this region. Even when they do so, there may be little discrimination in hue. Under a mixture of red, green and blue laser light, red, orange and purple flowers look very much the same hue.

There is also a psychological point. People who attend exhibitions of holography seem to be (at least at present) largely unaware of the presence or absence of color; a phenomenon that has been noticed with both cinema and television – people can seldom remember whether a particular old movie or TV program was in color or not. In holography it seems as if the three-dimensional nature of the image so overrides other visual stimuli that the unsophisticated viewer does not notice that the image is all-over green or yellow, merely that it seems to be suspended in space. However, just as in photography, it is probably only a question of time before color holograms become sufficiently commonplace for people to be surprised at seeing a monochrome image; even now there are indications of this. There is plainly a market in color holography for advertising purposes; no doubt we shall soon be seeing liquor bottles in what seems to be natural color and what is certainly three dimensions. 'Seems to be' is worth noting: colors may well be artificial; or the subject matter may have been specially prepared in off-color hues in anticipation of the effects of laser illumination, just as were photographic subjects in the early days of color photography, when it was usual to apply bluish or greenish make-up to compensate for the deficiencies of the material (see pp. 239–40). However, there is more to color imagery than just getting the colors right. In the early days of the color negative–positive process in photography, enthusiasts agonized over the difficulties of getting really good reproduction of reds and magentas, and of producing greens that didn't look bluish, completely forgetting that for over a hundred years painters had been using any colors that took their fancy, correct or not, and the public liked what it saw – and bought the pictures. Even today, one of the best-selling reproductions of any painting is of Tretchikoff's portrait of an Oriental woman; not only is it unfinished, with an area of bare canvas, but the woman's face is emerald green. An important characteristic of both transmission and reflection holograms is their ability to produce highly-saturated colors, so why not exploit this ability? In this chapter the production of holograms in natural colors (or as natural as they can be made) will be considered first, and then the more exciting world of synthetic color, known as *pseudocolor*.

Difficulties Inherent in Natural-color Holography

Basically, there is not much difference between a natural-color hologram and any other hologram bearing three images, as far as information content is concerned. The main problem in both cases is the lowered diffraction efficiency that results from the available dynamic range of the emulsion having to be shared between the three recordings. This has led to some workers experimenting with dichromated gelatin sensitized to red light with the dye methylene blue (see p. 277) and with photopolymers, which can readily be sensitized to long wavelengths. These materials have a higher potential information content than silver halide, as they are grainless; but their sensitivity is very low in comparison. New silver-halide materials now being evolved may solve the problem.

A second difficulty is cross-talk. This is particularly acute in transmission holograms, where there are nine virtual images and nine real images, most of the spurious ones overlapping and degrading the genuine one. These spurious images are to some extent suppressed in thick emulsions by the Bragg condition. In a reflection hologram, fortunately, all the spurious images are suppressed, but a new problem appears: shrinkage of the emulsion in processing, which causes a shift in all hues towards shorter wavelengths.

A further problem is the need for more than one laser, and the associated difficulty of getting two or more laser beams to coincide precisely in space, which makes the setting-up of a single spatial filter for the combined reference beam almost impossible. The use of a Kr^+ laser, which can provide suitable wavelengths of 647 nm (red), 521 nm (green) and 476 nm (blue), would seem to overcome this problem, but in the past it has not been easy to obtain stable operation at the two shorter wavelengths, and the power is lower than at the red wavelength. It appears that much of the difficulty has been due to the etalons used, and recent developments suggest that before long, given sufficient care and skill, and the right kind of laser, it may be possible to manufacture an etalon that can be matched to several frequencies without retuning (see p. 37). However, for the time being, most of us will have to make do with three separate exposures.

Yet another problem comes from the holographic emulsion. To date, no material of sufficiently fine grain for reflection holography has become available with a sensitivity that is uniform over the whole visible spectrum, so there is at present no real advantage in achieving a color hologram in a single exposure. In addition, the use of two plates, one (say Agfa-Gevaert 8E75HD) for the red, and the other (8E56) for the green and blue images, gives better diffraction efficiency than would a single hologram containing all three records.

Practical set-ups for Color Reflection Holography

There are several possible ways of setting up the beam. In one arrangement a HeNe and an Ar$^+$ laser are arranged parallel to one another, with a 1:1 beamsplitter splitting both beams from opposite sides. In a second arrangement a mirror simultaneously reflects the HeNe beam and blocks the Ar$^+$ beam; removing it allows the Ar$^+$ beam through, while the HeNe beam is no longer reflected down the path. A third arrangement employs a dichroic beamsplitter which reflects red and transmits green and blue light. The arrangements are shown in Fig 18.1.

In order to obtain correct registration of the emulsions a glass plate of the same thickness as the holographic plates needs to be placed in the holder along with each plate. This can be a spoiled plate which has had the emulsion removed with ordinary

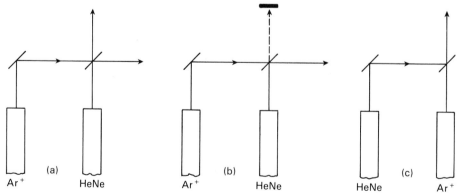

FIG 18.1 *Mixing light from two lasers. (a) A 1:1 beamsplitter produces two beams each containing half of the output from both lasers. (b) A mirror (not a beamsplitter) blocks the output of one laser; its removal allows light from the other laser to pass, while the original beam is no longer reflected down the path. (c) A dichroic filter reflects red light and transmits blue and green; thus both beams pass simultaneously along the same path.*

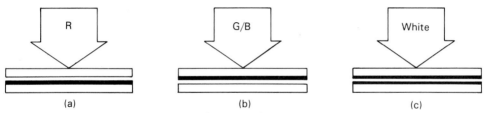

FIG 18.2 *When making color reflection holograms using more than one plate it is necessary to maintain accurate register by the use of a plain glass plate of appropriate thickness.* (a) *Shows the red exposure,* (b) *the green and blue exposure and* (c) *the reconstruction by white light.*

household bleach. The emulsion of the red-sensitive plate faces the object, with the glass in front, and the emulsion of the green/blue-sensitive plate faces away from the object, with the glass behind it, as shown in Fig 18.2.

It is not possible to use the Denisyuk configuration if you are using Agfa-Gevaert plates, as the 8E56 material has a transmittance of only 10% to green light. However, the green-sensitive Ilford material has a transmittance of 50% and is fully satisfactory for single-beam work, provided exposures are not more than 1–2 s. You can, of course, use either material for split-beam configurations.

The use of a HeNe laser for the red wavelength (633 nm) and the two main wavelengths of an Ar^+ laser (514 nm and 488 nm) for the green and blue wavelengths respectively gives a good range of colors (see Appendix 6), but the deeper blue line at 477 nm gives a better reproduction of blues. However, it has less than 30% of the intensity of the 514 nm line, whereas the 488 nm line has 70% of its intensity. Typical maximum power at 514 nm is 1–1.5 W, and as a HeNe laser suitable for holography can produce only about 25–50 mW you will need to do some balancing of the exposures. For large holograms the 647 nm line of the Kr^+ laser (about 0.3 W) is a better match for the Ar^+ beams.

As far as the table configuration is concerned, you can use any of the arrangements that are suitable for reflection holograms. In order to obtain the brightest results it is desirable to use a transfer technique, and you can do this by making three quite separate master plates, one for each wavelength, and transferring them one at a time onto two final-hologram plates set up as shown in Fig 18.2, for mounting emulsion to emulsion. A simpler method is to use the focused-image technique (see pp. 202–8). Whether you use a transfer or a focused-image technique, you will still find that you need to restrict the vertical parallax to about $15°$, using a broad slit over the master (about 25% to 30% of the height of the master), or (for a focused-image set-up) between the lens or mirror and the object, as in Figs 14.3 and 14.7 (pp. 203 and 206).

Color Rainbow Holograms

These are mastered and transferred in the same way as reflection transfers, but with a narrower slit (3–6 mm). Three spectra will be projected out into the field of view, but they will be displaced relative to one another in such a way that when viewed centrally the colors will be correct (Fig 18.3).

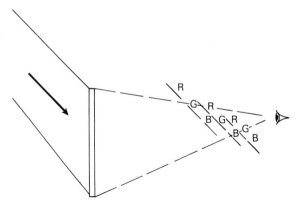

FIG 18.3 *In a correctly-made natural-color rainbow hologram, a spectrum will be created for each color. At the correct viewing height the spectra will overlap, projecting the true red, green and blue content of the image.*

For viewing of both reflection and transmission holograms the blue/green hologram should be in front of the red one and you need to bear this in mind when transferring holograms that are to be flipped for viewing.

Pseudocolor Reflection Holograms

When a reflection hologram is exposed with a HeNe laser, and subsequently processed by a method which removes material from the emulsion, the interference planes come closer together, so the Bragg condition is satisfied by a shorter wavelength. The reconstructed image becomes yellow, green or even blue, depending on the amount of material removed. A different way of achieving the same effect is to swell the material before exposure, then to process by a method that does not involve any removal of material from the emulsion. Washing removes the swelling agent. This technique results in a brighter image than shrinkage processing (Plate 25). A number of swelling agents have been used for this purpose, but the one most generally used at present is triethanolamine $(CH_2OHCH_2)_3N$. This is a clear viscous fluid which is somewhat difficult to dissolve in water, but once you have made up your solution you can use it repeatedly. Triethanolamine, usually abbreviated to TEA, has the property of *hypersensitizing* holographic emulsions, so that exposures can be reduced by a factor of about three compared with untreated emulsions. The effect lasts for about four hours, after which the emulsion begins to darken without any exposure; for this reason bleached holograms should not be treated with TEA to swell them after processing. (Sorbitol, another swelling agent, has a similar though less marked, effect.) For this purpose you should use parabenzoquinone (PBQ) $C_6H_4O_2$, or aluminum sulfate $Al_2(SO_4)_3 \cdot 16H_2O$ (see Appendix 6).

The most effective use for the pre-swelling technique is to provide synthesized color, or pseudocolor, in reflection holograms. Pseudocolor has been an artistic vehicle in white-light transmission holography for some years, and a number of artists have produced outstanding work in the medium (Plate 23), (methods for this are discussed later in the chapter). But it is much more recently that pseudocolor reflection holograms have appeared on the scene (Plate 24). They are dealt with first because they are basically simpler, although where complex subject matter is used there are likely to be color-registration problems unless the geometry of the table is adjusted carefully; this aspect is also discussed later.

Making a single-beam pseudocolor reflection hologram Pseudocolor holograms which need accurate register must have their image substantially in the plane of the hologram, and they thus require either a transfer or a focused-image system. However, if you use abstract subject matter such as brass wire or aluminum-foil shapes, registration is relatively unimportant and you can use a simple Denisyuk configuration. The method is as follows:

1. Pre-swell a test plate or film for 'red' by immersing it in plain de-ionized water for 5 min, then in a 2% solution of Kodak Photo-Flo for 1 min. (Photo-Flo is specified here because under these circumstances it has a hypersensitizing effect on the emulsion.) Remove the plate from the solution, place it emulsion side up on a piece of absorbent kitchen paper and wipe it with a blade squeegee until you can see no streaks when you examine it by the safelight. For 4×5 in plates push the squeegee away from you; for larger plates draw it towards you. Allow the emulsion to dry (if you are impatient you can use a hair dryer, though a film drying cabinet is preferable). Replace the kitchen paper with fresh sheets, put the plate down on the paper, emulsion side down, and polish the glass side with proprietary window-cleaning fluid; if you are using film be gentle with it. Complete the polishing by wiping 'downwards' (ie parallel to the plane of the reference beam), and examine the plate by looking through it at the safelight; ensure there is no dust or lint anywhere on it. Now set up the plate (use index-matching fluid in the usual way for a film) and allow 5 min settling time.
2. Make a normal exposure, and process the film as for reflection masters.
3. Check the brightness of the image against a known standard, and make more tests as necessary until the image is bright enough.
4. Place your best test result in a plate-holder, put the holder on a piece of white paper, and illuminate it with a small white light at the distance of the spatial filter. Place a red filter over the lamp and align the hologram so that it gives the brightest image. Mark its position. Now replace the red filter with a green one. Again adjust the plate-holder until you get the brightest image, and mark its position. Repeat with a blue filter. You now have the angles of incidence at which you will set the emulsion for the three exposures. The change in angle of incidence will usually be around $5-10°$.
5. Using a wax pencil, mark the top right-hand corner of a fresh plate and repeat Step 1. With film you can use the notch as a check.
6. Shoot the 'red' exposure, giving half the exposure you found correct for the test.
7. Remove the plate and replace the 'red' subject matter by the 'green' subject matter. Place the (dry) plate in a 15% solution of TEA for about 1 min. Remove it from the bath and squeegee and dry it as in Step 1. Move the plate-holder to the 'green' position, mount the plate and allow it to settle.

8 Shoot the 'green' exposure. Because of the hypersensitizing effect of the TEA, the exposure should be about 30% less than the 'red' exposure.
9 Remove the plate and replace the 'green' subject matter with the 'blue' subject matter. Place the (dry) plate in a 35% solution of TEA for about 1 min. Remove and squeegee it as in Step 1. Move the plate-holder to the 'blue' position, mount the plate and allow it to settle.
10 Shoot the 'blue' exposure, which should be about 60% less than the 'red' exposure.
11 Process the hologram using the method detailed on pp. 88–9 for making reflection master holograms.

The hues need not be confined to red, green and blue. Where the red and green images interpenetrate they will give yellow; red and blue will give magenta; blue and green will give cyan; and in any part where all three images interpenetrate the result will be white or neutral grey. By flipping the holographic plate through 180° between exposures the background part of the image can be made virtual and orthoscopic while the foreground is real and pseudoscopic.

If you feel the preliminary setting-up is going to be too tedious, forget it, and start the routine at Step 5. Increase the angle of incidence by about 7° after the first and again after the second exposures. You may find that not all the images are equally bright at the same angle, but the correspondence should not be far out. By using concentrations of TEA in between those given above you can obtain image components in oranges, yellows and blue-greens.

NOTE: This method is due to Peter Miller. John Kaufman favors exposing the 'blue' image first; this demands more working steps but leads to shorter overall exposure times. There does not seem to be any significant difference in image quality.

Pseudocolor reflection holograms by the transfer method   Once you have made a few Denisyuk multicolor images you will probably feel confident enough to move on to the next stage. This time the final hologram will be a transfer from three masters, preferably of the transmission type, which permits close control.

The first step is pre-visualization (see pp. 271–2). Determine the subject matter, and visualize the way the final image is to look. What color is to be where? What will be the lighting arrangement? and so on. Once you have decided exactly what you want, set up the subject matter and illuminate it appropriately. Let us suppose you want to make a hologram resembling Lon Moore's 'Still Life' of a red apple, a yellow pear and a green pepper, in a crystal bowl (Plate 26). This was an early example of a quasi-natural multicolor hologram, and Moore has given a description of how it was set up (1).

Moore's paper emphasizes the importance of correct registration, of the use of a collimated reference beam, also of not overloading the emulsion with information. The way to avoid this last problem is to use masks in the illuminating light beam, so that for the first master only the apple is illuminated, for the second one only the pear and for the third one only the pepper. For the transfers the plates can be masked down horizontally to about one-third of their height, which, as explained above, reduces the information content and increases the brightness of the image.

It is not necessary to use real fruit; it is not even desirable, as fruit is alive, and may move during the exposure. In addition, a green pepper is going to look very dark under red laser light. It is better to use artificial fruit sprayed white – glossy for the apple and pepper and semi-matt for the pear. If you want the pepper to appear darker you can spray it yellow-green or grey.

Having pre-visualized the image and set up the subject matter with appropriate illumination, your next task is to construct masks to go in the object illumination beam, and to try them out. Set up a plate-holder and collimated reference beam for transmission master holograms with a beam ratio of about 4 : 1 (looking straight up the beams). For an 8 × 10 in master hologram the subject matter should be about 175–200 mm (7.5–8 in) away from the plate, so that at the transfer stage the replay beam will not fall across the transfer emulsion. Now shoot the masters, changing the masks as appropriate between exposures. At this stage don't make any adjustment to the reference beam. Process all the masters at the same time and in the same solution if you can, using any method you have already found satisfactory for transmission masters. A reliable (if by now somewhat old-fashioned) method is D-19 developer followed by ordinary fixation and no bleach (see Appendix 10).

Now set up the table for making reflection holograms from transmission masters, using a collimating beam to replay the master holograms. To avoid possible errors in registration, try to ensure that any sharp-edged changes in hue are as close as possible to the plane of the hologram. The beam ratio should be about 1 : 1 with the meter pointed straight up the beams. You should restrict the reference beam to the area of the color for each transfer, as far as you can. From this point you should follow the instructions on pp. 265–6, reading '"red" master' for '"red" subject matter', etc.

Accurate color registration by geometry The change in the angle of incidence of the reference beam between the red and the green exposures, and between the green and blue exposures, is not critical, provided exact color registration is also not critical. However, if you are producing sharp-edged colors on a surface, as in John Kaufman's 'Rock Piece' (Plate 26), you do need to be exact, or the colored edges will drift apart as the viewer moves in one direction, and overlap on moving in the other direction. Strictly speaking, you should change the distance of the spatial filter from the hologram as well as the beam angle, but this is necessary only if you need very accurate color registration. The reason for the shift in reference source distance is that in changing from red to green, then to blue, the image is shifted in position and the scale is altered. More specifically, the blue image will be larger, farther away and nearer to the axis of the replay beam than the green image, and the same applies to the green image with respect to the red image. (Note that this is a parallel situation to the image formed by a lens with chromatic aberration, though the hues are reversed. This is because blue light passing through an uncorrected lens is refracted more than red light; a holographic 'lens' forms its image by diffraction, and blue light is diffracted less than red.) The mathematics of color table geometry have been worked out in detail by Stephen Benton (2). Steve McGrew (3) has found an elegant geometrical solution which is not at all difficult and leads straight to a plan of the table, it uses a number of simplifications as compared with Benton's more rigorous analysis, but is fully satisfactory provided collimated beams are used where appropriate. A version of the paper is reprinted in Appendix 7, by kind permission of its author.

Multicolor Transmission Holograms

Transmission holograms produce their colors in a somewhat different way, though the geometry involved has some close similarities to that of color reflection holography (see Appendix 7). As we have seen, every transmission hologram is a kind of diffraction grating. In particular, in vertical cross-section a rainbow hologram is a straightforward diffraction grating; if you look at people viewing such a hologram you can see a spectrum spread across the region of their eyes. If the viewer remains in the same position while the angle of the replay beam is changed, a different part of the spectrum will become aligned with the viewer's eyes, and the image will appear in a different hue. Thus the hue of the image depends on the angle of incidence of the replay beam with respect to that of the original reference beam.

Whether the viewer sees the image in a pure color, or in a desaturated color, or spanned by a spectrum, depends on the geometry of the table (see Fig 18.4). If the viewer's eye is in the vicinity of the real image of the slit, the image of the object will be in a pure saturated color, the hue of which passes through the whole visible spectrum over a vertical range of viewpoints of about 100 mm (4 in).

With a reference beam angle of 51° and a replay beam angle of 45° the hue will be green when the image is viewed from a central position coincident with the image of the slit; from points approximately 6° lower and higher, the hues will be blue and red respectively. If, therefore, a transfer hologram is made from several master holograms, with reference beams at slightly different angles for each, the images will appear in different hues.

In practice it is simplest to keep the master replay beam and the transfer reference beam fixed, and to move the position of the slit. If you move the master hologram, slit and all, in its own plane, you can obtain 'echo' images in different

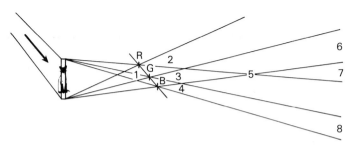

FIG 18.4 *Effect of viewing position on image color for a rainbow hologram. A viewer in position 1 will see blue fringes at the top and red fringes at the bottom. In position 2 the whole image will appear red, at 3 green and at 4 blue. At position 5 the image will be red at the top, green in the middle and blue at the bottom. At positions 6, 7 and 8 respectively only the bottom, middle or top of the image will be visible.*

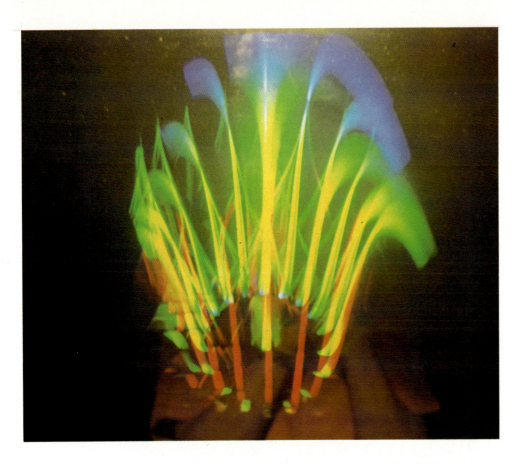

PLATE 24 *A multicolor white-light transmission hologram, part of the 'Yantra' series by Fred Unterseher (see p. 271). The colors in this type of hologram are obtained by dispersion of light, and must be viewed from the correct height or the wrong hues will be seen. (In* PLATE 16 *this effect is actually exploited).*

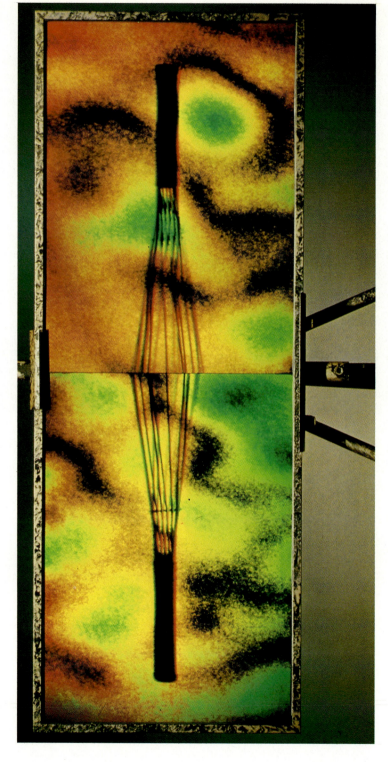

PLATE 25 *Expander*, by Michael Wenyon and Susan Gamble. *The colors have been produced in a random pattern by differential swelling of the emulsion before exposure (see p. 264). Photograph by Susan Gamble.*

(a)

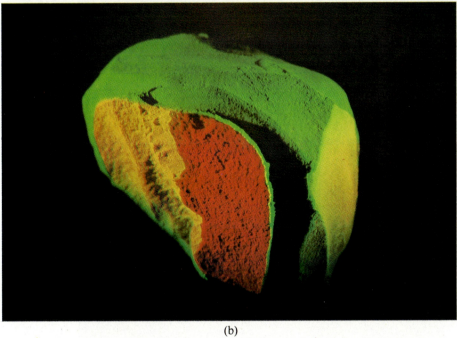

(b)

PLATE 26   (a) *'Still Life', a multicolor reflection hologram by Lon Moore. The pre-swelling has been controlled so as to produce what appear to be natural colors (see p. 266). (b) In contrast to Lon Moore's realism, John Kaufman's rock studies make use for creative purposes of deliberately false hues (see p. 267).*

PLATE 27 *Neutral-color reflection hologram by Richmond Holographic Studios, with acknowledgements to Tony Hopwood and Ilford plc. Pseudocolor swelling technique (see pp. 265–6) was used to produce the correct additive color mixture; correct registration of the red, green and blue images was achieved using a geometry similar to that of Appendix 7. Photograph by Tim Hawkins.*

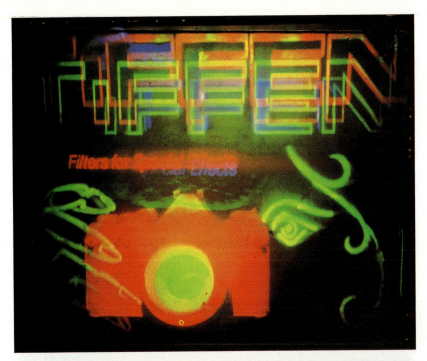

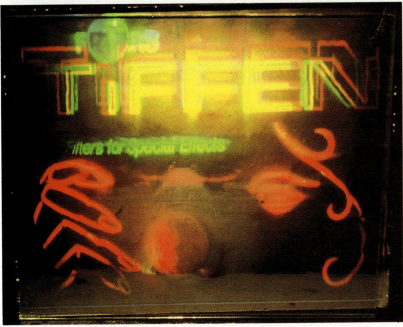

PLATE 28 *Embossed hologram by Jody Burns, sponsored by Tiffen Inc (see p. 285). Thirteen slit masters were used to make the original for this embossed hologram, which has multiple images appearing as the viewpoint changes in both horizontal and vertical directions. Two views are shown.*

PLATE 29 *Two further Stations of the Cross by the Sculptor Malcolm Woodward (see p. 304). Holograms by Advanced Holographics plc. Photography by the author. The strange atmosphere of the image straddling the glass plate is, of course, lost, as is the parallax; but something of the eerie light-quality of the original is retained.*

PLATE 30 *Birth of Adam*, a mixed-media piece by Jean Gilles consisting of a bronze sculpture and a rainbow hologram. At the most appropriate viewpoint (ie where the fingers of the two images touch) the hologram is very bright indeed; the photograph needed very careful camera positioning and separate lighting of the sculpture element (see p. 311).

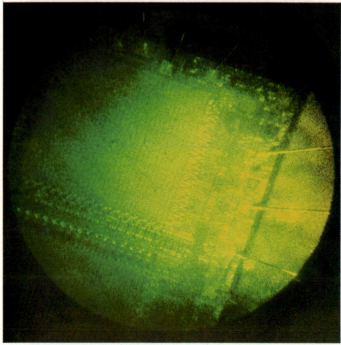

PLATE 31 *In the real-image reflection hologram of a microscope by Walter Spierings, it is possible to look through the image of the eyepiece and see the magnified image of the object (a microcircuit) that is on the microscope stage. (See p. 343.)*

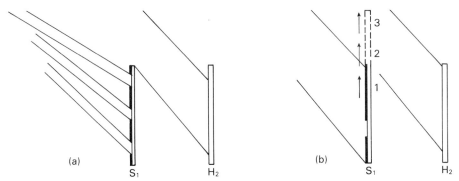

FIG 18.5 *Two ways of producing 'echo' images in different colors. In (a) the master hologram is masked to give three separate slits. The reference beam (expanded one-dimensionally) is swung to cover each in turn. In (b) a single slit is moved in plane between exposures.*

hues. This is very effective in holograms of abstract subjects such as cubes and spheres (Fig 18.5). You can do this by putting three slits over a single master hologram.

If the masters have been made with the subject matter at varying distances from the plate, you can make multiple-image final holograms which contain images in locations from well behind the final hologram to well in front of it, and which contain the whole spectrum of hues. Fig 18.6 is a schematic set-up which will give red, green and blue images, and intermediate hues if the angles are changed slightly. For holograms larger or smaller than 8×10 in, all the distances (but not the angles, of course) should be changed in proportion (Fig 18.6).

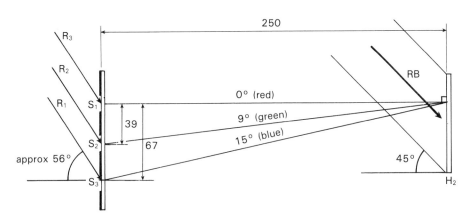

FIG 18.6 *A simple layout for multicolor rainbow holograms. Three slit master holograms (S_1, S_2 and S_3) (to appear red, green and blue respectively) are taped to a glass plate and exposed in succession. The angles must be fairly exact. If the master hologram–H_2 separation is 250 mm, separation of the centers of S_1 and S_2 should be 39 mm and of S_1 and S_3 67 mm. This layout does not give correct registration in depth or size, and is suitable for independent images in one hologram.* NOTE: *This drawing is not to scale.*

If you are using a one-dimensional beam expander with a collimating mirror you can make the three exposures separately. Mask one side of a piece of glass wide enough to accommodate all the masters with slits at the appropriate spacings, and mount the masters on the other side, using index-matching fluid. This should provide sufficient adhesion on its own, but you can make sure by using the springy plastic binding strips that are used for holding brochures together. Make the exposures one at a time, covering the transfer plate with a black plastic bag while you adjust the collimating mirror to move the slit to its next position to secure the ends of the masters. As long as you have a beam throw of not less than 1.5 m (5 ft) the change in angle of incidence will be negligible. Alternatively, you can move the relay mirror that is immediately behind your one-dimensional beam expander; this is the only possible method if you are using some kind of lens collimator. Each exposure makes a fresh set of fringes and a new hologram, so do not cut exposure times; give the full exposure to each*.

NOTE: If you compare the arrangement of Fig 18.6 with McGrew's geometry, you will see that in the Figure all the angles are correct, but that the slits are all at the same distance from the transfer plate. For most abstract creations this is of no importance, but for representative work where accurate registration is necessary, you should use the geometry given in McGrew's paper (3) or Benton's (2).

If you are able to use the simplified geometry of Fig 18.6, and are planning a run of copies, you will be able to save a good deal of time by covering all three slits in a single exposure. To do this you need to expand the beam at right angles to the slits, just wide enough to cover them all simultaneously. To do this, place a concave cylindrical lens of focal length -25 to -30 mm in the beam, close to the beam expander (a convex lens will also serve). The result will be an elliptical patch of light which will cover all the slits. For a cleaner beam you can put a $\times 10$ microscope objective in a spatial filter holder with a 50 μm pinhole immediately before the one-dimensional expander. A single exposure will suffice for all three images, and as the emulsion no longer gets an overdose of reference beam energy the image quality may be improved.

A great deal has been written about multicolor transmission holograms, and if you are going to take up this work seriously it would be well worth your while following up the references listed at the end of the Chapter. In particular, Suzanne St Cyr (4) has designed a worksheet for finding distances, angles etc, for both master and transfer holograms, derived from Benton's mathematics; a simplified version is reproduced (with permission) in Appendix 7. The full version of the worksheet, which is too lengthy for inclusion here, is given in Reference 4.

As with multicolor reflection holograms, for a multicolor transmission hologram the slits covering the master holograms should be displaced with respect not only to angle but also to distance. Alternatively, the three masters may be exposed in the same position, while the angles and distances of the reference beam sources are changed. This latter configuration is analogous to that of the multicolor

---

* Recent research suggests that exposure times should be increased by approximately 25% for each successive exposure, the final one being the 'calculated' exposure. This does not apply to pre-swollen plates. (See pp. 265–6).

reflection hologram, and is covered by McGrew's geometry, as is the displaced-slit configuration. You can find both these geometries in Appendix 7.

Remember: if you are playing with interpenetrating abstract designs there is no need to follow the geometry precisely; the methods given in the text above are sufficient (Plate 24). You need the exact geometry only where accurate registration of different colors is important. McGrew's geometry and St. Cyr's abridged worksheet are nearly always adequate, but for absolute precision you should refer to references 2 and 4.

Focused-image multicolor holograms   Using the table configuration shown in Fig 14.3 (p. 203) for one-step rainbow holograms, it is possible to produce multicolor results without the necessity of making three master holograms. This can save a good deal of time. In making these holograms it is necessary to achieve registration of the subject colors and at the same time to keep the real image of the slit in the same place. St Cyr (5) has studied the configuration in depth; although her paper is too long and complicated to be discussed here, the more dedicated color enthusiast will find a study of it rewarding.

A Note on 'Pre-visualization'

Anyone making holograms for display, particularly the complicated ones under discussion in this chapter, needs to spend some time thinking about the precise nature of the desired image before beginning to put together a table layout that will achieve the aim. John Kaufman (6) has compiled a checklist intended to help to reach the desired goal. The original paper goes into considerable detail; a summary of its main points is given below.

1 What an I trying to say?   At the highest aesthetic level, the basic concept may be inspired by a philosophical or political idea, a poem, a painting or a piece of music; or it may arise from purely holographic considerations, in terms of subtle interpenetrating shapes and colors. It may well spring from a found object, or from some experience such as a diving session on a reef, or even from a commission for advertising purposes that allows a fairly unrestricted choice of imagery.

2 What holographic techniques are appropriate to its visualization?   Should you use monochrome, achromatic or color? Reflection or transmission? Direct transfer or focused-image? Real, virtual, hologram-plane or all three? At this stage it is the holographer's knowledge of the details and limitations of the various processes that enable a suitable choice to be made. This leads to stage 3:

3 How do I link up concept and technique?   An internalized (or externalized, if easier) self-discussion over the way the various technical aspects of the process decided upon interact with the details of the concepts. At this stage the fine detail of both the concept and the technique become clear.

4 How do I go about realizing the concept?   The final stage is the detailed design of the table layout. This is the time for precision. By getting the geometry right (graph paper is essential) you will save an immense amount of time. This applies to even the simplest layout, for even with this you still need to know the length of throw required for your reference beam, what stray beams need carding-off, and whether you can match optical path lengths without the inelegance of optical sheepshanks, as well as the lighting necessary in order to obtain the optimum quality of subject illumination.

References

1 Moore L (1982) Pseudocolor reflection holography. *Proceedings of the International Symposium on Display Holography*, **1**, 163–70.
2 Benton S A (1982) The mathematical optics of white-light transmission holograms. *Proceedings of the International Symposium on Display Holography*, **1**, 5–14.
3 McGrew S (1982) A graphical method for calculating pseudocolor hologram recording geometries. *Proceedings of the International Symposium for Display Holography*, **1**, 171–83.
4 St Cyr S (1984) A holographer's worksheet for the Benton math. *Holosphere*, **12**, 4–13.
5 St Cyr S (1985) One-step pseudocolor WLT camera for artists. *Proceedings of the International Symposium for Display Holography*, **2**, 191–222.
6 Kaufman J (1982) Previsualization and pseudocolor image plane reflection holograms. *Proceedings of the International Symposium for Display Holography*, **1**, 195–207.

CHAPTER

19

Non-silver Processes

'Well, if I must, I must,' the King said, with a melancholy air, and after folding his arms and frowning at the cook, said, in a deep voice, 'What are tarts made of?'
'Pepper, mostly,' said the cook.
'Treacle,' said a sleepy voice behind her.
LEWIS CARROLL, *Alice in Wonderland*

In spite of all the advances in emulsion technology, there is still no grainless silver-halide emulsion. All silver-halide holographic images show *Rayleigh scatter*, which is inversely proportional to the fourth power of the wavelength (it is responsible for the blue color of the sky), and *Mie scatter*, which is independent of wavelength, and has only a very small effect. Rayleigh scatter causes a bluish background haze which can be a nuisance in display holograms; its effect is diminished by the yellow stain left by pyro-metol developer.

Part of the reason for the extraordinary sensitivity of silver halide to light lies in its crystalline structure. A very small number of photons striking a crystal at random is enough to produce a stable area of weakness (a latent image) which can be attacked by a developing agent. Crystal structure is thus an inherent element in the silver halide process. However, silver halide is not the only chemical substance that is sensitive to light energy. Iron complexes can also be broken down by light; if you are over fifty years old you will remember blueprints, which were negative white-on-blue copies of ink drawings. They were made using an iron compound as sensitizer, and were exposed to sunlight or mercury-vapor light and 'developed' by a simple wash in water. The process was rendered obsolete by diazo printing, which employs a variety of substances which react with ammonia vapor to give intensely-colored dyes, but are destroyed by short-wave radiation. The diazo process, which gives direct positives, works at the molecular level, and is effectively grainless. Apart from its use in making large copies of drawings, the diazo process is used mainly for microfiches and slides of graphics; but it has some potential in holography because of its extremely high *resolving power*.

Another light-sensitive element is chromium, in the form of the dichromate ion $Cr_2O_7^{2-}$. This has the property of rendering *colloids* such as albumen, gum arabic and, most notably, gelatin, insoluble when exposed to short-wave radiation. All three of these substances have been used in photography, but it is dichromated gelatin that has proved so far to be the most useful medium in holography.

Three other categories of photosensitive substances have also found use in holography: photoresists, photothermoplastics and photopolymers. Photoresists have been with us for a long time; they are substances which become insoluble (negative photoresists) or soluble (positive photoresists) when exposed to short-wave radiation. Their main application in industry is for making masks for printed-circuit boards and microchips. Although they are *binary recording media*, ie all-or-nothing, like photolithographic film, they are suitable for making masters for embossed holograms, and this is their main use in holography. Photothermoplastics become conductive to electricity when irradiated with visible light, and this property can be made use of in the production of real-time holograms. The material is indefinitely re-usable. Photopolymers change their chemical and physical structure, and thus their refractive index, on exposure to light and subsequent processing, and can be made sensitive to the whole visible spectrum. They can be used to make any type of hologram (unlike photoresists, which are suitable only for transmission holograms), and their application is the subject of active research.

Dichromated Gelatin: Principles

In photography, dichromated colloids have a long and honorable history. For over a century the medium for making photolithographic plates was dichromated albumen. Many classic photographs have been printed by the so-called gum-bichromate process ('bichromate' was an early misnomer for dichromate), and there are still many photographic enthusiasts who produce images in beautiful and subtle tones by means of this process. Dichromated gelatin, which becomes insoluble in warm water when exposed to light, is the basis of the Carbon and Carbro processes: the former produces photographic prints of extreme permanence and supreme richness of tone, and the latter was until the 1950s almost the only way of producing high-quality color prints (and a messy, unreliable, expensive and incredibly time-consuming process it was too!). With the advent of color negative–positive processes Carbro lapsed into obscurity, though there are still cells of activity around. But it was in holography that the dichromated-colloid principle was unexpectedly reincarnated.

Crudely speaking, gelatin is made from the parts of cows that people can't eat. If you take all the gristle, hoofs and bone and boil them for long enough, you end with old-fashioned carpenters' glue. If you then refine this further the final product is gelatin. Gelatin is a form of collagen, and has a number of remarkable properties, not the least of which is that it can absorb a very large amount of water and still remain more or less rigid. A table jelly is approximately 98% water and 2% gelatin. The molecular structure of collagen can change by a phenomenon called cross-linking, so that it becomes extremely tough; it not only resists penetration by water, but also holds water that has been previously absorbed, like a kind of three-dimensional fishing-net. Cross-linking occurs because the molecules of collagen are like long spirals, and parts of these spirals can become interlocked, rather like knitting, under certain stimuli. This is what happens in a dichromated-gelatin (DCG) hologram.

The principle of the method is simple: In the presence of dichromate ions, light energy stimulates cross-linking. Cross-linked gelatin becomes rigid, and has a refractive index which differs from that of unlinked gelatin when both are dried out forcibly. This is all that is necessary for the formation of a phase reflection hologram. The plate, coated with gelatin sensitized with dichromate, is given an exposure to light then 'developed' in a plain water bath which causes differential swelling of the exposed and unexposed areas, dehydrated in an alcohol bath and dried rapidly. In practice the method is by no means easy. The trouble is that as gelatin is a biological product its molecular structure is not fixed. Even if you buy the purest laboratory-standard gelatin, you will find enough difference from batch to batch to make it necessary to begin from scratch each time. There have been a great many papers on DCG holography, and most of them have proved unreliable in practice. The better ones are listed at the end of this chapter. The following account is due to Fred Unterseher (Saxby(1)), who has been consistently successful in producing DCG holograms commercially.

Making DCG holograms   Much of the literature on DCG refers to the use of fixed-out Kodak 649F plates. Although this is a good way of obtaining a uniformly-coated plate, it is a very expensive way of obtaining gelatin; in addition, the plates in question show considerable variation in quality. McGrew (2) gives a list of the conditions that can affect the quality and reconstruction characteristics of DCG holograms:

1 Initial thickness of the gelatin layer
2 Initial hardness of the gelatin
3 Concentration of the sensitizing dichromate bath
4 Drying conditions: temperature, humidity and time
5 Exposure duration, wavelength and energy
6 Time delay between exposure and processing
7 Alkalinity and temperature of the processing baths
8 Composition of the processing baths
9 Time spent in the processing baths
10 Recording geometry

The characteristics affected by these variables include fog, scatter, peak reconstruction wavelength, bandwidth and diffraction efficiency. McGrew notes that the initial hardness, the concentration of the sensitizer, the pre-exposure drying time and the total energy of the exposure all affect the reconstruction wavelength; the initial thickness of the emulsion affects the fog level; and the concentration of the sensitizer and post-exposure aging both affect the sensitivity.

DCG holograms have become standard for producing head-up displays (HUDs) for military aircraft; as a consequence much valuable information about the process has never been released, for security reasons. However, a good deal of know-how has been accumulated outside military establishments. The following method has consistently produced excellent results using a 15 mW HeCd laser, or an Ar^+ laser used at 488 nm.

Environment   The conditions for DCG holograms are rather more stringent than those for silver halide. You need a closely-controlled environment with a temperature of 24–26°C and a relative humidity of 35–40%. The environment must be absolutely free from dust, with air supplied by a laminar-flow air filter at slight over-pressure, and a fume extractor for the alcohol baths. The processing area should have a mixer control for providing water at constant temperature.

Emulsion preparation The gelatin to use is high-quality laboratory-grade gelatin of high *bloom strength* (resiliency). Prepare a solution of gelatin in de-ionized water, at a concentration of 50–200 g/l of solution (you will have to find the correct viscosity by trial and error). Warm the mixture slowly to about 50°C, stirring until the gelatin is completely dissolved. Now add 3 ml of a 9% solution of ammonium dichromate ((NH_4)$_2Cr_2O_7$) per gram of the original gelatin powder, and stir well. You should carry out this operation in subdued light, preferably with an amber safelight (Wratten OA or OB). Store the gelatin solution in brown glass bottles in a refrigerator at about 10°C, and allow it to ripen for about three weeks. Warm it up immediately before use.

Coating the plates Use 1.5 mm float glass (3 mm for large plates) and clean the plates in an ordinary domestic dishwasher using Calgon (sodium hexametaphosphate). You can also recycle spoilt plates by removing the emulsion with household bleach. The best way to coat a plate is by means of a *Mayer bar*. This is a piece of stainless steel rod about 12 mm in diameter, wound tightly with wire. You need a 24 SWG (Standard Wire Gauge) winding, which will give you a coating thickness of 7 μm. Take a dropper bottle of warm gelatin and spread a line of it across one end of the plate, then, as quickly as you can, pick up the Mayer bar and pull it across the plate. The plate must then stay perfectly level while the gelatin sets. At this stage it is very important not to let any speck of dust fall on the emulsion surface (as mentioned earlier, dandruff is the worst enemy, so wear protective head covering); any dust speck will suck up the gelatin through capillary action and form a sort of fish-eye. The coating process is tedious, and if you are going to coat a large number of plates you need to have a high threshold of boredom (and a personal stereo). Once the emulsion is coated it takes about an hour to mature. If the exposure is made too early the image will be noisy; if, on the other hand, it is left too long (over 24 h) the image will be dark green or blue. The optimum is 3–6 h.

Exposure and processing The required exposure may be anything from 25 to 250 mJ/cm^2, and you will have to find it by trial and error. The longer a plate is aged after coating, the bluer the image will be, and the more sensitive the emulsion. The processing steps are as follows:

1 Hardening: Kodak Rapid Fix with hardener, normal working strength, 10–30 s with agitation, the shorter times for more fully-aged plates.
2 Wash: 1–2 min in running water.
3 Wetting agent: About 10 s. Remove plate and shake off excess water.
4 Alcohol bath: Agitate vigorously in hot (43°C) propan-2-ol (isopropyl alcohol) for about 1 min.
5 Drying: Remove the plate from the bath and dry rapidly with a hairdryer.
6 Sealing: Cement a cover glass on the emulsion using optical-quality ultraviolet-

curing cement. Pour the cement on the cover glass in a dog-bone shape, place the plate on top of it and squeeze out any bubbles; cure the cement by exposure to an ultraviolet lamp or to sunlight.

NOTE: Handle the unsealed hologram with great care, as the emulsion is easily scratched, and the image will be rapidly degraded by the presence of any moisture from the fingers.

Color control in DCG holograms McGrew (2) has found that if several alcohol baths are used, with a concentration of alcohol rising from the first to the last, the reconstructed image is narrow-band, almost monochromatic. If, on the other hand, the hologram is immersed directly in undiluted alcohol, the image is wide-band, almost achromatic. This appears to be the result of sealing water in the innermost layers of the emulsion, so that the distance between fringe layers varies from the surface to the interior. Physicists sometimes refer to such a fringe structure as *chirped* (For an explanation of this term, refer to the Glossary). McGrew also notes that the image is more nearly monochromatic when the reference beam is incident on the glass side and more nearly achromatic when it is incident on the emulsion side.

DCG holograms with a HeNe laser The best laser for DCG holograms is undoubtedly a HeCd laser. At 442 nm the required exposure energy is only 3–6 mJ/cm^2, whereas with an Ar$^+$ laser at 488 nm it is 40–80 mJ/cm^2, and at 514 nm it is 200–400 mJ/cm^2. Unfortunately the coherence length of a 15 mW HeCd laser is only about 20 mm and Ar$^+$ lasers are very expensive (come to that, HeCd lasers are not exactly cheap, though they do not require any special installation). So, how about sensitizing DCG to red, so as to be able to expose the material with a HeNe laser? Well, if you fix Kodak 649F plates you keep the greenish dye, and this confers some slight red sensitivity to the emulsion, but not much. It is a better idea to add sensitizing dye to your own brew. There are two possibilities: methylene blue and methylene green. The main difficulty is that both these dyes tend to precipitate out of solution in the presence of dichromate. Kubota *et al* (3) suggest that a methylene blue solution at 1 mmol/l (milli*mole* per liter) should be mixed with ammonium dichromate solution in which the pH has been adjusted to 9 by the addition of ammonia. They also suggest the addition of EDTA.2NH$_4$ (ethylenediaminetetra-acetic acid, diammonium salt) to increase sensitivity. The late Andrejs Graube (4) suggested adding 1 mmol/l of solid methylene green, which is more readily soluble. It is perhaps worth noting that Graube's paper mutters darkly about exposures of several hours.

Photopolymers

Photopolymers are substances which polymerize when irradiated with light. Examples in everyday use are cyanoacrylate cement ('super-glue') and the ultraviolet-curing cement already mentioned in connection with the sealing of DCG holograms. Polymerization is a process whereby small molecules of a simple organic compound (the monomer) link up end to end, forming long chains. The physical properties of the polymer depend on the configuration of these chains – whether they are straight or coiled, whether they fit together to form crystals, the amount of cross-linking, and so on. The important property of a photopolymer is that on exposure to light and subsequent processing it changes its chemical and physical structure, and consequently its refractive index. The ideal photopolymer for holography would be transparent, flexible, sensitive to all visible wavelengths and completely stable after processing. It sounds as though this would be very difficult to achieve, and much research is currently being undertaken. The Polaroid Corporation recently brought out a material which appeared to possess all the desired properties, but early samples of the material gave images that eventually turned blue, and in some cases disappeared altogether. However, a more stable version has now appeared and is readily available. The material comprises a vinyl monomer and a photoinitiation system in a film-forming polymer matrix. The exposure for transmission holograms is about 5 mJ/cm^2 at a beam ratio of between 1:1 and 5:1 for transmission holograms, and about 30 mJ/cm^2 at a beam ratio of about 1:1 for reflection holograms. A single processing bath is used. Diffraction efficiency can be as high as 60% for a transmission hologram and 85% for a reflection hologram, and the signal-to-noise ratio is about 90:1 for the exposures which give the highest diffraction efficiency. These figures compare favorably even with those for DCG holograms. There seems little doubt that before long photopolymers will be playing an increasingly important part in display holography. Apart from their sensitivity to red light, photopolymer emulsions are consistent in their behavior, a quality absent from DCG and even (to a lesser extent) from silver halide emulsions (in which gelatin plays an active role).

Photothermoplastics

Photothermoplastics devices consist of a thin transparent layer of an electrical conductor such as indium oxide doped with tin on a quartz substrate, a thin (1.2 μm) layer of organic photoconductor such as polyvinylcarbazole sensitized with trinitrofluoronone, and a thermoplastic layer of thickness around 0.7 μm with a softening temperature of about 70°C (Fig 19.1).

Initially, the device is charged electrostatically; when it is exposed to light the charge leaks away in the exposed areas. The free surface of the thermoplastic material is recharged to an equipotential and the fringes are developed by passing an electric current through the indium oxide layer, which heats the plastic so that the surface forms corrugations in accordance with the electrostatic charge pattern. The hologram is erased by passing a further current of higher intensity or longer duration through the indium oxide layer. A holographic camera has been developed by Honeywell, and is described by Lee *at al* (5). Such a camera has obvious applications in industrial research, but would appear to have little use in display holography as the plates are very small (30 × 30 mm) and produce only laser-read transmission holograms.

Photoresists

Photoresists are proprietary substances which become either insoluble (negative) or soluble (positive) in an organic solvent after exposure to short-wave light. The

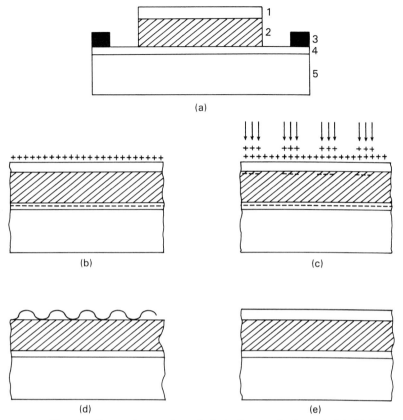

FIG 19.1 *Construction and operation of photothermoplastic material.* (a) *Shows the construction:* 1 *is the thermoplastic layer,* 0.3 µm *thick;* 2 *is the photoconductive layer,* 2 µm *thick;* 3 *are copper contacts;* 4 *is a transparent indium oxide conductor layer; and* 5 *is the glass substrate.* (b) *to* (e) *show the cycle of operations:* (b) *the material is given a surface positive charge;* (c) *exposure causes a redistribution of charges at the thermoplastic/photoconductive layer interface – this is followed by recharging of the thermoplastic surface to an equipotential;* (d) *when the thermoplastic material is heated by passing a current through the conductor the surface forms corrugation patterns in accordance with the electrostatic charge pattern;* (e) *a further heat pulse destroys the resistance of the thermoplastic material – the charge equalizes and surface tension restores flatness. The material is now ready for a further cycle.*

coating is typically 1 µm thick. The exposure is of the order of 10–100 mJ/cm^2; negative photoresists must be exposed through the base. In display holography the main use for photoresists is to make the master holograms from which stampers for embossed holograms are made, as the interference pattern is recorded as a relief. Photoresists cannot be used for making reflection holograms. They are sensitive to the 488 nm (blue) line of the Ar$^+$ laser and the 442 nm (blue-violet) line of the HeCd laser. The method of using photoresist material for masters for embossed holograms is dealt with in Chapter 20.

References

1 Saxby G (1984) Fred Unterseher on dichromated gelatin holograms. *British Journal of Photography*, 1984, 176–7.
2 McGrew S (1980) Color control in dichromated gelatin reflection holograms. *Proceedings of the SPIE*, **215**, 24–31.
3 Kubota T, Ose T, Sasaki M and Honda K (1976) Hologram formation with red light in methylene blue sensitized dichromated gelatin. *Applied Optics*, **15**, 556–8.
4 Graube A (1978) Dye-sensitized dichromated gelatin for holographic optical element formation. *Photographic Science and Engineering*, **22**, 37–41.
5 Lee T C, Skogen J, Schulze R, Bernal E, Lin J, Daehlin T and Campbell T (1980) Automated thermoplastic holographic camera development. *Proceedings of the SPIE*, **215**, 192–6.

CHAPTER
20

Embossed Holograms

> 'The thing can be done,' said the Butcher, 'I think.
> The thing must be done, I am sure
> The thing shall be done! Bring me paper and ink,
> The best there is time to procure.'
> LEWIS CARROLL, *The Hunting of the Snark*

The heading quotation gives something of the flavor of the hunt for a method of producing holograms mechanically rather than optically. The solution was indicated by the development of techniques in the electronics and sound-recording industries which led to the production of ultrahigh-resolution masks for making microcircuits and to CD and videodisks. Using these techniques it became possible to produce a hologram with its primary fringes in relief, and to replicate this relief pattern by a mechanical embossing process. The basic steps for producing embossed holograms are shown in Fig 20.1.

Any kind of hologram, with the exception of a reflection hologram, can be used as an original – even an existing embossed hologram, provided it contains only one image, though this is at about the same level of effectiveness as copying an audio

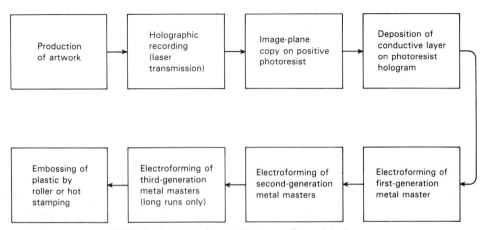

FIG 20.1   *Basic steps for producing embossed holograms.*

How the cover hologram was made

When David Pizzanelli of See 3 first suggested to the author that there should be an embossed hologram on the cover of the book, they both felt that the image should be related to the Siamese cat that forms the subject in most of the diagrams and many of the photographs of holograms that appear within it. It was at first intended to base the hologram on a realization of an Escher woodcut, but Kevin Baumber, who was responsible for the art direction, was not happy about the result, and suggested instead a hologram of a live cat, making use of the pulse-laser studio at the Royal College of Art. As this studio is in great demand, it was necessary to move quickly once official approval had been given: after many agitated telephone calls a small cream kitten was discovered in a pet-shop in Streatham. It proved to be ideal, and David and his business partner Jonathan Ross transported the kitten to the RCA, where Mike Burridge, Tutor in Holography, was waiting with Kevin to supervise the making of the master hologram.

David had carefully calculated the distance and tilt angle to compensate for the shift in wavelength when the master would be transferred in See 3's lab using an argon laser. In the event, a very lively subject equipped with razor-sharp claws made any measurements out of the question, and it was only by intuition and good luck that the geometry turned out to be correct. Persuading the subject to utter a bellow on cue was also far from easy, but eventually everybody (including the kitten) got it right, and the image was captured for posterity.

Back at See 3's headquarters, calculations were complicated by the decision to surround the hologram with an art-deco border of Laserfoil diffraction graphics, as it was necessary for the kitten's image to appear in the correct color relative to those in the border.

At a meeting with the author and publisher the other elements of the cover design were decided: overall color, layout and typeface to complement the silver square of the hologram. The cover would also need to be recessed to reduce the risk of the hologram's becoming scuffed.

Once the artwork had been put together, Jonathan made the separation masks, each with apertures appropriate to a particular Laserfoil color. David could now begin work on the embossing master, using a glass plate coated with photoresist. Once this was made, the nickel shims were grown and sent to Jayco Holographics, who took care of the embossing, metallizing, adhesive-backing and die-cutting of the hologram ready for application to the cover.

See 3 (Holograms) Ltd, 4 Macaulay Road, London SW4 0QX.
Jayco Holographics Ltd, 29/43 Sydney Road, Watford, Herts WD1 7PY.
Royal College of Art, Kensington Gore, London SW7 2EU.

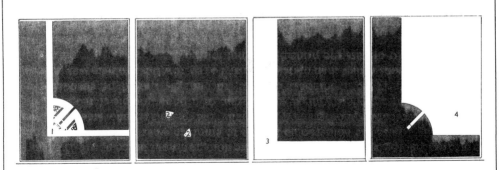

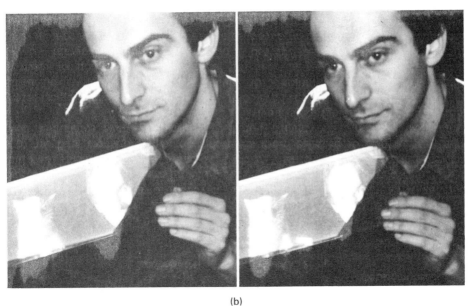

FIG 20.2 (a) *Set of artwork masks used to expose the various elements of the 'Cat's Whiskers' hologram. The figure shows detail of the bottom left corner. The white areas indicate the different colors or elements: reading from left to right, 1 red, 2 blue, 3 green, 4 kitten.*
(b) *David Pizzanelli with pulse-laser transmission hologram. Stereophotography by Caroline Taylor.*

recording by placing a microphone in front of a loudspeaker. At each stage in the process there is a certain loss of information, so it is important that the original should be of the highest possible quality.

Production of Originals

Full-aperture holograms are not really bright enough for this type of work, and they offer only restricted depth. Even where the images are shallow, as in *2D–3D images*, or where very tiny objects are used (eg for the mini-holograms used as security devices on credit cards), slit optics are necessary.

Companies which specialize in making up 2D–3D and similar quasi-holographic images require a camera-ready drawing of the image, and a key sketch of the colors required (usually red, yellow, green and blue). From this they will prepare a set of separation masks which will be used for the hues required (Fig 20.2). If you want full depth you will have to pre-visualize in terms of a rainbow master hologram; but even here you will need to remember that the embossed hologram is likely to be viewed, for the most part, under unsuitable forms of illumination; it is therefore advisable to arrange for the important parts of the image to be close to a single plane, which will become the plane of the final embossed hologram. If the final hologram is to be multi-image, you will need a separate model for each image, except where an image (such as a company logo) is to be repeated at different distances and/or in a different hue.

Holographic Recording of the Initial Image

Master holograms for embossed work are made on photoresist material, which records the primary holographic fringes in relief. This is a difficult task requiring a powerful ion laser and exceptionally stable conditions. There are a number of small companies which specialize in this work (see Appendix 12). If you are going to have the photoresist master made by such a specialist company from your own hologram you should supply them with both a slit master and a final transfer hologram. In general they will work from the slit master, using the final transfer as a guide. If possible, arrange the geometry of the slit master so that a single-beam transfer is possible. If, however, your set-up is a complicated one involving a number of slits, the final transfer may be used for the resist master. McGrew (1) suggests making an 'inter-master' which is recorded between the slit master and the final hologram plane, using a reference beam originating from the same point as the reference beam which is to be used to make the image-plane hologram. Otherwise, make slit or full-aperture holograms for transfer with a collimated reference beam, as in such cases the beam used for reconstruction will be collimated. Cross holograms (holographic stereograms) are resist-mastered by contact-printing, as are cylindrical holograms, but flat Benton stereograms are much better for this purpose.

There is no special requirement for the master hologram except high quality, but this is vital. Any noise caused by fog, intermodulation, light scatter or non-linear recording will be aggravated in the embossing shim, so you need to produce the brightest, clearest image possible. If the subject matter is contrasty, the beam ratio may need to be 8:1 or higher. It may be advisable to use a developer with a physical action, such as Kodak LX-24 or DX-80, followed by straight fixing and no bleach (see Appendix 10). Specialist companies usually require masters made with the reference beam at a specific angle, and they may ask for a master tuned to a specific wavelength. Whether this is so or not, it is important that the noise level should be as low as possible; the less damage that is done to the fringe structure in the processing the better. Short exposures and high table stability are mandatory. For multi-slit transfers use the method devised by McGrew (2) (see Appendix 7) or Benton (3) to get the positions of the slits and reference sources right; use the multiple-slit geometry rather than the multiple reference beam system.

Making the Photoresist Master Hologram

If you have a sufficiently powerful Ar^+ laser or a HeCd laser, you may be eager to learn how to make your own photoresist master holograms. You can make contact copies using an Ar^+ laser at 488 nm or, better, 458 nm, or a HeCd laser at 442 nm. For most projection transfer configurations you will need a Fabry-Pérot etalon with an Ar^+ laser, and this will reduce the output power at 458 nm to about 120 mW; the HeCd laser, without an etalon, has sufficient coherence length for this purpose and its light is much more actinic, so the lower power (around 15 mW) does not matter too much.

Most photoresist holograms are recorded on Shipley AZ1350J, which is the only commercially-available positive photoresist with sufficient resolving power to produce high-quality fringes. Unfortunately, it has very low sensitivity even to violet light. The nominal sensitivity at 458 nm is around 250 mJ/cm^2, but the material suffers from low-level reciprocity failure, and according to Burns (4) the energy must be delivered within 30 min exposure duration and preferably within 10 min. In the same paper Burns suggests that the maximum size master that is obtainable from a 5 W Ar^+ laser is 150 mm (6 in) square, using a single 3 mm wide slit master hologram 400 mm long, and bleached. An 18 W Ar^+ laser could expose an 8×10 in image from a maximum of two slits simultaneously. Burns's well-known Tiffen hologram (Plate 28) was made with no fewer than thirteen slits, and plainly this could not be recorded from slit masters directly; complicated holograms of this type must be made as contact copies. Burns suggests the use of an overhead-beam copying table, avoiding the need for a plate-holder.

When making transfer holograms directly onto photoresist it is necessary to have absolute stability, and unless you are using a purpose-built transfer frame you should consider installing a fringe stabilizer. This is a device which senses the position of the primary fringes and signals any change to a servomechanism which restores the position (see Appendix 9).

It is to be hoped that before long a more sensitive photoresist will be developed. The main user of photoresists is the electronics industry, for producing microelectronics circuits, and for this purpose it is possible to use powerful sources of ultraviolet radiation, to which the resist is much more sensitive. Holography represents only a tiny fraction of the market for photoresists, so it seems as if the initiative will have to come from the holographers themselves, as it has already done in the case of photopolymers.

The developer for AZ1350J is Shipley 303A, usually diluted about 1 : 6 with de-ionized water; development time is somewhat less than a minute.

Coating your own plates is not in general recommended, though you are welcome to try it. The glass plates need some kind of anti-halation coating: Burns recommends a low-reflection chrome layer between the glass and the resist, but suggests trying matt black spray paint on the back as a possible alternative. The coating should be about 1 μm thick, mirror-smooth and free from surface flaws. Coated plates of sizes up to 325 mm square are commercially available. The most uniform coating appears to be produced by a whirler, but even so it is quite difficult to get plates made by this method that are completely free from radial striations. Dip-coated plates are not sufficiently uniform.

Electroformation of metal masters The first stage is to make the surface of the photoresist master electrically conductive. This can be done with a proprietary silver spray, using a double-nozzle spraygun to mix the two reagents. Coverage must be complete, with no pinholes, and the layer must be uniform in thickness, no more than a few tens of nanometers. Vacuum deposition is a simple method which works well for small holograms, but the equipment is expensive. The so-called electroless nickel process, described by Burns (4), has much to commend it, as it provides a nickel-to-nickel bond between the initial conductive layer and the subsequent electroformed metal.

The electroforming process is standard, using nickel sulfamate in a plating tank. As with all plating operations, such variables as temperature, current density, pH and concentration of solution demand tight control. The photoresist is held in a conductive metal jig, the whole forming the cathode; the anode is a bar of pure nickel. The jig needs to be some 25% larger than the image area, as otherwise the metal master will be thicker at the edges than at the center; this may result in voids in the central area of the embossed hologram. The minimum thickness of metal required is about 0.05 mm (0.002 in) but it can be thicker and more rigid if the bath is well maintained and operated. The metal master is carefully separated from the photoresist, and rinsed in a solvent to remove any residual resist material. It is then 'passivated' in dilute potassium dichromate solution for several minutes and then rinsed in de-ionized water before being placed in a new jig. The passivation is necessary in order to allow the duplicate electroform to be separated after forming. Up to ten of these second-generation shims can be made from a single metal master, and this will allow the production of well over 20 000 embossed holograms. For very long runs each second-generation metal replica is used to make ten further stamper shims. It is not considered good policy to go beyond three generations of electroforms, as each generation contributes to a certain amount of degradation of the final image.

Successive generations of metal masters (see Fig 20.1) are always orthoscopic but are alternately reversed right to left. (In this respect there is an error in Burns's paper, which states that images are alternately orthoscopic and pseudoscopic.) As the production hologram is mounted downwards to protect the embossed surface (it is only some 200 nm deep and is very delicate) the final shim needs to be the right way round (orthoptic). This means that if there are two generations of metal masters, the second being the stamping shim, the first generation must be pseudoptic and the photoresist master orthoptic. The master hologram is thus made in the usual manner for rainbow holograms, or it may be made as a contact copy of a reconstructed silver-halide virtual image. When the processed master is flipped, the image will be orthoptic when the photoresist surface is next to the viewer. If, on the other hand, a third-generation master shim is to be used for stamping, in order to permit a longer run of embossed holograms, the photoresist image must be pseudoptic. One way of achieving this would be to use a negative photoresist and expose it through the back of the plate. Burns (4) recommends a method of transfer that uses a 45° mirror; in practice this does not work well, and it is better to make a further generation. Another technique is simply to make a silver-halide transfer, and to contact-print the projected-slit image so that when viewed through the photoresist the image will be pseudoptic.

The Embossing Process

There are three methods of producing embossed holograms. The first is known as *flat-bed embossing*. This is slow, and it is used mainly for test pressings in-house rather than for production runs, though it has been used for short runs where very high quality is required, or for unusual materials that do not lend themselves to high-speed runs. The process is similar to that used for pressing audiodisks: two flat plates are each heated and then cooled while pressure is maintained on the stamper and the plastic sheet that will become the hologram. Several layers of platens, stampers and plastic sheets can be used to increase the rate of production. Flat-bed embossing also requires rigid stampers up to 1 mm thick and very uniform in thickness.

The second process is used mainly for the bumper-sticker, cornflakes-packet and candy giveaway market. It is known as *roll embossing*. The shims are affixed to rollers, and the holograms are hot-rolled in a continuous process. The usual material is polyvinylchloride (PVC) or polyester. The latter produces better-quality and more durable results, but it is difficult to produce large images owing to the difficulty in maintaining sufficient pressure uniformly across large rollers; also, because of the high temperatures and pressures required, the shims wear out much more quickly. After pressing, the embossed side is aluminized: this serves the double purpose of protecting the delicate surface and turning the transmission hologram into a reflection hologram. It is also usual to add a self-adhesive backing and protective paper, so that the hologram can be mounted on a flat surface; the backing also helps to protect the metallization, and minimizes the likelihood of creasing and folding in transit. The method is fast: a throughput of more than 30 m/min is possible.

The third method, known as *hot-foil blocking* or *hot stamping*, has become the standard method of applying holograms to materials such as paper, banknotes and credit cards. It results in a flush surface which is impossible to remove without damage, and it is therefore very useful in security applications. Hot-foil is well-suited to the graphics industry as it provides a system whereby a hologram can be printed directly onto card using conventional hot-foil equipment as used for putting metallized etc surfaces onto card. These machines need some modification in order to get the holographic image correctly positioned, but small hand-operated presses are readily available. However, large companies such as American Banknote and Bradbury Wilkinson employ huge expensive presses with throughputs of up to 5000 cycles/h.

In the hot-foil process the making of the hologram is separate from the mounting. Conventional hot-transfer foil is used to print gold and other metallic colors

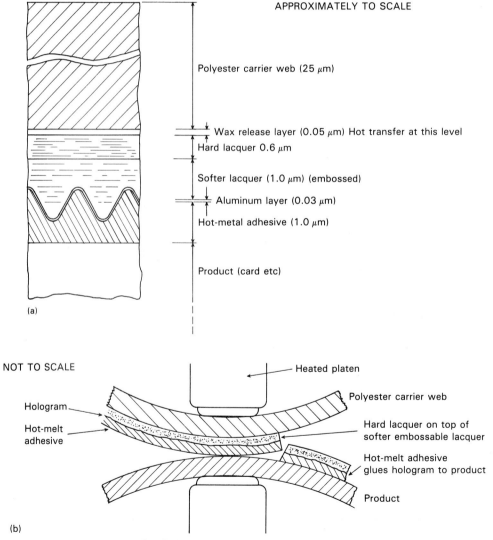

FIG 20.3 *Hot-transfer foil process. (a) Shows the construction of embossed thermoplastic film. (b) shows the hot-stamping process.*

onto book covers, diaries, business cards, etc. For some years it has replaced the old skilled craft of laying expensive gold leaf on book spines and other surfaces. The process consists of transferring an extremely thin layer of metallized lacquer from a support layer of polyester onto the product by means of a heated platen. Holographic hot foil is produced by embossing the lacquer while it is supported by the polyester carrier layer. This lacquer layer is extremely thin, about 1.6 μm, only about three times the depth of indentation, and the aluminum coating is even thinner, about 30 nm.

The way in which the transfer is accomplished is shown in Fig 20.3. Between the polyester carrier web and the hologram there is a wax release layer about 50 nm thick, and below the hologram there is a layer of hot-melt adhesive about 1 μm thick. When a hot die is brought down onto the sandwich, the release layer and the adhesive layer both soften, and the hologram adheres to the product and the carrier web separates from it. As the hologram is now underneath the lacquer layer it is protected from damage. As the total thickness of the hologram is less than 3 μm, the surface of the hologram is effectively flush with the surface of the product, and it cannot be removed without destroying it.

References

1 McGrew S (1982) Custom embossed holograms. *Proceedings of the International Symposium on Display Holography*, **1**, 185–8.
2 McGrew S (1982) A graphical method for calculating pseudocolor hologram recording geometries. *Proceedings of the International Symposium on Display Holography*, **1**, 171–83.
3 Benton S A (1982) The mathematical optics of white-light transmission holograms, *Proceedings of the International Symposium on Display Holography*, **1**, 5–14.
4 Burns J R (1984) Large format embossed holograms. *Proceedings of the SPIE*, **523**, 7–14.

CHAPTER

21

Display Techniques

> The shop seemed to be full of all manner of curious things – but the oddest part of it all was that, wherever she looked hard at any shelf, to make out exactly what it had on it, that particular shelf was always quite empty, though the others round it were crowded as full as they could hold.
> LEWIS CARROLL, *Through the Looking Glass*

Most holographers carry one or two holograms around with them all the time. They don't want to be caught out by somebody who asks them 'What *is* a hologram? What does a hologram look like?' They know that the large majority of the public still has no real idea of what a hologram is, even though most of them have credit cards with holograms printed on them. Of course, merely carrying a hologram is not enough; in these days of universal shadowless lighting you need a pocket torch too, and you certainly need an appropriate line of patter. The sort of hologram to carry around is a reflection hologram with very little depth – say, of a coin or small trinket – which will give a tolerable reconstruction under almost any lighting conditions. If you are giving a talk to a photographic club or similar gathering, you will probably be using a slide projector to show slides of holograms, or using an overhead projector to display diagrams. Either type of projector provides an excellent light source for showing reflection holograms, provided you hold them at the correct angle. If you have a mirror you can also show rainbow holograms, though this is more difficult (Fig 21.1).

The Basic Types of Hologram and their Display

There are four principal types of hologram for display purposes:

1 Laser transmission The image appears in a single hue, and shows laser speckle. The illumination is usually from the side, and always from behind. The image is dramatically deep and the resolution very high; transfer holograms can put real images up to a meter or more in front of the plate if it is large enough. The snag is that to get the full benefit of these properties you have to use a laser to illuminate the hologram. If the hologram is large you will need a powerful laser, which means, apart from the initial expense, that you will need to employ some-

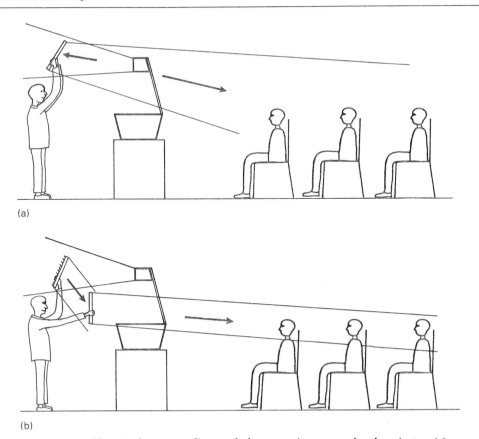

FIG 21.1 *How to show an audience a hologram using an overhead projector.* (a) *a reflection hologram;* (b) *a white-light transmission hologram. Align the mirror and hologram so that the projected spectrum falls across the audience's eyes.*

one to give it regular attention to keep the spatial filter aligned and the laser itself in good running order. In some countries you may find it difficult or impossible to get permission to use such a laser in a public place. In such a case you will have to use a mercury-arc source with a narrow-band filter, and you will lose some of the definition and depth that makes this type of hologram so dramatic. (However, you will also lose most of the speckle.) For these reasons laser transmission holograms have virtually disappeared from exhibitions, except for a few that are of historical interest (Plates 9 and 10).

2 *White-light transmission* The most frequently-seen white-light transmission hologram is the rainbow hologram which, as we have seen, produces an image in pure spectral colors, though special geometries can produce achromatic images. As a rule, creative white-light transmission holograms exploit the ability of this technique to produce bright saturated colors, and as these depend on critical positioning of the illuminating light, such holograms can be quite difficult to illuminate. The viewing angle is somewhat limited in a horizontal direction and

severely limited vertically; nevertheless, it is possible to produce a depth of image as great as that of a laser transmission hologram, and with a comparatively simple light source. This is situated behind and above (occasionally below) the hologram, which thus has to be suspended in the room space.

3 Reflection As far as display is concerned, this is the most good-natured type of hologram. Provided the image depth is not more than about 300 mm, an ordinary commercial spotlight will provide adequate illumination. The hologram can be hung on the wall like a picture, and the angle of incidence of the replay beam is not critical. There is full horizontal and vertical parallax, though in image-plane transfer holograms some of the vertical parallax may be sacrificed in order to increase the image brightness.

4 Holographic stereograms Until recently these have always been Cross holograms made on 240 mm width film wrapped in a 120° curve (sometimes 360°) with a radius of curvature of 400 mm. More recently, flat stereograms have begun to appear and, most recently, large stereograms which may be of computer-generated imagery. All of these are white-light transmission holograms. Cross holograms are illuminated from below by a small-filament tungsten bulb that is usually built into the display unit; the newer, flat stereograms are more often illuminated from above than from below.

General Display

Reflection holograms The easiest way to display a reflection hologram in domestic surroundings is simply to hang it on a wall with a ceiling spotlight playing on it, just as if it were a picture. You can mount a plate hologram flush on hardboard, using chrome clips, with a ring at the back about two-thirds of the way up, to hang on a hook or masonry nail. A film hologram can be mounted behind glass or acrylic sheet. If you hang the hologram a little above average eye level, the ring support will allow a small amount of forward tilt; this is important, as it means that people of short stature will be able to see the image by standing a little farther away. If you hang a hologram flush with the wall, with the light right for tall people to view it, small people will not be able to see the image, and vice versa.

Of course, you cannot put the spotlight anywhere on the ceiling: it has to be directly in front of the hologram and at the correct angle: it should also cover the hologram without wasted light, so the distance has to be right too. The usual distance is about 1.5 m (5 ft), and the usual angle of incidence is about 45°. You will need help from one or two friends when you are hanging the hologram and spotlight. Get them to hold the light at ceiling height and to move the light around until the contrast looks highest when you are square on to the hologram (Fig 21.2). (NOTE: This is not necessarily the angle that gives maximum image intensity.) Mark this position on the ceiling and fix the lamp. You can now make small adjustments to the tilt of the hologram with a wooden spacer. Once you are satisfied that you have the best image you can fix the spacer with Blu-tack; don't fix it permanently, as you may want to change the set-up at some later date.

DISPLAY TECHNIQUES   293

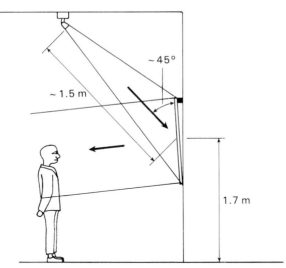

FIG 21.2 *Hanging a reflection hologram. Dimensions and angles are approximate and must be optimized by trial and error. It is preferable to have the hologram a little above average eye level and angled slightly down (see text).*

Again, you can't use just any old spotlight. Some have filaments and reflectors which do not match, and give distorted, blurred and even double images. Test a spotlight by shining it on a light-colored surface and placing your hand in the beam a foot or so from the surface. If the shadow of your hand is reasonably sharp, the lamp should be satisfactory. A selection of suppliers of spotlights is given in Appendix 12. You can often obtain a more dramatic lighting effect by filtering the illumination in the hue of the replayed image. This does not lead to any loss of brightness, and often improves the sharpness of the image, as well as suppressing much of the scattered light. You can mount your reflection hologram elsewhere, of course. It can equally well form part of a coffee-table arrangement, or a mobile. There are many ingenious ways of displaying reflection holograms, as a trip round any up-to-date exhibition will show.

Transmission holograms Transmission holograms are much more difficult to hang satisfactorily. As a general rule they need a clear meter of space behind them; in order not to take up too much space they are best mounted in an alcove or a corner. The range of satisfactory viewing distances is rather small, and if there is not much room the optimum display position is often below eye level (Fig 21.3).

If space is restricted you can reduce the lamp distance by combining a wide-angle beam with a magnifying lens or a Fresnel lens sheet (Fig 21.4). You can also fold the beam with a mirror (Fig 21.5).

Get a friend to hold a card in front of the hologram about 1 m away from it, while you adjust the tilt of the hologram until the spectrum projected on the card is at its brightest. Move the card backward and forward to find the point at which the spectrum appears to have the purest colors. This should be at about eye level

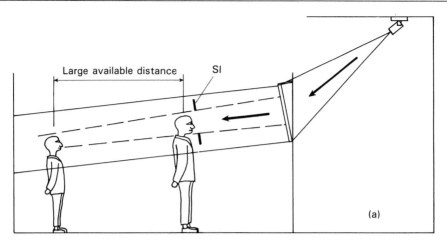

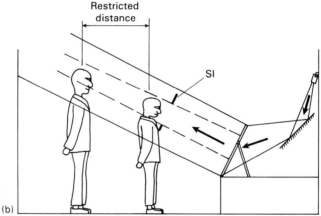

FIG 21.3 *Displaying white-light transmission holograms. (a) When there is plenty of room, hang the hologram facing slightly downwards, so that both tall and short people can find a viewpoint within the slit angle. (b) Where space is restricted, mount the hologram below eye level and use a folded replay beam if necessary. SI is the real image of the master hologram slit.*

FIG 21.4 *Use of a lens or Fresnel lens sheet to shorten the replay beam throw.*

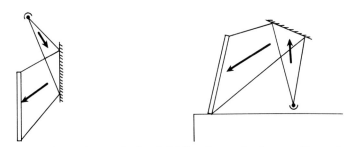

FIG 21.5 *Alternative methods of folding the replay beam with a mirror.*

for a person of average height. If not, move the hologram and the light source together, until it is at eye level. If the hologram has been made with the correct geometry the brightest spectrum will appear when it is formed on a line perpendicular to the plate.

The area behind the hologram should ideally be matt black. If there are any light-catching objects behind the hologram, they will show through and cause a distraction.

The illuminating lamp for a white-light transmission hologram can be very simple. A 25 W car headlamp bulb with a straight filament end-on to the hologram is adequate. Slit transfer holograms are very bright, because all the light forming the image passes through the image of the slit, and the pupil of the eye intercepts a much larger portion of the beam than with full-aperture holograms.

Where there is no space to hang a white-light transmission hologram it is usual to back it with a mirror (it need not be a front-surface mirror) and to illuminate it from the front, as with a reflection hologram. If you do this you should bear in mind that the lighting geometry is as critical as it is for rear-lighting, and that anything in front of the hologram, including yourself, will be reflected in the mirror. Stephen Benton (1) has designed a mirror system which avoids the second of these problems (see Box).

A novel reflection mount for transmission holograms

The usual configuration for a mirror-backed transmission hologram is as shown in Fig 21.6; this has the disadvantage that the image of the room is reflected in the mirror.

Benton's design is based on an angled mirror. The general principle is shown in Fig 21.7.

The mount can be made thinner if the space between the hologram and the mirror is filled with acrylic material, as the mirror angle then needs to be only $28°$. Room light entering the prism suffers total internal reflection at the back of the hologram and is absorbed by the matt black surface (Fig 21.8).

Of course, acrylic material of this bulk is expensive and difficult to find, and although the space could equally well be filled with a liquid such as glycerol or liquid paraffin, such liquids are notorious for their ability to find the tiniest crack and escape through it. Benton suggests fabricating a Fresnel prism with alternate

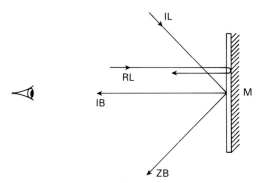

FIG 21.6 *When a white-light transmission hologram is backed with a mirror* M *for display, the illuminating light* IL *produces an image beam* IB *which is reflected towards the viewer, and a zero-order beam* ZB *which is reflected towards the floor. Unwanted room light* RL *is reflected straight back, degrading the image.*

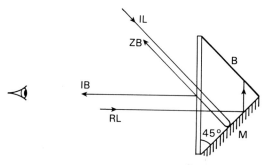

FIG 21.7 *If the hologram is mounted with a* 45° *mirror* M *behind it,* ZB *is reflected towards the ceiling and* IB *towards the viewer as before, but room light* RL *is reflected up to the black baffle* B *and absorbed. For this configuration the hologram geometry requires reconstruction from below.*

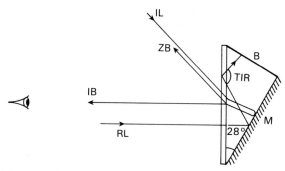

FIG 21.8 *If the space between the hologram and the mirror is filled with acrylic material* (RI > 1.5) *the angle needs to be only about* 28° *and the depth of the mount is reduced correspondingly. Stray room light is reflected back from the mirror at an angle greater than the critical angle and suffers total internal reflection* (TIR) *at the hologram surface. The viewer sees the image against a deep black background.*

DISPLAY TECHNIQUES

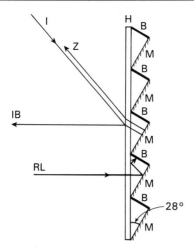

FIG 21.9 *Fresnel prism/mirror light trap based on geometry of* FIG 21.8.

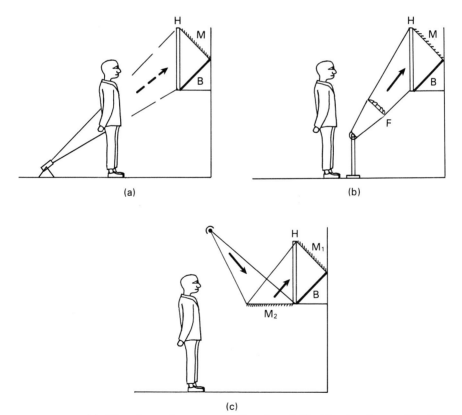

FIG 21.10 (a) *The viewer interrupts a floor-based illumination beam.* (b) *A Fresnel lens enables the source to be in front of the viewer.* (c) *A horizontal mirror allows ceiling illumination.*

blackened and mirrored facets (Fig 21.9). Such a device has been manufactured experimentally by the Polaroid Corporation, though at the time of writing it is not available commercially. However, the simple mirror mount of Fig 21.7 takes up only a little more room, and works well.

As shown, the device requires a hologram in which the replay beam comes from below. By inverting the geometry it can be used for existing holograms in which the illuminating beam comes from above, but this raises the difficulty of installing a floor-level replay light: the viewer's body interrupts the beam. Two ways of overcoming this problem are shown in Fig 21.10.

Cross holograms, as mentioned above, are usually illuminated from below by their own small-filament light source situated at the center of curvature of the cylinder and designed as an integral part of the stand. Other types of stereogram should be treated as if they were ordinary white-light transmission holograms and illuminated by one of the methods described above.

Displays to Accompany Lectures and Informal Presentations

Anybody who needs to give talks about holography needs to be able to display some holograms in order to be convincing. In a brief talk during a conference on more general matters it may be sufficient to hold up a hologram in the beam of a slide projector or overhead projector (Fig 21.1). If you are giving a more extended lecture on holography to a lay audience, you will need some sort of portable exhibition containing examples of what you are talking about. You can find much of what you need in any chemistry lab. Burette stands with heavy cast-iron bases can support the lamps with their clamps. The lights can be cheap single-slide projectors, which cost less than many spotlights and give a broader beam; four of them are sufficient to illuminate as many as forty 4 × 5 in holograms. You can lay reflection and embossed holograms directly on the bench, which you should cover with black card to prevent glare. Bring a few thin laths with you, in case some of the specimens need a bit of extra tilt. White-light transmission holograms can be taped to a piece of acrylic sheet held by clamps mounted on retort stands. To demonstrate laser transmission holograms, slit masters and Fourier-transform holograms you need a laser and beam expander, a dark corner and plenty of black card, and a roll of masking tape. To produce a satisfactory real image under these circumstances you need a 5 mW laser, which you can mount on a steel rod or a triangular optical bench together with its spatial filter; the whole can be clamped to a burette stand. It is always a good idea to take a low-power laser along with you; people who go to lectures on holography expect to see a laser beam and are disappointed if they don't. You do not need more than one laser beam, however; virtual-image and Fourier-transform holograms reconstruct perfectly well with a sodium lamp with its aperture masked down by kitchen foil with a circular aperture 5–10 mm in diameter.

It is a good idea to include in your briefing some form of words such as: 'Please don't hesitate to pick up the holograms if you want to, but make sure you put them back in exactly the same position they were in before', and have a notice saying something similar displayed prominently. Of course, whether you allow this depends on the nature of your audience. In some venues you need a heavy sheet of glass over the lot if you are going to have any holograms at all to take home.

Exhibitions

If you are organizing a full-scale public exhibition of holography, you are in a different situation altogether. Anything from thirty to two hundred holograms must be illuminated correctly, and in addition arranged so that individual pieces can be viewed as the visitor walks round, without the constant necessity to get closer or move farther away, or to duck and weave or, worst of all, to bang heads with a neighbor. There are other considerations too, such as insurance, packing holograms for shipping, installation devices, illumination, ambient lighting, music, safety considerations, maintenance and the way to take the exhibition down after it has ended. Rosemary (Posy) Jackson, sometime Director of the New York Museum of Holography, has written a paper (2) which deals in detail with these matters and gives examples of forms appropriate for exhibitors (loan forms, installation data, etc). The paper is too long to quote in full, but the points that are relevant to this book are summarized here, with the author's permission.

Packing a hologram for shipping Whenever possible, send images on film. Where plates are unavoidable, first tape the plate between foam polystyrene sheets (interleave if several plates are to be shipped). Wrap each corner with cardboard or bubble sheet. Wrap the entire piece in bubble sheet with the bubbles inside. With large holograms, roll newspaper into a long cylinder and wrap round the piece first. Then place the hologram in a cardboard box, skewed, not straight, with at least 75 mm of polystyrene foam pellets in all directions, packed tightly. Now pack the box in the same way inside another box and put at least two labels on opposite sides indicating both receiver and sender. Use 'Fragile' and 'Glass' notices. (Jackson suggests using a suitcase lined with high-impact foam as a sensible and simpler alternative to the series of boxes.) Finally, insure the hologram for the full amount you would expect to get for it if it were sold.

Light sources for display

1 *Lasers* Use the lowest power you can, preferably 5 mW or less (in the United States the maximum permitted is 5 mW). In order to get the most out of the illumination, arrange the display in a darkened area of the exhibition, and expand the beam so that it barely fills the area of the plate. If you are going to have even illumination you must use a spatial filter, preferably fixed directly to the laser head. It is wise to invite an official from your local Health and Safety Authority to inspect your laser display; this will cover you against any complaint from a member of the public who has misapprehensions about lasers. In the United States you need to file details of your display system with the Department of Health, Education and Welfare well in advance. You can obtain details of laser display regulations from the Department.

2 *Arc lamps* For large laser-read holograms you will need a 0.2–0.5 kW arc-lamp unit equipped with a narrow-band filter. Compact-source xenon (CSX) and compact-source iodine (CSI) lamps are the only really satisfactory sources for this purpose: CSX lamps have a source area of about 6 mm^2; the source area of the CSI lamp is even less. The radiation from these sources generates ozone gas, and with the more powerful lamps you may find that safety regulations compel you to install extraction equipment. Narrow-band filters usually have a bandwidth of about 5 nm, which is sufficiently low to provide good reconstruction, without too much loss in light intensity. You can obtain filters centered on a number of regions in the spectrum, the most useful being orange-red (for HeNe) and green and blue (for Ar$^+$). You can also use a low-pressure sodium lamp without a filter for small transmission holograms, using a short-focus lens and a 3 mm diameter pinhole to improve the spatial coherence of the somewhat large source area. A filtered mercury-vapor lamp is suitable for green or violet illumination.

3 *White-light illumination* Low-voltage straight-filament lamps are best for reflection or white-light transmission holograms. The best illumination for reflection holograms is a small-source tungsten-halogen lamp with built-in reflector; a number of models are made by firms listed in Appendix 12. Sources with collimators are better still, and for large holograms or arrays of holograms you can use a 2 KW photographic spotlight. Transmission holograms need a less powerful source if they are not to dazzle the viewer; the beam geometry should approximate to that calculated when the hologram was made (see Appendix 7). A straight-filament car headlamp bulb is usually adequate; the filament should lie along the axis of the beam. It may, exceptionally, be allowed to be vertical, but it should never be horizontally across the beam axis; and it must not sag. A vertical-filament lamp is suitable for stereograms and white-light cylindrical holograms. You may sometimes find that reflection holograms made using a wide-band processing technique show blur in the out-of-plane regions due to color dispersion. In such cases it is a good idea to eliminate the color that is causing the blur by filtration of the illuminating light (though you need to bear in mind the exhibitor's wishes in this respect).

Finishing techniques If you are submitting holograms to an exhibition, have them adequately mounted and/or framed first, otherwise the exhibition organizers will have to do this for you. The result may then not be what you would have wished, and you will not be popular with the organizers, who have already quite enough to do without finishing off your work for you. You can frame holograms in the same way as exhibition photographs, using sunken or flush mounts. You can also laminate them to glass or acrylic sheet. To do this take about 25 g (1 oz) of plain cooking gelatin and dissolve it in 250 ml ($\frac{1}{2}$ pint) of boiling water; just before it begins to set pour an extended puddle of it onto the pre-cleaned cover glass. Lay the film or plate on the glass and press it down to exclude all air (use a roller squeegee for film). Wipe off all the surplus at once, making sure you leave no trace on the outer surfaces. The job is best done with the glass and hologram at the same temperature (about 30–40°C) as the gelatin. For a more permanent mount you can use ultraviolet curing cement as described for DCG holograms (pp. 276–7), but Jackson points out that the archival stability of these cements has not been satisfactorily established yet. For holograms which are to be attached by screws or cables, four drilled holes will provide suitable points for attachment. If you sandwich a film transmission hologram between two acrylic sheets, you can use hollow pop rivets in the holes; these will hold the sandwich together and provide support points. If the acrylic sheet becomes scuffed, you can clean it with ordinary metal polish.

Installation of exhibits Structurally, the soundest method of hanging holograms is to mount them on walls. Beware of hanging them on nails or hooks: at best they will be knocked awry; at worst they will be stolen. There are several types of fixing screw that will enable you to adjust the angle of the hologram. Transmission holograms of all types look best mounted in a wall if there is space for backlighting; as this is not often possible, you may have to build alcoves for them instead. A simple method is to hang them from the ceiling using 22 SWG pre-stretched brass wire. Don't be tempted to use monofilament nylon, which can be severed by a carelessly-waved cigarette, and think hard before you decide on floor-to-ceiling cables, except for the biggest white-light transmission holograms. If such cables are tripped over they can cause injury to spectators and damage to both the holograms and the fabric of the building. Mark such fixings off with brightly-colored poles to stop spectators from squeezing between the holograms. Professional camera tripods are an excellent way of mounting large white-light transmission holograms, but they must be fixed firmly to the floor. Horizontally-oriented holograms on tables or the floor need a cordon round them, otherwise when people bend over them things may fall out of their top pockets and damage the exhibits.

Floor plan When designing the floor plan you must bear in mind that the flow of traffic will be limited by the narrowest section of the gallery; its width should nowhere be less than about two meters. Exits must comply with safety regulations. Establish a consistent eye-level for all the exhibits, at a constant distance. This is very important: people should not be expected to have to bob and weave. Jackson suggests that if exhibits are mounted flush with the wall they should be at a height of 1.63 m (64 in), and suggests that plastic drinks bottle crates should be provided for children and very short people to stand on. (Experience has shown that drinks containers are not really safe to stand on, and that it is better to have wall-mounted holograms above eye level and tilted slightly down, and transmission holograms below eye level and tilted up, as suggested earlier; this avoids the problem altogether.) Jackson says: 'A good floor plan takes into account the traffic flow of the gallery, the installation requirements of the holograms on exhibition, the "feel" of the work, and the aesthetics and intellectual flow of images round the room.' There should be no viewing overlap and no stray reflections. In drawing up a plan for an exhibition, each exhibit needs to be given a diagram showing the plan view complete with viewing distance, viewing angle and position of illuminating lamp (Fig 21.11). Draw these to scale on card for each hologram, cut out the areas and arrange them on a plan of the exhibition area drawn to the same scale.

In making the arrangement you should try to get about a meter of free space between viewing areas (not between holograms), as otherwise people will bump

302 PRACTICAL DISPLAY HOLOGRAPHY

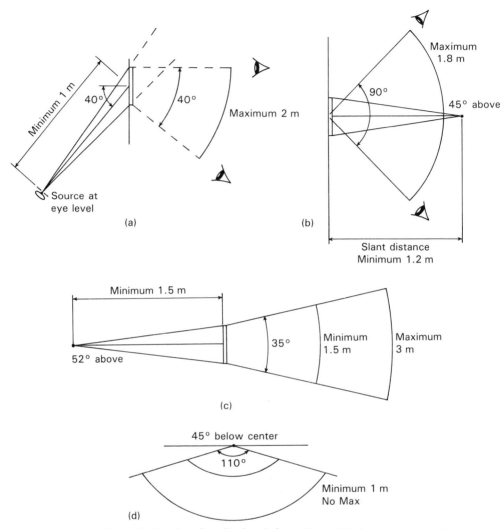

FIG 21.11 *Typical sketches for display information.* (a) *Laser transmission hologram;* (b) *reflection hologram;* (c) *white-light transmission hologram;* (d) *holographic stereogram.*

heads. Restrict each light to the area of a single hologram or group of holograms, otherwise you will produce ghost images in neighboring holograms.

Relevant information It should no longer be necessary to provide a display of graphics showing the principles of holography, any more than it is necessary in an exhibition of photography to provide an explanation of how a camera works. Those who already know do not need to be told yet again; those who do not know probably do not want to know, and in any case can find out from the catalog or the exhibition bookshop. It is much more important to have a label beside each exhibit indicating the name of the artist, the type of hologram and date it was made, any sponsorship, ownership (if relevant) and any production credits. These credits must be in a legible typeface such as Helvetica or Folio, *not* ordinary typescript such as Elite or Pica (which are all but illegible from a meter away, even when enlarged), with initial capitals where appropriate, in a minimum of 16 point (ie 4 mm height for capitals and lower-case letters with ascenders).

Environment The ambient lighting, decoration, plants, furniture, etc, must support the exhibition and not distract from it. In this connection, one of the most distracting elements can be music. In the early days of holographic exhibitions the accompanying music was invariably loud and weird. Holography has long outgrown its space-age image, and deserves better. To the author's knowledge nobody has ever complained at an exhibition that the music was too quiet. If you must have music, keep it neutral, and don't keep playing the same tape over and over again. Remember that many people will be coming back to the exhibition for a second and third time, and will not be favorably impressed by evidence that you appear to be able to afford only one tape. Ensure that safety regulations are observed, staff have unobstructed views of the public, and there is supervised space for leaving such hazardous impedimenta as backpacks, umbrellas and large briefcases. Make sure small children are properly supervised, and discourage smoking. Clean all glass surfaces with a proprietary window cleaner daily (spray the cloth, not the hologram, and make sure that your last wipe of the polishing cloth is downwards, not across. Keep a good stock of replacement lamps; failing to replace a lamp promptly is a guarantee that the artist who made the hologram it was illuminating will appear in the gallery within minutes to see how his or her piece is looking. Finally, when you demount the exhibition, take down the holograms first; check them; wrap them up carefully; and put them in a safe place before you do anything else.

References

1 Benton S A (1984) Reflection mounts for transmission holograms: format compatibility issues. *Proceedings of the SPIE*, **462**, 2–4.
2 Jackson R (1982) Exhibition techniques and materials for holography. *Proceedings of the International Symposium on Display Holography*, **1**, 215–38.

CHAPTER
22
--- ✸ ---
The Photography of Holograms

'It seems very pretty,' she said, 'but it's rather hard to understand.'
LEWIS CARROLL, *Alice in Wonderland*

Making successful photographs of holograms is not particularly easy. Black-and-white prints show up the laser speckle effect (which exists even in a white-light-viewable hologram), and in many display holograms nowadays the color is very important, so the use of color transparency material is essential except where the photograph is required for a publication that does not use color. On occasion a really good projected slide can show up the qualities of a hologram better than a view of the hologram itself, especially if the original is small, shallow and not very bright (as was often the case in the early days of display holography).

One of the difficulties associated with the photography of representational holograms is that they tend to turn out looking like grainy, not-quite-sharp photographs of the original subject matter, so that one might as well have simply photographed the subject rather than the holographic image. This can be seen clearly in the photographs of the Stations of the Cross, made for Coventry Cathedral by Advanced Holographics plc, the sculptures having been specially made by Malcolm Woodward (Plate 29 and Fig 22.1). Although these superb sculptures were made with the medium of holography in mind, and although the penetration by the holographic image of the glass plate produces an other-wordly impression that overrides any *trompe-l'oeil* effect, in a photograph the glass plate simply disappears and we are left with what appears to be merely a photograph of a sculpture, and a grainy one at that. Nevertheless, photographs of holographic images are often needed (this book, for example, would be a great deal less attractive without them), and it is surprising that even professional photographers are often unaware of the special problems associated with the photography of holographic images. You can take photographs of holograms with any camera, of course: just as you can make a hologram of anything you can see, including the optical image formed by a camera lens, so you can take a photograph of anything you can see, including a holographic image. But there are some difficulties. For a start, not every camera is really suitable. There are two types of camera that *are* suitable: a single-lens reflex camera, and a small studio camera. The former should be equipped with through-the-lens metering

FIG 22.1 *In a black-and-white photograph of a representative hologram such as the Stations of the Cross, it is impossible to tell whether the photographs are of the holograms or of the sculptures themselves (cf* PLATE 29*). Sculpture by Malcolm Woodward. Holograms by Advanced Holographics plc. The above pictures and those in* PLATE 29 *are photographs of the holograms, made by the author.*

and manual focusing (autofocus systems are fooled by the glass surface of the hologram); the latter should have lens and back movements.

Reflection Holograms

When photographing reflection holograms it can be difficult to prevent stray reflections and other bright objects from getting into the photograph unless you take suitable precautions. If you are unable to take the hologram away and set it up in a properly-designed surrounding, the best way to prevent such artifacts from getting into the picture is to use a piece of black velvet about 150 cm (5 ft) square, with a hem sewn or stapled at the top and a bamboo cane thrust through this to hold the cloth out straight, and a slot cut at the center through which you poke your camera lens. Hold the velvet sheet and check through the camera viewfinder that it does not cast any shadows on the hologram (Fig 22.2). For the best results, you need a tripod that will extend to at least 1.5 m (5 ft), higher if possible.

The main problems in photographing reflection holograms are depth of field and speckle. These require opposite treatments: if you are to get sufficient depth of field to render a deep image sharp all over, you need a fairly small lens aperture, but if you want to minimize speckle you need the largest lens aperture you can

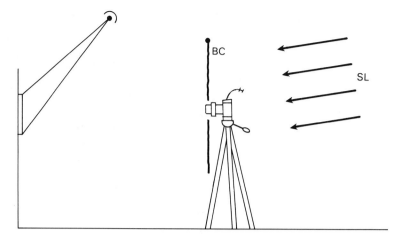

FIG 22.2 *A black cloth* (BC) *with a hole for the lens eliminates reflections from stray light* (SL).

manage. The answer is to use the longest focal length you can, and the largest aperture that will give you sufficient depth of field. The way to check depth of field accurately is as follows:

1 Focus on the nearest point of the holographic image. Note the point on the focusing scale.
2 Focus on the farthest point of the image that is visible. Note the point on the focusing scale.
3 Adjust the focus by moving the focusing ring on the lens until the 'focus' mark on the lens barrel is equidistant from the two marks (Fig 22.3).
4 Set the lens aperture to that indicated by the limits (ie read off the f/no as indicated on the lens barrel opposite these two points), and make your exposure.

In restricted spaces the use of a zoom lens makes photography very much easier, as you are able to frame the hologram in the viewfinder without the necessity of moving the whole camera and tripod back and forward. A zoom range of focal lengths from 80–200 mm is about the optimum, with an extension ring for close focusing for the occasional very small hologram. If you do use a zoom lens try to

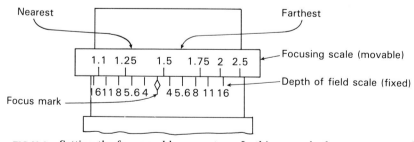

FIG 22.3 *Setting the focus and lens aperture. In this example the nearest part of the image was at about* 1.3 m *and the farthest part was at about* 1.65 m. *These two distances are bracketed across the focus mark and the indicated aperture (in this case* f/5.6*) chosen.*

avoid using the extremes of focal length, as there is usually a measure of *distortion* at these extremes.

For most holograms the exposure indicated by your meter, or the automatic exposure indication, will be correct. However, if the image is on a dark background you should give only half the indicated exposure, as a general rule; conversely, if it is on a bright background, give double the indicated exposure. If your camera has exposure-control override, use the $\times \frac{1}{2}$ and $\times 2$ overrides respectively; if it does not have such controls, set the film-speed indicator to half the film speed (ASA or ISO arithmetic) for a light-ground image or twice the film speed for a dark-ground image. If you have plenty of film 'bracket' your exposures a further notch up or down.

If you bracket your exposures in this manner, you may find with some rather contrasty images that one of your transparencies shows a correct exposure in one part of the image and under-exposure in another part, while another shows correct exposure in the second part and over-exposure in the first. In this case you can get a much better result (if you are able to go back and re-shoot) by using the sort of 'dodging' and 'burning-in' techniques with the replay beam that photographic printers use with an enlarger when making top-quality prints. The technique is simple, especially if you are one of those people who are good at making animals with the shadow of their hands: you simply hold back the light in the brighter regions of the image for part of the exposure, using the shadow of your hands, or a piece of paper torn roughly to the shape you want. If you are used to doing this in the photographic darkroom, you should find it easy enough. Remember, though, that with a color slide film, *more* exposure means a *lighter* result.

Although color transparencies are the most generally useful variety of photograph, you may want to use black-and-white or color-negative film, for prints. In this case you can do the dodging in the darkroom, of course, and your main concern is to obtain adequate exposure in the shadows; you can usually rely on your exposure control with this type of film, though it is still a good idea to give extra exposure to subjects with bright backgrounds. The reason this is necessary is that exposure meters are calibrated for an average grey, and when confronted by a subject that is unusually bright they simply turn down the exposure.

Choosing a film In general, you should use a film with a speed of around ISO 125/22°. If you are making transparencies use a daylight-balanced film, and select a make that combines accurate color reproduction with high resolving power. Ektachrome 100 is among the best (most of the color plates in this book were shot using it), though some photographers prefer Kodachrome 25, which gives high contrast and color saturation, at the expense of some latitude in exposure. Film balanced for tungsten-filament lighting seems to give rather bluish results that are less lifelike than daylight-balanced film.

When you are setting up the camera, get the hologram to fill the frame as far as possible. Most amateur 35 mm cameras have viewfinders which are pessimistic, ie you get more in the picture than you see in the finder; typically 5% all round, so you can afford to crowd the image a little. The aspect ratio (ratio of width to height) of a 35 mm format is 1.5:1, but that of most holograms is lower, about 1.25:1. This does not matter with a negative, as it can be masked when making the print;

but with a color slide film you should arrange it in the viewfinder to fill three sides. When the slides come back from being processed, mask off the odd edge with blocking-out medium or black gummed strip.

If you want to photograph reflection holograms at an exhibition and you don't want to be bothered with a black cloth, you can still get very passable photographs using a flash. It needs to be a dedicated flash, ie it is interconnected to the exposure control mechanism of the camera, and it must have an extension lead. Connect the flash to the camera via the extension lead, and hold the flash-head in the replay beam, as far up the beam as you can. Examine the image through the finder, align the flash – the shadow of the flash-head should be a clean rectangle and not a hexagon – and operate the shutter. Stray reflections are swamped by the flash. With this technique you do not need a tripod, but you do still need to control the exposure for dark or light backgrounds by using either the override control or the film speed adjustment.

With care, you can get very good photographs of holograms that are on public exhibition; a number of the plates in this book were made this way. However, if you can, it is always much easier to photograph them in your own studio, where you can get the replay condition exactly right. The ideal 'studio' is a walk-in cupboard which you have sprayed matt black all over, with an open door and a large matt black screen behind you in the room. You can take as long as you like to adjust the replay light. It should be adjusted for maximum contrast, not maximum brightness (which often produces a noisier image). In the case of a broad-band hologram, you can place filters over the replay light to cut down on the dispersion colors; a fairly narrow-band filter (such as a cheap interference filter of the right hue) on the camera lens is even better. You can also ensure that the surface of the hologram is perfectly clean. As mentioned earlier, it is a good idea to put a proprietary window-cleaning fluid on a cloth and polish the glass, finishing the polishing in a downward direction; this is the only sure way to avoid streaks being visible in the transparency.

Transmission Holograms

Transmission holograms require a different technique. Laser-viewable holograms are particularly difficult to photograph successfully owing to the presence of obtrusive speckle and to the very high contrast they often possess. As mentioned above, the speckle size is larger and more obtrusive for small lens apertures, and in order to minimize the effect it is important to use the largest possible lens aperture. Also, if the image is being replayed using HeNe light, the red color may fool your exposure meter, so bracket the exposure both upwards and downwards. Ar^+ illumination seems to give less trouble, and filtered mercury or other arc lighting suppresses the speckle to some extent. Rainbow holograms are less subject to speckle effects, which is fortunate as they often possess considerable depth, and it may be necessary to employ quite a small aperture. For all types of white-light transmission holograms the distance and elevation of the camera is important. As a general rule the camera should be at a distance of two or three times the width of the hologram, which will be approximately the position of the plane of the real image of the slit, or a little farther away. If the camera is too close the image will appear red at the bottom and blue at the top; if it is too far away the image will be blue at the bottom and red at the top. In either case, in a multicolor hologram the hues will be wrong. When the camera lens is in the plane of the real image of the slit the hue of the image as seen through the viewfinder will be uniform. Once you have found this position, adjust the height of the camera until the overall hue is a yellowish green. (This is where you may need the full height of your tripod.) With multicolor and multiple images it is important to have the camera at the correct height, otherwise the transparency may give a quite misleading impression of the real appearance of the holographic image. It is, of course, important to examine the image from exactly the same viewpoint as the camera lens, and this is where the single-lens reflex camera scores: everything you see through the finder is exactly as it will appear in the final result (plus a little extra space round the edges in most amateur cameras).

All rainbow hologram images are astigmatic, with the vertical aspect of the image being in the plane of the plate regardless of where the horizontal aspect may be. You can test the truth of this by turning your head sideways while looking at a rainbow hologram: the image will suddenly flatten itself onto the surface of the plate. This may result in distortion of the image unless the camera is the correct distance from the hologram. There will in fact be a small amount of astigmatic distortion at the plane of uniform image hue, as the zero-distortion point or 'intended viewing distance' is the position of the *horizontally*-focused real image of the master hologram slit, which is some distance farther from the hologram than the uniform-hue plane, which is at its *vertical* focus. Nevertheless, the most satisfactory result seems to be from the latter position. The use of a yellow-green interference filter makes it easy to find this plane with some precision.

The new generation of holographic stereograms has a reconstruction geometry that is very similar to that of a straightforward rainbow hologram, and the same restrictions apply to the photography of these images. However, it is very difficult to obtain a satisfactory photograph of a Cross hologram. This is partly because the vertical aspect of the image is some 150 mm nearer to the viewer than the horizontal aspect, and partly because of the mismatch of the printing and viewing optics, which produce a further helping of vertical/horizontal mismatch of scale. The only way to get anything like a satisfactory photograph of a Cross hologram is to use a very long-focus lens and set up the camera as far as possible from the hologram; otherwise the image will be stretched vertically to such an extent that your slides will look as though the painter El Greco had a hand in their production.

One advantage of transmission holograms, at least when photographed in situ and not in your studio, is that they are usually bright enough to permit hand-held exposures at about $\frac{1}{15}-\frac{1}{30}$ s (reflection holograms usually demand exposures of a second or more). But beware of casual photographs without a black cloth: you will find your own self-portrait reflected in every hologram. If a transmission hologram is hung in the middle of the room you will need to take particular care not to record unwanted objects or people behind it; you will find it an advantage to co-opt the services of a friend to hold a second black cloth a meter or so behind the hologram. Flash exposures by the method described above for reflection holograms are possible, but you will need a flash extension lead at least 3 m long, and a friend to hold the flash for you. If you have been commissioned to take photographs for an exhibition catalog, or for publicity for the artist, or if, indeed, the holograms in question are your own, the only way to be totally fair to the artists concerned, as well as to yourself, is to photograph transmission holograms in your own studio, where you can take your time, and experiment with distances and angles until you are satisfied that you have caught the best possible image.

Viewpoint and Parallax

Although they operate on different principles, a photograph and a hologram have much in common, visually speaking. The important difference, of course, is the three-dimensional nature of the holographic image, and a single photograph cannot show this. Although holographic reproductions have already begun to appear in books, there is as yet no way of printing them directly onto the page. A further problem is that whereas slides do make a fairly good job of reproducing the saturated colors of multicolor holograms, the available printers' inks fall far short of perfection in this respect; good as the color plates in this book may be, they do not stand comparison with the transparencies from which they were made, which are themselves one important remove from the original holographic images.

One partial way round the problem of three-dimensional imagery is to make photographs in stereoscopic pairs. When these are examined through a stereoscopic viewer, or projected through polarizing filters onto a reflective screen and viewed through polarizing spectacles, the impression of depth can be very strong. If you have a large portfolio of holograms which you are unwilling to carry around to potential clients, or if you want to send sample transparencies to the organizers of a forthcoming exhibition of holography, a set of stereo pairs are the next best thing. It is therefore a good idea to take stereoscopic pairs of any hologram you photograph. All that is necessary is to move the camera horizontally through a distance of about 60–70 mm ($2\frac{1}{2}$–$2\frac{3}{4}$ in) between exposures; not more, or the perspective will seem unnatural. It is possible to obtain jigs in which the camera is slid between two endstops, between the exposures; it is not difficult to build such a jig yourself. There are also specialist companies which make stereoscopic cameras, usually by modifying two identical 35 mm cameras into Siamese twins. These, as you might expect, are very expensive.

One difficulty about making stereoscopic pairs is that you may find it necessary to toe in the two camera axes to face the center of the hologram, resulting in 'keystoning' (the rectangular shapes of the plates appear as trapezia). The resulting distortion can cause problems with stereoscopic fusion of the two images. Often, too, in single-shot photography, the best viewpoint may be well to one side of the hologram, or above or below it. This is where the studio camera with its lens and back movements comes into its own. It does not matter if the viewpoint is skewed: as long as the lens panel and the camera back are parallel to the surface of the hologram its shape will be a rectangle on the film. A 35 mm camera with a perspective-control ('shift') lens will achieve the same thing. Unfortunately, shift lenses are of unacceptably short focal length, unless you use them in combination with a $2\times$ teleconverter lens, which will give you an effective focal length of about 70 mm.

Unusual Holograms

To display the qualities of animated or trick holograms which when viewed from different angles show several images popping in and out, you will need to take as many shots as will cover all the variants. Holographic stereograms need a minimum of three exposures. Transmission holograms of abstract material may need up to fifteen exposures, three rows of five at different heights, in order to do any sort of justice to the originals. Mixed-media pieces demand very careful positioning and sometimes separate lighting for the hologram and the other material (see Plate 30).

Occasionally the frame of the hologram is an integral part of the overall composition, and you may find that in such cases the hologram is simply not bright enough to record well without the border material being over-exposed. In such cases you simply have to take two exposures, one for the hologram and one for the border. If the result is intended for reproduction, the printer will be able to make a montage of the two. Whenever you take a photograph of a mixed-media piece, always take a second shot from the same viewpoint, showing the hologram alone. If the photograph is to appear in print this may be the only way to do the artist justice.

Presenting Slides

Every lecture seems to have at least one slide the wrong way round, or upside down. Those who are familiar with the operation of Murphy's law point out that there are eight possible ways of loading a slide into a slide tray, and seven of them are wrong; perhaps it is surprising in view of the odds that so many slides do get shown the right way up. In fact it should not be difficult to get things right every time. With all standard projectors the slide has to go into the projector gate upside down but right way round, if you are looking through the slide at the screen, and it should be marked so that this can be done without reference to the material to be projected. The international standard for marking slides is a spot in the bottom left corner of the mount when the slide is viewed correctly on a light table. When projecting, the spot goes at the top right of the side away from the screen. Without such a spot, Murphy's dice are heavily loaded against you.

Copyright

All works of art are automatically copyright, and if you make a replica without permission (except in some very narrowly-defined circumstances) you are breaking the law. This certainly applies to making a holographic copy of someone else's hologram. Taking a photograph of almost any copyright material is likely to be illegal, in the strictest sense of the word, though many people do so, for private or educational purposes. It is when they do so for profit that they are likely to be in legal trouble. If you photograph somebody's hologram at a private exhibition without permission, you are likely to be unpopular, at least with the artist, and most exhibitions do not allow the photography of exhibits without such permission. The proper procedure is to contact the artist first, and make an arrangement that is satisfactory to both of you. As a photographer you are probably aware that the copyright of a photograph is not even necessarily yours just because you took it; it belongs to the person who commissioned the photography (whether or not payment entered into the contract) unless the contract states otherwise. So if you do enter into an agreement, make sure the ownership of the copyright is clarified. This may seem unnecessarily formal, but most agreements between holographers and photographers concern publicity material, and may involve reproduction fees, etc. On the other hand if *you* are the holographer (and if you are reading this book you probably are!), why not learn to take photographs, and be your own publicity photographer, and protect your work at the same time? After all, you are the best person to judge whether any given photograph does justice to your work. What is more, photography, as a parallel art form with much in common with holography, may give you fresh insights into holography itself.

PART
III
HOLOGRAPHY EARNS ITS KEEP

CHAPTER

23

---※---

Holography in Measurement and Displacement Analysis

'While you're refreshing yourself,' said the Queen, 'I'll just take the measurements.' And she took a ribbon out of her pocket, marked in inches, and began measuring the ground and sticking little pegs in here and there.
     LEWIS CARROLL, *Through the Looking Glass*

The potential of holography in measurement science was appreciated early in its history. In photography a discipline known as *photogrammetry* had existed since the First World War; this was a method of making accurate maps from aerial photographs, and an important technique that developed from it was the measurement of distances out of the object plane (ie heights and depths) by the comparison of overlapping adjacent pairs of photographs using a stereoscope. The techniques of photogrammetry soon became applied to the measurement of other things, including the human body. The opportunity provided by holography for direct rather than indirect measurement on a three-dimensional image seemed almost too good to be true, though the techniques of *hologrammetry* have proved to be more difficult than was at first anticipated. However, a technique known as *holographic interferometry* provides a method for very precise measurement of small displacements, and it is to this technique that the greater part of this chapter is devoted. A different approach, using laser light but not holography, is known as *speckle interferometry*; although outside the scope of this book as a technique, it complements holography in many ways, and a description of the principle is included.

Conventional Measuring Techniques

Nils Abramson (1) describes a laser transmission hologram as a 'window with a memory', and points out that all the measurements that could be made through an ordinary window could also be made through a holographic plate of the same size. Provided the glass plate is of high optical quality, and there has been no lateral displacement of any part of the emulsion during or after processing, and the reconstruction geometry is identical with the exposing geometry, the precision of measurement that is possible is equal to the diffraction-limited resolution, which is the same as the *objective speckle size*. Unfortunately, the inherent resolution cannot be realized with the virtual image, owing to the small size of the pupil of the eye, which renders the *subjective speckle size* much larger. However, reversing the hologram in the (collimated) reconstruction beam solves this problem. A photographic emulsion placed in the image space, exposed and processed, will record in sharp focus everything in the plane of the emulsion that can be seen from any viewpoint in the virtual-image reconstruction. Tozer and Webster (2) show that with a 1 × 1 m hologram a photographic emulsion 1 m from the hologram plate can record an image with a resolution of 1 μm; for an 8 × 10 in plate the resolution would be 6–10 μm. It is important that the lasers used for both taking and reconstruction should be absolutely stable and of exactly the same wavelength; the latter requirement raises difficulties when a pulse ruby laser is used for making the hologram, as the (CW) reconstruction laser has to be tuned to precisely 694 nm.

Such applications as microscopy and the analysis of aerosol particles also belong to this general category, but they are more appropriate to the material of Chapter 24, and are described there.

Holographic Interferometry

This book began by defining a hologram as the record of the interaction of two beams of coherent light which are mutually correlated, in the form of a microscopic pattern of interference fringes. In this and subsequent chapters these fringes will be referred to as *primary fringes*. However, much larger fringe patterns, known as *secondary fringes*, appear on a hologram whenever the object or the plate has been moved by a small distance while being exposed. They are a moiré pattern generated by two sets of primary fringes that have been displaced relative to one another. The effect was first documented by Powell and Stetson in 1965 (3), who showed the value of secondary fringes in the analysis of vibration. Since then, holographic interferometry (sometimes shortened to *holometry*) has become the most important technique in applied holography, being used in stress and vibration analysis, quality control and non-destructive testing (NDT) and as an investigative tool in the preservation and restoration of works of art.

Real-time interferometry If you make a laser transmission hologram of an object, and leave the object in position while you process the plate, you can return the processed hologram to its holder and, provided nothing in the system has moved, the object will coincide precisely with its own image. But if you squeeze the object between your thumb and forefinger, or warm it with a hairdryer, a pattern of secondary fringes sweeps across the object as you view it through the hologram. The more the object is distorted, the closer together are these fringes. Because they are being produced in real time they are known as *live fringe*s; what is happening is that the wavefront from the object itself is interfering with the wavefront of the object re-created by the hologram. The reconstructed wavefront shows the object as it was originally, and the live-object wavefront shows the object as it is now, after distortion. Each whole secondary fringe represents one wavelength of distortion, in a direction approximately along the line of sight. Thus holographic interferometry can be an extraordinarily sensitive method of visualizing small amounts of movement (Fig 23.1).

FIG 23.1 *An example of live fringes. If the object remains in situ while the hologram is displayed, any distortion results in the generation of secondary fringes which contour the distortion. In this case, the bolts were tightened after the exposure. Hologram by P A Storey. Photograph courtesy of Rolls Royce plc, Derby.*

Interferometry using partially-coherent light has long been used in checking the accuracy of lenses and optical mirrors, but it works only with optically-polished surfaces. Holographic interferometry works equally well with rough surfaces, which may be up to a meter or more in size, and situated well away from the holographic recording equipment. In practice it is quite difficult to ensure that the emulsion does not become distorted in processing. The amount of force exerted on the base by the drying emulsion is considerable and can result in the distortion of even a glass plate. One method used to overcome the problem, as well as the problem of repositioning the plate precisely in its holder after processing, is the *liquid gate*. The plate is held in a cell formed by a pair of optical flats, which is initially filled with de-ionized water. The hologram is exposed and processed in situ. Abramson (4) suggests suspending a plate-holder by a clamp, and bringing a processing tank up from beneath, thus allowing a dry-to-dry system without removing the plate.

Usually, a few fringes will be seen even when the object has not been disturbed, and these can often be brought down to just one fringe by a very small amount of repositioning of the plate. This single broad dark fringe is sometimes called a 'fluffed-out fringe field'. If there are no fringes visible at all, it is likely that the object or plate has a very large repositioning error. In such cases you may be able to see a double outline. As you adjust the outlines and they begin to come into register, the fringes will appear, and you can then begin to minimize them. Fringes that are caused by translational movement obscure those that are formed by deformation; the former are parallel lines, and once you have got rid of these you can see the deformation pattern. The principle of this, as applied to *double-exposure holographic interferometry*, is dealt with later under the heading of *sandwich holography*.

Live fringes are generally used in practice as a run-up to a more permanent form of holographic interferometry such as those described below. By making a few test loads on a component you will be able to see what are the loads that will result in fringes that are easy to count and evaluate. For this it is unnecessary to eliminate all fringes at the outset. On loading the component, each new fringe represents a movement of one wavelength along a line bisecting the object and reference beams (to a first approximation, the out-of-plane direction).

At one time, this method showed considerable promise for quality control of very accurately-machined components: all that was necessary was to make a hologram of the pattern, then to introduce production components into the image space, one at a time. However, methods based on *speckle interferometry* (see pp. 327–9) have superseded holography in this respect, and a speckle camera is now available (5).

Double-exposure interferometry   To obtain a more permanent record of inter-

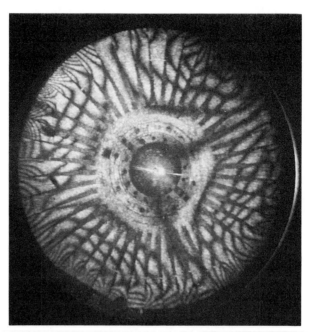

FIG 23.2   *Double-pulse focused-image shear interferogram of a fan rotating at high speed. An optical derotator (see text) eliminates translational fringes. Hologram by P A Storey. Photograph courtesy of Rolls Royce plc. Derby.*

ference patterns produced in this manner, the method of double-exposure interferometry is used. With this technique the holographic plate is not processed immediately after the primary exposure but remains in its holder while the component in question is subjected to loading, temperature change, etc. A second holographic exposure is then made. A permanent holographic interferogram results; it can be used for making measurements or kept for record purposes. This is known as a *frozen-fringe technique*.

Holographic shear interferometry Since the advent of the Q-switched laser, with a typical pulse duration of 25 ns, it has been possible to make holograms of fast-moving objects. This can be extended to rapidly-rotating components such as turbofans, using a technique known as *holographic shear interferometry*, though it is necessary to rotate the holographic plate in synchrony with the component or, more simply, to use an *optical de-rotator* (Fig 23.2 and Box). Holographic shear interferometry is useful in the analysis of irregularly-vibrating components which are in a condition of rapid translation.

Principle of derotation

A pulse laser with a typical pulse duration of 25 ns can produce a good-quality hologram of an object that is moving quite fast (about 4 m/s). Picosecond lasers exist that can easily stop a bullet. But when an attempt is made to produce a double-

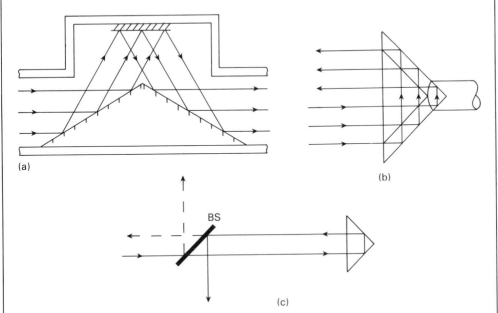

FIG 23.3 *Inverting prisms for derotation. (a) Mirror version of Abbé prism, with entire optical path in air, mounted in hollow shaft. (b) Roof prism mounted on end of solid shaft. (c) Shows losses occurring at the beamsplitter BS in a roof-prism system.*

pulse interferogram of a rapidly moving object, the translational displacement is too great for any useful information about vibration to be derived from the hologram.

As many engineering designs are concerned with the vibrations of fan, compressor, turbine etc blades it is necessary to find some method of stabilizing the fringes on the hologram plate. The obvious solution would be to rotate the plate in synchrony with the fan but this is usually out of the question, for apart from the mechanical problems the reference beam would have to rotate too. A practical solution is to use an optical derotator. This is a prismatic device which inverts the image by multiple reflection; the two devices most often used are the *Abbé prism* and the *roof prism* (Fig 23.3).

When the prism is rotated the image also rotates, in the same direction and twice as fast. If such a prism is placed in the optical path of an image-forming device such as a camera lens, and rotated in the direction opposite to that of the fan etc, but at half its rotational speed, the image remains stationary. The Abbé-prism system does not need a beamsplitter, and thus wastes less light than the roof-prism system, but it needs a cage mounting, whereas the roof prism can be mounted on a simple shaft. For lightness, either type of prism can be made up of front-surface mirrors rather than total-internal-reflection surfaces.

Time-averaged interferometry One of the simpler forms of holographic interferogram is of an object which is translationally stationary, but which is vibrating in a stable manner. This technique, which can often be carried out using a very simple Denisyuk set-up, is known as *time-averaged interferometry*. Although a vibrating object seems to be in continuous motion, it is instantaneously stationary at the two extremes of its movement. You can see this if you clamp a metal rule or a hacksaw blade in a vise and look at it from the side while you twang it with your finger. You can clearly see the extremes of movement, which look like two separate blades; the rest is blurred out. Although the time spent at the extremes is infinitesimal, the time spent in the immediate neighborhood of the extremes, say 0.1 of a wavelength, is a substantial proportion of the cycle. So if you want to investigate the modes of break-up of a loudspeaker cone, or the vibration pattern of a guitar belly or even a car silencer, all you need to do is to set it into steady vibration with a suitable transducer and make a straightforward Denisyuk reflection hologram of it. Let us look at the way this would work for a small loudspeaker cone (Fig 23.4).

Successive fringes denote an out-of-plane movement equivalent to one-quarter of a wavelength for each fringe, not a whole wavelength as you might guess. This is because (a) you are dealing with a reflection hologram and the path difference is folded and (b) the fringes are measuring a peak-to-peak excursion, not a zero-to-peak excursion: each of these factors halves the distance. (In practice, an excursion of only a few wavelengths produces quite a loud sound. It is necessary to increase the input as the pitch rises: at 9 kHz you need a powerful squeal to get good fringes.) One of the snags of time-averaged interferometry is that the contrast of the fringes falls off rapidly as the fringe order gets higher; this does not occur with other types of holographic interferometry. However, this is one of the simplest techniques;

FIG 23.4 *Time-average holographic interferograms of a small loudspeaker, made on the single-beam frame described on pp. 96–9. The antinodes (points of maximum excursion) of the cone are contoured by fringes. Holograms by Jane Dunnicliffe.*

it is used successfully in the design and testing of many manufactured objects that undergo steady vibrations – not necessarily musical ones.

When used for measurement purposes it should be noted that adjacent fringes represent a displacement of slightly *more* than one-quarter of a wavelength. This is because the fringes are recorded in terms of the *average* distance over which the vibrating component dwells longest; this is slightly less than the total excursion (Fig 23.5).

Strobed interferometry One of the limitations of time-averaged holographic interferometry is that it records only a comparison between the two extreme positions of the vibrating object. It is unsuitable for unstable vibrations and does not show the relative phases of the various modes of vibration. It would be helpful to

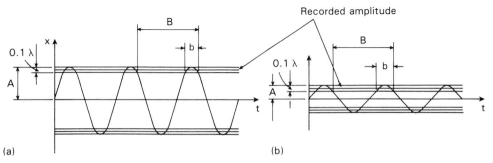

FIG 23.5 *Time-averaged interferometry.* (a) *Wide-excursion;* (b) *narrow excursion. The apparent position of the fringe is at the mid-point of the* 0.1 λ *width at the ends of the excursion* A *and is slightly less than* A. *The relative length of time dwelt within the* 0.1 λ *width* (b/B) *is a function of* A, *and is greater for lower amplitudes.*

be able to compare the extreme positions of the vibrating surface with the position of the surface at rest, particularly if the vibration may possibly not be symmetrical with respect to this position. It would be better still to be able to compare *any* position with the rest position. This can be achieved using a frozen-fringe technique with a double-pulse laser: any two positions can be compared by varying the interval and the timing of the impulses, or an exposure can be made with the surface stationary followed by a second one while it is vibrating, at a point triggered by the voltage output of the transducer. A CW laser can be used to produce live fringes by coupling it to a stroboscopic device or *chopper* based on a Pockels cell and having the same frequency as the vibration; it is possible, though tedious and not altogether reliable, to make frozen-fringe stroboscopic holograms by this method. In general, a pulse laser is more satisfactory for this, and the process lends itself well to the powerful techniques of sandwich holography (see pp. 325–7).

Visualization of fluid flows by holographic interferometry In aerodynamic and ballistic research, an important requirement is the visualization and analysis of shockwaves and vortex phenomena. For some purposes a shadowgram made with a high-intensity spark suffices, but the method is not very sensitive and does not lend itself to mathematical analysis. *Schlieren photography* involves a beam of collimated light that is passed through a chamber and focused on a knife edge before being expanded again to be recorded on film. Any density gradient in the chamber implies a gradient in the refractive index, so any such region causes light passing through it to be refracted upwards or downwards, respectively clearing or fouling the knife-edge and resulting in a bright or dark line in the image. Strips of color filter may be substituted for the knife edge. There are two difficulties associated with

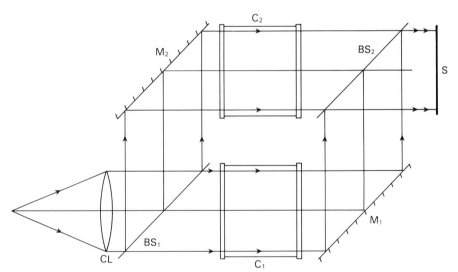

FIG 23.6 *Principle of Mach-Zehnder interferometer. The optical paths via the chambers C_1 (undisturbed) and C_2 (containing test material) produce fringes which are examined at S. In the holographic version there is a single route; exposures are made first without, then with, the test material or event.*

schlieren photography: first, analysis can be carried out only in a direction perpendicular to the knife blade; secondly, what is recorded is a density gradient, not the density itself, so that mathematical analysis, though possible, is still difficult. Holographic double-exposure interferometry solves both problems. The principle used is that of the *Mach-Zehnder interferometer* (Fig 23.6), but the holographic version does not require large high-precision optical elements.

Using a pulse laser, a holographic exposure is made of the chamber alone, filled with air or some other gas as appropriate, followed by a second exposure recording the event itself. Any variation in density of the gas in the chamber will alter the optical path length through it, causing interference between the two holographic images. The result not only shows internal detail which does not appear in schlieren photographs, but shows it in three dimensions. Real-time techniques are also possible: thermal convection can be observed directly, and phenomena such as vortex streets can be examined in stills or by means of cinematography or video recording (Fig 23.7).

Interferograms in difficult conditions If you have a table with full vibration isolation and heavy optical equipment, it is possible to make interferograms with long exposures, but sometimes conditions are less suitable than they might be. In particular, if the air within a one-meter space warms up by as little as $1\,^\circ C$ there will be a difference in optical path length equivalent to about two wavelengths. Abramson (6) recommends a single-beam transmission set-up, with the geometry calculated on the basis of his *holodiagram* (see Appendix 4), ensuring that the reference beam passes through almost the same volume of air as the object beam (Fig 23.8).

FIG 23.7 *Visualization of fluid flow patterns in a labyrinth seal. Hologram by R J Parker. Photograph courtesy of Rolls Royce plc, Derby.*

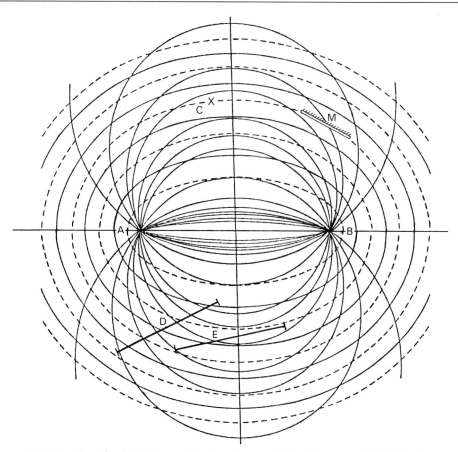

FIG 23.8 *Use of a holodiagram for deep images (after Abramson (6)). A is the reference source and B the holographic plate. The reference mirror is at M. The ellipsoids represent the loci of equal distances from A and B, in steps of 10 cm. An object lying along the ellipsoid marked by the cross C will be within the coherence length (say 20 cm) of the laser. Object D lies across a range of six ellipsoids (60 cm) and will not be imaged at its extremes; a small shift to E brings it within the coherence length.*

Hans Bjelkhagen (7) recommends a somewhat similar arrangement for in-situ holography, and details several set-ups.

NOTE: While Bjelkhagen's paper is sound, his diagrams show mirrors and plates at impossible angles, and if you want to use his set-ups you will need to arrange beams to fall on the objects and plates at much smaller angles of incidence than are indicated in his diagrams.

When working with a reference beam produced by a mirror close to the object, the angle between the beams is small and the primary fringes are large. In extreme cases of instability the mirror can be fixed to the object. As such a mirror will be so close to the object itself, the spurious real image can overlap the genuine virtual image, where the object is large, and a way of avoiding this is to aim the mirror off, and divert the beam to the hologram by a second mirror (Fig 23.9).

David Rowley (8, 9) has modified a 35 mm camera to produce focused-image holograms. Owing to the great reduction in scale the images are to all intents two-dimensional. By placing a narrow-band filter on the camera lens the system can be used under normal lighting conditions for double-exposure interferometry, using a 5 mW HeNe laser and an ordinary photographic tripod (Figs 23.10 and 23.11).

Sandwich holography In ordinary double-exposure interferograms there is no way of telling whether the out-of-plane movement was towards or away from the viewer, because the plate does not remember which of the two exposures was made first. In live-fringe holography it is easy to find out simply by pushing the object with a finger and seeing whether the number of fringes increases or decreases, but this works only as long as the holographic set-up remains unchanged.

A technique known as sandwich holography provides a solution and gives added versatility to the double-exposure set-up. The principle of sandwich holography is that the two exposures are made on separate plates, which are then combined in an accurately-adjustable plate-holder. The system permits the exposure of many plates with the object under different loading conditions; when processed, the plates can be viewed in any combinations of pairs. One important advantage is that any rigid-body motions of the object between exposures can be compensated for, so that only the fringes caused by deformations remain. However, the method is complicated by the necessity to reposition the plates with high accuracy, and it is necessary to have a plate-holder that can do this. The most effective plate-holder for

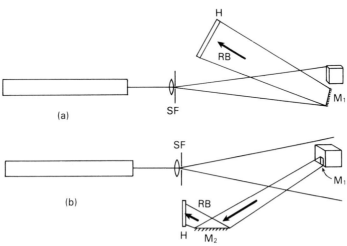

FIG 23.9 *Single-beam geometry for objects in difficult conditions. (a) The mirror M_1 is close to the object, on the same base. (b) A modification to place spurious images well out of the way of the genuine image. Mirror M_1 is fixed to the object directing the beam onto the relay mirror M_2 to increase the angle between the object and reference beams (after Abramson (6)).*

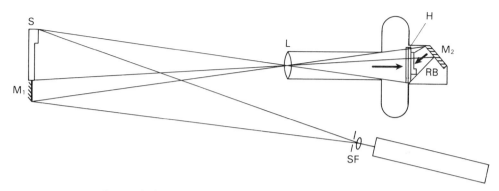

FIG 23.10   35 mm *holographic camera (after Rowley (8, 9)). The optical image of the specimen* S *is focused close to the hologram plane* H *by the camera lens* L. *This also focuses the reference beam which passes through the edge of the frame and is reflected from the mirror* M_2 *in an extension of the camera back to fall on the rear side of the film. A more recent version of the camera (FIG 23.11) uses fiber optics to lead the reference beam to the film from the front.*

this purpose is one that holds the plate in position by gravity alone, with a slight tilt back and to one side, with pins or ball-bearings as supports (Fig 23.12).

As there are always two plates in a sandwich hologram there must be two plates for the exposure. In all cases the emulsion faces the object. The first plate from the second exposure should be placed in front of the back plate of the first exposure; Abramson (10) recommends using two unexposed plates for this rather than one unexposed plate and a plain glass plate, as there will then be the possibility of combining any pair of holographic records. In case the two plates from a single exposure

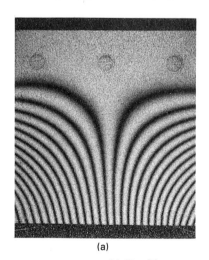
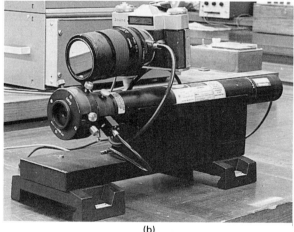

(a)   (b)

FIG 23.11   (a) *Double-exposure holographic interferometry (cantilever flat plate) with* (b) *a modified amateur* 35 mm *camera. Photographs by Ken Topley. Courtesy David Rowley, Department of Mechanical Engineering, Loughborough University of Technology.*

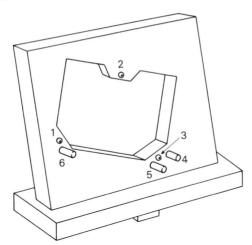

FIG 23.12 *Gravity plate-holder (after Abramson (6)). 1, 2, and 3 are ball-bearings; 4, 5, and 6 are pins; the plate is thus fixed in all six degrees of freedom.*

are mixed up in processing, the correct positions can be found by simply remounting them: if they are in the right order there should be at most one or two fringes, but if they are in the wrong order a set of concentric circular fringes will appear. Any sandwich combination can be bonded with cyanoacrylate ('super-glue') using a few tiny drops on two opposite corners and pressing the center of the front plate gently onto the rear one in the plate-holder for a few seconds. Evaluation must be by a beam of the same divergence as the reference beam; ie the hologram must be at the same distance from the spatial filter.

It should be possible to tilt a sandwich hologram in such a way that no fringes are visible on the reference surface (a piece of metal included in the object space that does not suffer any distortion). In order to discover whether the specimen has been bent towards or away from the hologram at a particular point it is necessary only to tilt the sandwich until the fringes disappear at that point. The sandwich plane will then be tilted in the same direction as the object plane, but with a magnification of about 2000 times the tilt, making it easy to see. With a little practice this method can be used hand-held to scan the surface for more complex distortions.

Evaluation of fringe patterns for both out-of-plane and in-plane motion is a much less tricky business with sandwich holography than with double-exposure or time-averaged holography. A very complete account of the technique appears in Abramson (10).

Speckle and speckle interferometry

Whenever a diffusely-scattering surface is illuminated by a laser the surface appears to be covered in a random pattern of bright and dark speckles. This is called laser speckle, and occurs because neighboring microscopic elements making up the surface produce optical paths that are randomly different and may differ in length by several wavelengths. At any point, therefore, diffracted waves are arriving from many of these elements simultaneously, and as they are all highly correlated their instantaneous amplitudes add algebraically. However, as the phases are random they may provide at any point either constructive or destructive interference overall. At points where the aggregate amplitude is non-zero there will be a bright speckle; where it is zero there will be a dark speckle. In the original or objective speckle all spatial frequencies are present, but when the illuminated surface is imaged by a camera or on a video screen (or, for that matter, by a hologram), the size of the speckles is determined by the f/no of the imaging device; it is then known as subjective speckle, and its size is larger for smaller apertures, and vice versa (Fig 23.13).

If an object illuminated by laser light is photographed, the speckles will appear on the negative. If a double exposure is made, and the object is distorted in some manner between the two exposures, the speckles will be displaced in proportion to the in-plane distortion, and contour fringes will appear in the photograph. The resolution is not as high as in holography, but the process is much simpler and does not need special emulsions. By defocusing the object (usually by setting the lens to infinity so that the emulsion is in the Fourier-transform plane), fringes due to out-of-plane motion can be observed, because the speckles move when the object is spun, ie rotated out of plane. This can be appreciated from the fact that if you focus your eyes behind a laser-illuminated surface and move our head sideways the speckles move in the same direction, whereas if you focus your eyes in front of the surface and move your head then they move in the opposite direction.

Speckle photography has been largely replaced by *electronic speckle pattern interferometry* (*ESPI*). This is a technique which permits the speckle-pattern correlation fringes to be seen on a monitor screen. In double-exposure work the first image is stored and the second subtracted from it; thus the areas where the speckle pattern remains stationary (correlated) will give a zero signal (dark) while uncorrelated areas will give a signal that may be positive or negative. The signal is processed by rectifying so that both the negative- and positive-going signals show as bright fringes, and high-pass filtering removes low-frequency noise and improves the

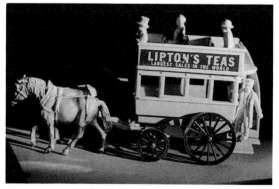

FIG 23.13 *An object illuminated by laser light and photographed at* (a) f/2.8, (b) f/16. *Notice the increase in speckle size as the aperture size is decreased. Due to the reproduction process the speckle is less obtrusive than in the original photographs.*

clarity of the fringes. In time-averaged and double-pulse speckle interferometry, the comparatively slow fading of the camera tube image (typically about 0.1 s) necessitates addition rather than subtraction of the signals. In this case areas of maximum correlation have maximum contrast and areas of minimum correlation have minimum contrast. However, the average intensity is constant. DC filtering and rectifying produces an image in which the fringe maxima correspond with the fringe minima that would have been obtained by subtraction methods, and vice versa. The visibility of the fringes is not as good as with the subtraction method, but a storage tube is not required. ESPI is a rapidly expanding technique which has been extended to the solution of many complex problems concerning displacement and the measurement of shape, and its techniques are inherently much simpler than either speckle photography or holography. The standard work is by Jones and Wykes (11); some other recent important papers are listed at the end of the chapter (12, 13).

Holographic Contouring with Two Wavelengths

One of the difficulties associated with holographic interferometry is its extreme sensitivity; many of the movements we should like to investigate are so large that the fringes would be too close together to be resolved. In particular, this is a problem with time-averaged interferograms, where the contrast of the fringes falls off rapidly as they become closer together. The techniques used in speckle interferometry (see Box above) overcome this problem, as the fringe contour interval is determined by the geometry of the system; this is why speckle techniques have in some areas superseded holography. However, there are techniques which can extend the scope of holography by increasing the contour interval: these depend on the use of two wavelengths rather than a single wavelength. The most important use of the

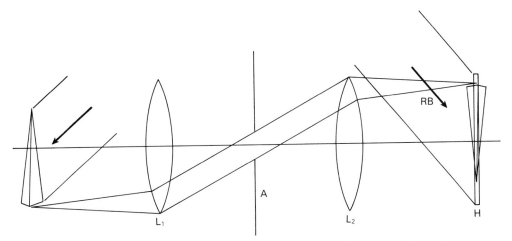

FIG 23.14 *Optical system for two-wavelength holographic contouring (after Zelenka and Varner (14)). A is a circular aperture and the object illumination and reference beams make equal and opposite angles.*

holographic contouring technique does not measure distortion, but the contours of a static object. The system uses an image of the object focused by a *telecentric lens system*, ie a symmetrical system with the object at the front principal focus of the first lens and the image at the rear principal focus of the second lens, the lenses being spaced two focal lengths apart (Fig 23.14). The system is described in detail in Zelenka and Varner (14).

The illumination is collimated. A first exposure is made using one wavelength, then a second exposure using a slightly different wavelength. After processing, the hologram is illuminated with just one of the two wavelengths. Because of the confocal set-up there is no in-plane (lateral) displacement, but there is out-of-plane (axial) displacement of one image relative to the other, and contour fringes will be seen, the contour interval being greater the closer the wavelengths are; for example, using an Ar^+ laser at 51 and 477 nm the interval is 10 μm. With a dye laser giving a continuous range of wavelengths the contour interval can be chosen for any value required.

If only one wavelength is available, the object can be immersed in liquids of slightly differing refractive index (such as sugar solutions of different concentra-

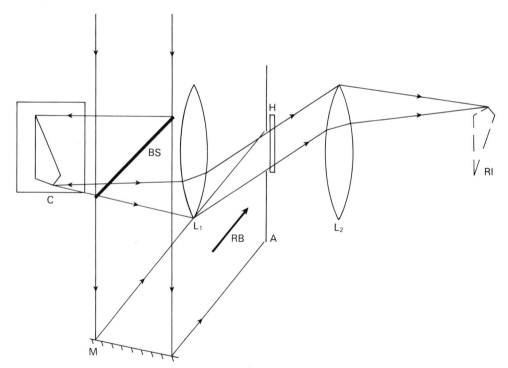

FIG 23.15 *Two-refractive-index contouring systems. C is a cell containing liquid surrounding the object; after the first exposure the liquid is replaced by another liquid of differing refractive index before a second exposure. Note that the hologram is situated at the aperture (A) in this configuration. Reconstruction is by the same beam as made the hologram in either one of the liquids, forming a real image which shows the contours (RI) (after Zelenka and Varner (15)).*

tions) for the two successive exposures (Fig 23.15). This technique is also due to Zelenka and Varner (15); both procedures are discussed by Hariharan (16).

Summary of Applications

This summary is not intended to be exhaustive. In some cases there are rival non-holographic techniques: speckle techniques and schlieren photography have already been mentioned, and several photographic methods exist for contouring. However, in many cases the holographic and non-holographic methods are complementary, each having its own special merits.

1 Direct measurement
1. Measurement of deformation: Comparison of measurements of components over a long period, eg measurements designed to detect for creep and fatigue on the servicing of components subjected to prolonged heat and/or stress, such as parts of a nuclear reactor, or re-usable components of spacecraft.
2. Measurement of components in a hostile environment: For example, in-situ examination of fuel pins in a nuclear reactor (see above) or underwater, as in offshore oil rigs.
3. Medical data in three dimensions: For example dental casts, records of computerized tomography scans (see pp. 350–1).
4. Microscopy: Usually for living objects such as bacteria (see pp. 343–4).
5. Particle counting: Either by off-axis holography (eg for fuel-injection systems) or by in-line holography (for very small particles) (see pp. 343–5).
6. Contouring: Measuring the accuracy of, for example, shallow stampings or optical components.

2 Interferometric measurements
1. Real-time, or live-fringe: Pre-testing of components before using frozen-fringe methods; quality control of components produced within very narrow tolerances where speckle methods are inappropriate.
2. Double-exposure (CW): Measurement of distortion in stressed components. Can be used, with some difficulty, in longitudinal studies of creep and wear on repeatedly-stressed components such as springs and tappets. Routine non-destructive testing on remold tires. Investigations concerning the preservation and restoration of works of art.
3. Time-averaged: Investigation of the modes of vibration of objects such as loudspeakers, turbine blades, car silencers and musical instruments.
4. Double-pulse: A subset of double-exposure interferometry appropriate to moving objects, with or without image motion compensation via an optical device such as a derotator (see pp. 319–20). Used for measurements on moving fan blades, human muscular action, fluid flow patterns.
5. Stroboscopic: For deeper investigations into areas covered by 3 and 4 above. Can be real-time.
6. Sandwich: Most useful where there is enough translational movement in a double-exposure hologram to mask small distortions, eg in large machines where several components have large distortions in different directions. The technique allows each part to be examined individually.

References

1 Abramson N (1981) *The Making and Evaluation of Holograms*, p 70. Academic Press.
2 Tozer B A and Webster J M (1981) Holography as a measuring tool. *Central Electricity Generating Board (CEGB) Research*, Jan 1981, 3–11.
3 Powell R L and Stetson K A (1965) Interferometric vibration analysis by wavefront reconstruction. *Journal of the Optical Society of America*, **55**, 1543–8.
4 Abramson N (1981) *The Making and Evaluation of Holograms*, p 77. Academic Press.
5 'Vidispec' electro-optics test equipment marketed by Ealing Electro-optics Inc. The HP4000 ESPI system manufactured by Newport Research Corp.
6 Abramson N (1981) *The Making and Evaluation of Holograms*, pp 225–6. Academic Press.
7 Bjelkhagen H (1977) Experiences with large-scale reflection and transmission holograms. *Proceedings of the SPIE*, **120**, 122–6
8 Rowley D M (1980) A 35 mm holographic camera. *Journal of Photographic Science*, **28**, 198–202.
9 Rowley D M (1983) The use of a fibre-optic reference beam in a focused-image holographic interferometer. *Optics and Laser Technology*, Aug 1983, 194–8.
10 Abramson N (1981) *The Making and Evaluation of Holograms*, pp 206–67. Academic Press.
11 Jones R and Wykes C (1983) *Holographic and Speckle Interferometry*, pp 165–97. Cambridge University Press.
12 Accardo G, De Santis P, Gori F, Guattari G and Webster J M (1985) The use of speckle interferometry in the study of large works of art. *Journal of Photographic Science*, **33**, 174–82.
13 Pickering C J D and Halliwell N A (1985) Laser speckle photography: preprocessing of fringe pattern data to improve dynamic range. *Journal of Photographic Science*, **33**, 183–6.
14 Zelenka J S and Varner J R (1968) New method for generating depth contours holographically. *Applied Optics*, **7**, 2107–10.
15 Zelenka J S and Varner J R (1969) Multiple-index holographic contouring. *Applied Optics*, **8**, 1431–4.
16 Hariharan P (1984) *Optical Holography*, pp 246–51. Cambridge University Press.

CHAPTER
24
Data Storage, Processing and Retrieval

'I'm afraid I am, Sir,' said Alice. 'I can't remember things as I used – and I don't keep the same size for ten minutes together!'
'Can't remember *what* things?' said the Caterpillar.
'Well, I've tried to say "How Doth the Little Busy Bee," but it all came different!'
LEWIS CARROLL, *Alice in Wonderland*

Judging by the number of papers published on the subject, a first impression might well be that the storage, processing and retrieval of data was the most important area in holography. A study of the papers themselves, however, revealing as it does a great deal of repetition, and often a paucity of content disguised under nigh-impenetrable thickets of mathematics, might raise the suspicion that the papers were being generated for their own sakes rather than as pieces of genuine research. Both views are to some extent justified. This chapter is an attempt to cut through the arcane language and describe what is going on. For the reader who wishes to pursue the subject in greater depth there is a list of the more important papers at the end of the chapter.

Data Storage

To consider data storage as a specialized aspect of holography might seem to be a redundant exercise. After all, data storage is the whole purpose of holography, just as it is with photography, or even typewriting. But holography has some rather special aspects. For a start, its storage capacity is very high indeed: an ordinary holographic emulsion can store roughly ten times the information that can be put onto a photographic microfiche. Secondly, although in a holographic 'microfiche' the information is to some extent localized, in that a page of information can be displayed by directing an unspread laser beam at a chosen area, within that area the information is *not* localized, so that a small scratch or dust speck, which would wipe out a large amount of information from a photograph, has little effect.

There has long been talk of information storage in three-dimensional holographic formats, and in theory it should be possible to store and retrieve vast quantities of information in light-sensitive crystals such as lithium niobate, but this does not seem to have had the success predicted for it (Burke *et al*, (1)).

Data Processing

In photography, data processing consists mainly of such techniques as contrast enhancement, tone separation and enhancement of fine detail by unsharp-masking techniques. These, and more sophisticated processing techniques, can be carried out much more easily if the input signal is one-dimensional, ie it comes in the form of an electronic signal that is used to build up a picture sequentially, as in a television picture. It is the processing of such pictures that has led to the splendid images of objects in space that we have had over the past few years. Such processing is known as *linear* or *sequential processing*, and it is the conventional way in which a computer works – one instruction at a time. The most recent generation of computers is being designed to perform a large number of operations at the same time. This is known as *parallel processing*. As holography operates in two and even three dimensions, holographic information should be the ultimate in parallel processing.

Spatial Filtering with Fourier-Transform (FT) Holograms The area where holographic information processing works best is in spatial filtering. We have met this concept before, in cleaning up laser beams (pp. 99–100), but the use of Fourier-transform (usually abbreviated to 'FT') holograms in spatial filtering is more subtle. The Fourier approach to image formation, introduced on pp. 12–5, is dealt with in more detail in Appendix 3, and if you are uncertain about the concepts underlying this approach, it would be a good idea to study Appendix 3 before tackling the rest of this chapter. The central thesis of the Fourier approach is that any source of light waves (ie any object illuminated by a beam of light) produces a complex wavefront that can be synthesized precisely by a large number of plane wavefronts of various amplitudes and phases traveling in fixed directions. These wavefronts behave as if they were the waves diffracted by a large number of diffraction gratings at various angles to one another and of various spatial frequencies. The gratings with the highest spatial frequencies carry the information relating to the finest details of the object, and these wavefronts diverge at the greatest angles (this is why closing down the aperture of a camera lens lowers its capacity for recording fine detail). If a large lens is placed in the path of these wavefronts with the object (the source) at its front focal plane, the wavefronts will be converged to pairs of points at its rear focal plane; these pairs of points are symmetrically placed with respect to the center of the field, and their position and intensity provide unique information about the object. The electromagnetic field at the rear focal plane is the optical Fourier transform of the field in the front focal plane (ie at the object), and if a further lens is placed so that the Fourier-transform plane of the first lens is in its front focal plane, a real inverted image of the object will appear in its rear focal plane.

The optical set-up (Fig 24.1(a)) may appear obvious to anyone who has 'O' level physics or has studied the rudiments of photographic optics; what is less obvious is what happens at the rear focal plane of the first lens. Using coherent light to illuminate the object, the pattern indicating the existence of the optical Fourier transform (Fig 24.1(b)) can be clearly seen. For a simple object such as a ruled grating it is just a row of spots; for a two-dimensional object such as a piece of copper mesh it is a more complicated (but still predictable) two-dimensional array of

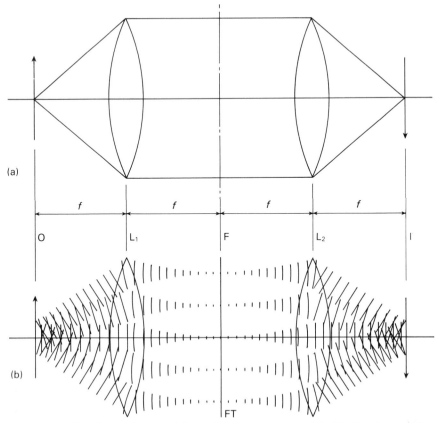

FIG 24.1 *Forming an image with two lenses.* (a) *Ray model;* (b) *Fourier model. Examples of the patterns observed in the* FT *plane can be seen in* FIG A3.23.

spots. For a very complicated object such as a photographic transparency it is more like a miniature version of the Milky Way, but an examination reveals a rotational symmetry (the type of symmetry shown by the letter **S**). It is made up of pairs of points, each carrying all the information about a particular spatial frequency present in the object wavefront.

The important thing is that if this wavefront has been further disturbed in any way, eg by being passed through an uneven piece of glass, the information about the disturbance is also present, and in a form that can be disentangled from the genuine information about the object. To see how this happens requires a familiarity with the Fourier-transform *convolution theorem*, which is discussed in more detail in Appendix 3. Broadly speaking when two functions are multiplied together in *x*-space (the space of the object) their transforms in *frequency space* (the space of the Fourier transform) are *convolved* and vice versa. Now, when some kind of function (say an Airy diffraction pattern) is convolved with another function (say an array of points) one function is 'dealt out' to the other function. Thus, in astronomy, the Airy pattern of a telescope objective is convolved with the field of stars: each star is imaged as an Airy pattern. In the frequency space of the focal plane of an optical Fourier-transform set-up, however, functions that have been convolved in *x*-space have their transforms multiplied. Thus if we could be clever enough to put into the rear focal plane of such a system a filter that *divided* the product of the transforms by the transform of the unwanted function (the Airy pattern), and then reconstructed the original function by a further identical set-up, we could also achieve *deconvolution* of the two functions, and be able to recover the original function (the star field) uncontaminated by the telescope function (the Airy pattern). This may not seem too important for stars (astronomers have found other, less complicated ways of doing the same thing), but it is quite important in other situations, eg where an important photograph has been blurred through camera shake or has accidentally been exposed out of focus. It is particularly important in electron microscopy, where aberrations of the magnetic lenses introduce artifacts into images, rendering their structure unintelligible.

Provided you know the exact nature of the 'blur' function it is in principle easy to *deconvolve* it. The first part – finding the blur function – can be quite simple. For example, the camera-shake function is a short straight line representing the actual movement of a point on the negative during the exposure. All points on the image will be convolved with this function, ie they will all appear to be straight lines of the same length oriented in the same direction. Similarly, the blur function of an out-of-focus lens is a small disk, and all points on the image are convolved with this disk. Electron microscope errors are more complicated patterns, but are still calculable. Other, more crude artifacts such as TV raster lines and half-tone screen dots, and even quasi-random noise, can be specified in two dimensions. Having found the 'unwanted' function, the next step is to find its two-dimensional Fourier transform. The FT of the combined function must be divided by this, which is the same as multiplying it by the reciprocal of the transform of the 'blur' function. The only difficulty here is that if the reciprocal becomes higher than 1 its value has to remain at 1 (you can't have a transmittance greater than 1). All this can be achieved quite simply by photographic methods. In fact, a photographic technique known as 'unsharp masking' was used for cleaning up woolly images long before anyone had heard of optical Fourier transforms. It is easy to correct for amplitude errors, which involve only the contrast of fine detail. Those errors which involve a shift in the image (phase errors) cannot be corrected using ordinary light detectors, including photographic emulsions, because these do not record phase – the very problem that began this book. As we have seen, holography *can* do this.

FT Holograms in Data Processing

Principle of the FT hologram If the history of photographic optics had been a little different, the first hologram might well have been a Fourier-transform hologram, for such a hologram can be made with ordinary amateur film and only partially-coherent light. The sudden insight of Denis Gabor which resulted in the original concept of the hologram did not take place until the late 1940s. Several years earlier, an equally sudden and radical insight by the physicist E W H Selwyn had founded what was to become the concept of the *optical transfer function*, and this was eventually to revolutionize the whole way of thinking about the design of optical devices.* Like Gabor's ideas, which had to await the arrival of the laser, Selwyn's ideas were forced to lie fallow awaiting the arrival of the digital computer. (There is another parallel between the two cases: the small amount of research that ensued directly from these discoveries remained in both cases under security wraps.) And even after Selwyn's discovery of the Fourier relationship between the point spread function and the optical transfer function had become the basis of modern lens design, the idea that a camera lens produced its image via a double Fourier transform was not physically demonstrated until after the arrival of the laser. By then Selwyn was dead. The first FT hologram was made by George Stroke in the mid-1960s.

As we have seen, light-sensitive devices record not amplitude and phase, but time-averaged intensity; all phase information is lost. This is where holography comes in. As long as a wavefront can be compared with a reference wavefront, both amplitude and phase can be recorded. Provided we supply an off-axis reference beam we can record a hologram in the Fourier-transform plane (the rear focal plane) as easily as anywhere else, and we can re-launch the wavefront by illuminating the processed hologram with the original reference beam (Fig 24.2). There are several other equivalent geometries, the simplest practical one having been originally described by Stroke (2), and called by him a 'lensless' FT hologram (pp. 210–1).

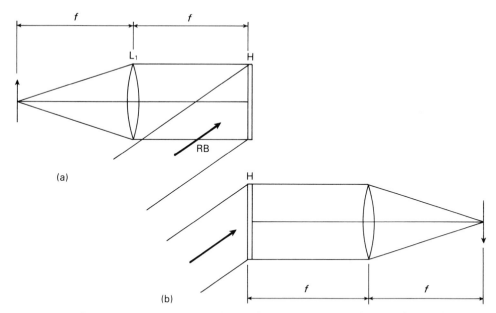

FIG 24.2 (a) *Holographic recording of an optical Fourier transform.* (b) *Reconstruction of the object beam beyond the hologram, giving an inverted image (the zero-order beam and the second image have been omitted for clarity – cf Figs 14.10/11.*

---

* P M Duffieux has also been credited with this insight.

The hologram itself acts like a lens, and if it is illuminated by a beam of collimated light from a laser both the erect real image and the inverted virtual image will be at infinity, and can be focused by a lens on the side of the hologram opposite to the reconstruction beam, just as in Fig 14.11 (p. 210).

This type of hologram is quite easy to make. Although for a true FT hologram the object should be flat, in practice this does not matter provided its distance from the hologram is large compared with its depth. The reference source should be close to the object and in the same plane. According to Stroke *et al* (3), holograms made by this technique do not need holographic emulsions as the primary fringes have a large spacing, and the way the information about the wavefront is coded means that the resolution of the image is determined by the size of the hologram and not by its photographic resolution. (see Fig 14.4)

To replay an FT hologram you simply place it in an expanded laser beam (or other quasi-monochromatic light) and look straight through the hologram at the light source. You will see two images, one each side of the central spot, in the same plane, one inverted and (if you have used a solid object) pseudoscopic. If you rotate the hologram the images rotate with it, but if you move the hologram in its own plane the images stay fixed in space. This property has led to suggestions that FT hologram geometry might be used as a basis for holographic movies.

FT holograms in spatial and correlation filtering The FT hologram has some very interesting applications in image processing. The first of these is in image de-blurring. Hariharan (4) shows that a suitable filter can be made by superimposing a negative of the diffraction pattern of the blur function in the rear focal plane of a lens (the blurred transparency is in the front focal plane), developed to a *gamma* of 2 (see Appendix 2, p. 370) superimposed on an FT hologram made of the same blur function and mounted in the same plane. This takes care of both amplitude and phase, and removes the blur function from the reconstituted image by *deconvolution*. Stroke *et al* (5) discuss the effectiveness of optical versus digital scanning processes, and decide that whatever the flexibility of the latter (they are routinely used for processing images from cameras on spacecraft) the capacity of holography for parallel processing puts it well ahead.

A further use for FT holograms is as matched or *correlation filters* for identifying features in photographic transparencies which may be of anything from Biblical quotations to fingerprints. An FT hologram of the character to be identified is placed in the FT plane of the set-up of Fig 14.13, and the transparency is positioned in the front focal plane. Any correlations between the two FTs will show as bright spots in the image plane, in the appropriate positions (Van ver Lugt *et al* (6)).

Computer-generated holograms

You may have seen pictures created with a typewriter keyboard. Nowadays it is fairly easy to construct these if you have a home computer with a word-processing facility; but originally these were created entirely on keyboards by teleprinter operators during the Second World War. The method of designing such pictures was to break them up into letter-sized rectangles which would then have one or more letters typed into each to achieve the desired level of blackness. *Computer-generated holograms* (CGHs) have a number of characteristics in common with typewriter pictures. They, too, attempt to reproduce the structure of an image (in this case the fringe structure of a hologram) by breaking it down into rectangular elements, called *pixels* (short for 'picture elements'), and they have similar limits as to their structure and tonal grading. The rules for designing a CGH are, of course, more complicated than the rules-of-thumb of the teleprinter operators: they are quite difficult mathematical rules which require a large database to store the result of their application to a particular case. The calculations involved in designing a three-dimensional Fresnel hologram are so complicated that we may never have the ability to produce an ordinary transmission hologram of, say, an apple, by drawing its holographic fringe structure by computer; the problem is of the same order of difficulty as the writing of a distinguished musical performance of Schoenberg's Variations for Orchestra directly onto a CD master; the most we can manage at present are some rather trivial noises. In holography, the realistic database to use is the FT hologram. Fortunately, there are applications where artificially-generated FT holograms can be very useful.

Applications for Fourier-transform CGHs Lee (7) has itemized five main areas where GGHs are of demonstrable use:

1 Three-dimensional displays. Although the near-field wavefront of an object beam is very complicated, it can nevertheless be computed if the object is thought of as being an array of point sources of varying intensity and distance. Each point source produces a wavefront at the plane of the hologram which can be calculated; these can be summed and the result plotted in terms of intensity. A more practical approach starts with a description of the three-dimensional object, from which the computer calculates the two-dimensional aspect from a large number of viewing angles. These are then recorded on motion-picture film to produce a holographic stereogram. Although the process does not involve writing the hologram itself, it is certainly computer-generated holography, and merits inclusion here because of its growing importance in display holography. However, Yatagai (8) describes a method of converting the projection images into FT holograms which are plotted by a printer, reduced photographically and arranged in the same order as the projection images. Each eye sees a holographically-reconstructed image from a different direction, giving a stereoscopic image. Although this method does not give anything like the spectacular results achievable by direct photography of a VDU screen, it can be used to display computer-generated three-dimensional images in real time.
2 Optical data processing. The insertion of a mask in the FT plane of a lens can modify the image of a transparency as described above – simple examples are the elimination of high spatial frequencies with a restricting aperture to give an image with soft edges or the elimination of low spatial frequencies by an opaque central spot in the FT plane giving images that resemble line drawings. At a slightly more complicated level we can find the location in the FT plane of the FTs of unwanted artifacts such as TV rasters and remove them by means of carefully-

designed masks. In the case of more sophisticated requirements, computer-generated holograms (CGHs) can provide an answer. The edges in a low-contrast image can be enhanced by operating on the amplitude transmittance function of the system (this is the coherent-light equivalent of the *modulation transfer function*). Image deblurring (pp. 334–6) can be accomplished by using a CGH of the FT of the blur function, which is usually a line (for image movement) or a disk (for incorrect focus), the FTs of which are well known. Such CGHs can restore the detail of the original object surprisingly well.

3 Matched filtering. The application of this technique to data obtained from synthetic-aperture radar, though important, falls outside the scope of this book. Lee (7) deals with it in some detail. Another important application is in interferometry, in particular in the testing of optical components. The traditional technique is to place a pattern directly on the surface of the component and to examine the Newton's fringes, and, by repeated working of the surface, to reduce their number to an acceptable minimum. There are devices such as the *Twyman-Green interferometer* which simplify the testing (Fig 24.3).

The reference waves in these interferometers are usually just plane waves, perhaps with a little tilt to produce broad zebra-stripe fringes. Sometimes, as in the testing of spherical optical components, the reference wavefronts are spherical. But in the testing of aspherical components the residual aberrations are so large that the interference pattern does not easily yield useful data on fabri-

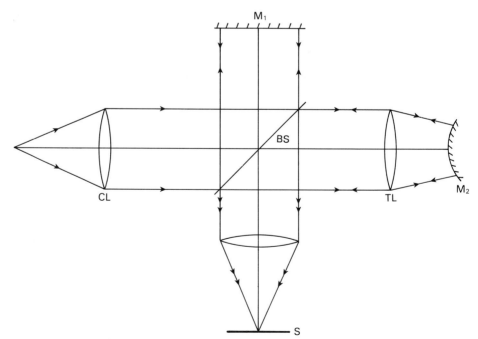

FIG 24.3 *Twyman-Green interferometer set up to test a lens* TL. *Except for the layout of the test arm with its retroreflective mirror* M_2 *the principle is similar to that of the Michelson interferometer. The interference pattern at* S *contours any optical inadequacies in the lens under test.*

cation errors. If a CGH is used to set up a special reference wavefront matched to the surface in question, the fringe pattern will be reduced to parallel bars showing only the errors in figuring. There are several variations of the system, which is discussed in detail by Lee (11).

4 Optical data storage and random phase coding. Information (usually in one-dimensional form) can be collected and stored as a series of FT holograms which may be partially or completely digitized. This area is the subjective of active research, and Lee (11) discusses the limitations of the process. There is much promise in the field of laser machining and other processes amenable to numerical control.

5 Laser-beam scanning. Holographically-recorded gratings (see Fig 25.1) can scan at high speed with low mechanical tolerances compared with conventional optics, and do not need a focusing lens.

Lee (9) is an updating of his earlier extensive publication (7), giving some new methods for making CGHs and discussing their application for non-optical wavelengths. Loomis (10) has contributed a useful paper on the design and writing of CGHs for optical testing purposes.

Techniques for making CGHs At present there are four general categories of technique for making CGHs: the results are known respectively as detour-phase holograms, modified off-axis reference beam holograms, kinoforms and computer-generated interferograms.

1 Detour-phase holograms. These were first described by Brown and Lohmann (11) in 1966, and are the oldest type of CGH. In order to simplify generation of the hologram the transmittance of the hologram is binary, ie at every point it is either transparent or opaque. It is capable of recording both amplitude and phase, and does not explicitly use an off-axis reference.

The desired wavefront, denoted by its two-dimensional wave equation, is first sampled at equally-spaced intervals using standard sampling theory. The paper base on which the hologram is to be plotted is divided into square cells in which the transmittance is coded by the height and width of a black rectangle within the cell. On photoreduction the rectangle becomes transparent on an opaque ground (Fig 24.4).

Several variations of the technique have been introduced to reduce noise, and these are described in the same paper. Later workers such as Haskell (12, 13), Lee (7, 9) and Burckhardt (14) describe methods which further subdivide the cells, resulting in better control of transmittance and phase.

2 Modified off-axis reference beam holograms. In 1967 Burch (15) showed a way of simplifying the amplitude-transmittance equation for a CGH. The result is a hologram with variable transmittance which is recorded directly onto film using the optical system of a *microdensitometer*, with a spot about 20 μm in diameter. Huang and Prasada (16) and Lee (7) describe somewhat different methods of achieving a similar result.

3 Kinoforms. These were first described in 1967 by Lesem *et al* (17), and have received a good deal of attention from later workers. They differ from the types described above in being phase holograms, and are recorded on film from a

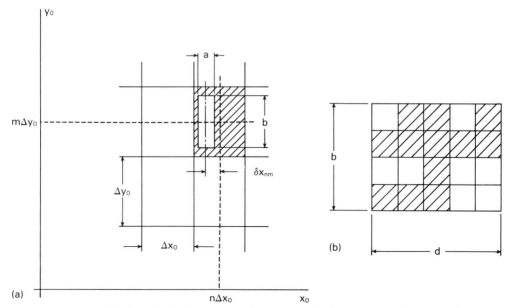

FIG 24.4 (a) *Typical cell in a binary detour-phase hologram, after Brown and Lohmann (11). The symbols represent the various coordinates defining the size and position of the components.* (b) *Typical cell in a generalized detour-phase hologram (after Haskell and Culver (12)). This method provides finer levels of amplitude and phase, and reduces noise.*

VDU display. The emulsion is processed by a bleach technique which leaves a relief pattern on the film. Chu *et al* (18) have extended the principle of the kinoform. By using Kodachrome film they were able to record both phase and amplitude. To make the kinoform, which they called ROACH (referenceless on-axis complex hologram), the film is first exposed to the intensity-variation pattern through a red filter, then to the phase-variation pattern through a cyan filter. During processing of a Kodachrome transparency the silver is removed, leaving a positive photographic image in dye. The red-sensitive layer bears a cyan image, which controls the red light passing through it. The green- and blue-sensitive layers finish as magenta and yellow respectively, both of these dyes being transparent to red light. There is a marked relief effect in the processed transparency. If the processed film is illuminated by light from a HeNe laser, the cyan layer modulates the amplitude and the other two layers modulate the phase.

4 Computer-generated interferograms. When wavefronts that contain only phase data are recorded as image-plane holograms they resemble interferograms. This was first pointed out in 1968 by Bryngdahl and Lohmann (19). The non-linearity of the emulsion response can be exploited to produce a hologram that is approximately binary. Following through the mathematics (Lee (7)) discloses that this type of hologram can record both amplitude and phase information without the approximations of the detour phase hologram. There are a number of methods

of recording phase and amplitude separately, and these are discussed in Lee (9, 20).

Although CGHs are undoubtedly exciting – one could visualize a technologically-enlightened artist of the future printing out a Fourier or even a Fresnel hologram directly from the imagination, just as did those wartime artists of the teleprinter keyboard – one should not become over-excited about their potential. Most of the authorities consulted when this chapter was being prepared seemed agreed that CGHs were probably more useful for generating PhD theses than useful imagery. However, there are two areas where CGHs are clearly valuable: in the testing of aspherical optical components; and in the production of holographic optical elements (HOEs). The latter are dealt with in Chapter 25.

Data Retrieval

Holographic micrography   One of the most important properties of holographic recordings is that everything that can be seen within the window of the hologram is recorded, and can be retrieved. There is a hologram by Walter Spierings which gives a dramatic example of this. It is a real-image hologram of a microscope. If the viewer moves up to the position of the image of the eyepiece, a pinpoint of light expands to fill the field of view, and suddenly there is a full-field image of a microcircuit, as seen through the microscope itself. Spierings's imagery is not in itself intended to be a piece of serious scientific importance; in spite of its dramatic impact it tells us little about the holographic storage and retrieval process. However, it gives valuable clues as to how holography might be used in microscopy (Plate 31).

If an ordinary camera is used to produce a photographic replica of the optical image seen in a microscope, only a single plane of the subject will appear in exact focus, because of the limited depth of field. If the subject matter is, say, a drop of water containing living micro-organisms, it will never be possible to make a satisfactory three-dimensional image by taking optical slices, owing to the movements of the micro-organisms. *Holomicrography*, on the other hand, can record the whole drop at a single instant, using a pulse laser. The processed hologram can then be placed back in its original position illuminated by the original reference beam, and the microscope can be focused right through the depth of the hologram, to examine any plane within the original drop.

In practice it is difficult to get a reference beam (or even a holographic plate) onto the microscope stage, but in fact the hologram can be recorded anywhere between the stage and the eye with identical results. The image can also be played back to form a real image at the microscope stage; this can then be examined by a conjugate microscope. The best method to date seems to be to generate a real image of unit magnification by placing two objectives back to back (Hefliger *et al*) (21). The various possible systems are shown schematically in Fig 24.5.

Particle counting   An even simpler method can be used for particle counting and analysis. The method is similar in geometrical principle to Gabor's original in-line

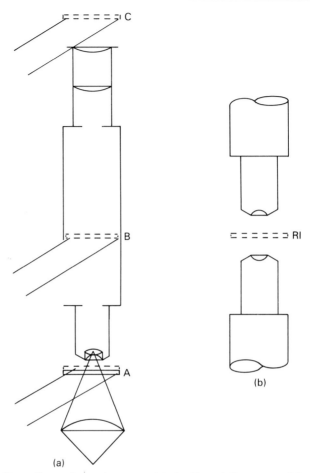

FIG 24.5 *Recording a holomicrogram* (a) A, B and C *are possible recording positions. For positions* A *and* B *the reconstruction can be viewed through the microscope; for position* C *the image is projected back through the microscope to the stage, where it can be viewed* (b) *by means of a second microscope focused on the other side.*

holography; however, the difficulties Gabor found because of the spurious real image do not occur, as this effectively adds only a constant to the genuine image term, and thus does not have a significant effect on the image. The technique is known as far-field or Fraünhofer holography. The layout is straightforward: the particle chamber is illuminated by a pulsed beam of collimated laser light. The particles form a characteristic diffraction pattern which forms the object wavefront and interferes with the undisturbed background, which acts as a collinear reference beam. For a point object, the interference pattern forms a Gabor zone plate which produces a real image of the point when reconstructed. In practice a relay lens is used to produce a real image which may if necessary be magnified. A point-by-point

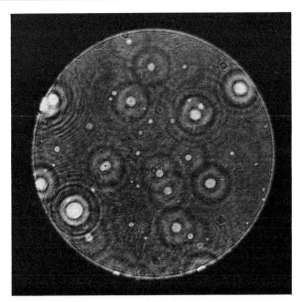

FIG 24.6 *In-line hologram of ragweed pollen. Although convolved with their own diffraction pattern, the images of the grains can be seen in three dimensions, and many useful measurements are possible. Hologram and photography by Paul Dunn. From Cartwright S L, Dunn P and Thompson B J (1980) Partical sizing using far-field holography; new developments. Optical Engineering, 19, 727–33.*

analysis of larger objects shows that results are still valid, though lower spatial frequencies may be lost in the geometric shadow (i.e. the objects show black, with higher-order fringes round them) (Fig 24.6).

Thompson and Dunn (22, 23) discuss the use of Fraunhöfer holography in the measurement of atmospheric fog particles, studies of cloud chambers, measurements of aerosol particles, flow diagnostics (with double-pulse illumination) and measurements of fibers. An important application of in-line holography to drop size measurement in dense fuel sprays in given by Jones *et al* (24) and Webster *et al* (25), with particular reference to combustion systems. Fraunhöfer holography has also proved useful in studies of cavitation. It is possible that achromatic analysis may soon be achieved.

References

1 Burke W J, Staebler B L, Phillips S W and Alphonse G A (1978) Volume phase storage in ferroelectric crystals. *Optical Engineering*, **17**, 308–16.
2 Stroke G W (1969) Lensless Fourier-transform method for optical holography. *Applied Physics Letters*, **6**, 201–3.
3 Stroke G W, Brumm D and Funkhouser A (1965) Three-dimensional holography with 'lensless' Fourier-transform holograms and coarse P/N Polaroid film. *Journal of the Optical Society of America*, **55**, 1327–8.

4 Hariharan P (1984) *Optical Holography*, pp 200–1. Cambridge University Press.
5 Stroke G W, Halioua M, Thon F and Willasch D H (1977) Image improvement and three-dimensional reconstructions using holographic image processing. *Proceedings of the Institute of Electrical and Electronic Engineers*, **65**, 39–62.
6 Van der Lugt A, Rotz F B, and Klooster A Jr (1965) Character reading by optical filtering. In *Optical and Electro-optical Information Processing*, ed J T Tippett, pp 125–41. MIT Press.
7 Lee W-H (1978) In *Progress in Optics XVI*, ed E Wolf, pp 120–232. North-Holland.
8 Yatagai T (1976) Stereoscopic approach to 3-D display using computer-generated holograms. *Applied Optics*, **15**, 2722–29.
9 Lee W-H (1980) Recent developments in computer-generated holograms. *Proceedings of the SPIE*, **215**, 52–8.
10 Loomis J S (1980) Computer-generated holography and optical testing. *Proceedings of the SPIE*, **215**, 59–69.
11 Brown B R and Lohmann A W (1966) Complex spatial filtering with binary masks. *IBM Journal of Research and Development*, **13** 160–7.
12 Haskell R E and Culver B C (1972) New coding techniques for computer-generated holograms. *Applied Optics* **11**, 2712–4.
13 Haskell R E (1973) Computer-generated binary holograms with minimum quantization errors. *Journal of the Optical Society of America*, **63**, 504.
14 Burckhardt C B (1970) A simplification of Lee's method of generating holograms by computer. *Applied Optics*, **9**, 1949.
15 Burch J J (1967) A computer algorithm for the synthesis of spatial frequency filters. *Proceedings of the Institute of Electrical and Electronic Engineers*, **55**, 599–601.
16 Huang T S and Prasada B (1966) *MIT/RLE Quarterly Progress Report*, **81**, 199.
17 Lesem L B, Hirsch P M and Jordan J A (1969) The kinoform: a new wavefront reconstruction device. *IBM Journal of Research and Development*, **13**, 150–5.
18 Chu D C, Fienup J R and Goodman J W (1973) Multi-emulsion, on-axis, computer-generated holograms. *Applied Optics*, **12**, 1386–8.
19 Bryngdahl O and Lohmann A W (1968) Interferograms are image holograms. *Journal of the Optical Society of America*, **58**, 141–2.
20 Lee W-H (1979) Binary computer-generated holograms. *Applied Optics*, **18**, 3661–9.
21 Hefliger L O, Stewart G L and Booth G R (1978) Holographic motion pictures of microscopic plankton. *Applied Optics*, **17**, 951–4.
22 Thompson B J and Dunn P (1980) Advances in far-field holography: theory and applications. *Proceedings of the SPIE*, **215**, 102–11.
23 Dunn P and Thompson B J (1982) Object shape, fringe visibility and resolution in far-field holography. *Optical Engineering*, **21**, 327–32.
24 Jones A R, Sargeant M, Davis C R and Denham R O (1978) Application of in-line holography to drop size measurement in dense fuel sprays. *Applied Optics*, **17**, 3.
25 Webster J M, Wright R P and Archbold E G (1976) *Combustion and Flame*, **27**, 395.

CHAPTER
25
Other Applications of Holography

'But oh!' thought Alice, suddenly jumping up, 'if I don't make haste, I shall have to go back through the looking-glass, before I've seen what the rest of the house is like.'
LEWIS CARROLL, *Through the Looking Glass*

In many books on technological subjects, the last chapter seems to be a kind of ragbag of partly-baked ideas. Indeed, Collier *et al* (1) end their excellent text with 'Three topics in search of a Chapter'. This chapter also contains a number of topics, but they are by no means necessarily partly-baked; some are restricted for reasons of commercial or national security, and others have been held up by patents squabbles. Nevertheless, progress continues to be made in all these areas.

Holographic Optical Elements

The principles and production of HOEs for use as collimators, beamsplitters, etc for making holograms has already been described (pp. 228–34). However, HOEs have far wider applications in industry, and for military purposes. For these, HOEs have so far invariably been made using dichromated gelatin (DCG), in spite of its variability, as no non-biological material with comparable qualities has yet been developed. HOEs have a number of advantages over conventional optics:

1 They are only as thick as a photographic emulsion and base. This means that a HOE can be put into a place where there would be insufficient room for a lens or optical mirror system.
2 The optical surface does not need to be oriented at an angle that obeys the usual lens laws. It is the direction of the interference planes and their spacing that determines the direction of the outgoing beam. We can thus produce a mirror that sits at an angle of, say, 45°, yet reflects light straight back along its path, or a lens that not only focuses a beam of light but also turns it 60° off axis.
3 HOEs do not even need laser beams for their manufacture: they can be drawn by a computer and photographed down to the appropriate size. In this way it is possible to produce corrector plates for lens aberrations of any type simply by feeding the appropriate instructions into a computer programmed to draw HOEs.
4 They are comparatively cheap to produce. Although the making of the master

FIG 25.1 *Bar-code scanner by IBM used at a supermarket checkout. The complicated optics needed to produce five or more directions of scanning simultaneously are now being produced holographically. Photograph by Richard Turpin.*

hologram can take up a large amount of human and computer time, and although the preparation of the plates demands rigorous clean-room conditions, the resulting HOEs cost far less than glass optics.

5 They are wavelength-selective. Where HOEs are required to substitute for conventional optics in producing images using incoherent light, this can be a drawback. However, the narrow-band selectivity that can be obtained with HOEs can be turned to advantage. For example, in *head-up displays* (HUDs) used in combat aircraft, it is important that the display should be as bright as possible, while still allowing maximum light to be transmitted through it from outside the aircraft. The conventional HUD optics use a projector which projects an image to infinity, with a half-silvered mirror close to the windscreen. Typically, by day the image is of instrument and weapon-sight readings, and by night a view of the terrain ahead suitably amplified by an image intensifier. The trouble is that the half-silvered mirror is in effect a 1 : 1 beamsplitter, and so the view ahead is dimmed by the mirror. But a HOE that is tuned to the color of the display can be made to reflect 95% of it, while maintaining 100% transmittance to all other wavelengths. This means that the pilot has a bright clear view at all times, and at night can make use of available visible cues as to the terrain, as well as seeing the HUD image brightly superposed. In addition, the HUD can be tailored to fit

the available space, whatever the angle of the windscreen; it can also be curved, if necessary, and in practice has a much wider angle of view than would be possible using conventional optics. HUDs making use of HOEs are becoming standard in combat aircraft, and it seems likely that they will be used increasingly on civil aircraft too. Small versions of HUDs have also been made to fit inside pilots' helmets.

HOEs are finding increasing applications replacing conventional optics for such complicated optical components as bar-code scanners for checkouts in supermarkets. It is not easy to draw a bar code over a light beam sufficiently accurately for the code to be recorded correctly every time, but if a network of beams moving in different directions scans a stationary bar code, one or other of the beams will scan the code bars in the correct direction whatever the orientation of the goods, and the code will be recognized. Originally this was done by a rotating array of mirrors, but these are expensive to produce, whereas HOEs can be replicated with ease (Fig 25.1).

Holography in Biology and Medicine

The main uses of holography in biology are in microscopy, which has already been discussed (pp. 343–4). Smith and Empson (2) have designed a holographic comparison microscope for the identification of small fossils. The problem of speckle is overcome by making a hologram of the specimen in the form of a number of miniholograms arranged in a circle, which is then scanned at high speed by a rotating replay beam. This microscope uses its stereoscopic facility to compare the specimen with a hologram of a known microfossil; this avoids the necessity of working in museums with valuable specimens which are not permitted to be carried away to sites. Heflinger *et al* (3) have described an improved holographic ciné camera which produces motion-picture holograms of tiny marine animals. The camera can be used as a projector to produce a real image which can be examined in depth, frame by frame, using a microscope. Briones *et al* (4) have analysed the resolution of the system compared with that of white-light microscopy.

Holography has often been used as a model for certain types of brain function, and people unfamiliar with the philosophical concept of a model as a metaphor which helps the achievement of insights into a particular phenomenon (we have had plenty of them in this book) can easily be misled into thinking that memories are literally recorded as holograms somewhere in the brain. The model is in effect a very oblique metaphor for the fact that memories for specific events do not seem to be strongly localized in the brain, so that the destruction of parts of the cerebral cortex does not lead to the loss of specific memories but only to a slight dimming of memories in general. There is also evidence to the contrary, however, and the whole model is very shaky. On the other hand, quasi-holographic images do seem to form a part of the sensory systems of some animals. Dolphins and related species use an acoustic method of navigation which has strong affinities with synthetic-aperture radar; and bats probably perceive objects in a similar manner. Pollen *et al* (5) put a persuasive case for the operation of the visual cortex of the brain in terms of Fourier holography.

In the field of medicine, holography is becoming important in three main areas, two being described by Higgins (6). The first is the three-dimensional record. The main use for this so far is in dental records. For legal, medical and forensic purposes it is necessary to store dental casts for up to twelve years, and as all these have to be boxed there is an enormous storage problem. By making simple Denisyuk holograms of each cast the storage problem is minimized. At the same time it becomes comparatively easy to make measurements on the reconstructed image using ordinary calipers, and the cast can be viewed as an impression simply by flipping it in the replay beam. Casts made at intervals can be directly compared by placing one hologram over the other and adjusting for the best register.

The second important field for holography is computerized axial tomography (CAT). This is a technique for displaying X-ray cross-sections of the body such as the pelvic area, thorax and, skull, and is used in the diagnosis of tumors and other lesions. The patient is moved into the scanner a few millimeters at a time, and successive slices of the head or body are displayed on a VDU screen. The images are stored initially in a computer memory, and can be processed to eliminate or emphasize detail. The difficulty in practice is that as only one slice can be displayed on the VDU at a time, the diagnostician has to mentally assemble all the successive visual information into a three-dimensional picture. In the holographic build-up of the image in depth, up to twenty slices are needed; these are recorded on 35 mm film, and the resulting transparencies are placed in succession in a projector to produce images on a ground glass screen. These are used to make transmission or reflection holograms on a single plate, by moving the ground glass between exposures by a distance corresponding to the distance between successive slices. The final hologram

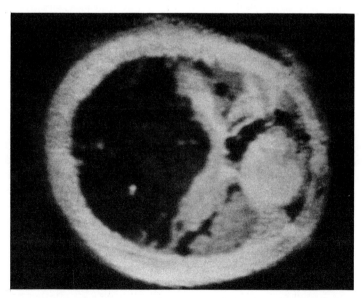

FIG 25.2 *A hologram synthesized from a number of photographs of cranial CAT scans (see text). Photograph by Sam Higgins. Courtesy of Royal Sussex County Hospital.*

contains images of all the slices in their correct position in space, and when viewed appears with full parallax. The image does not appear solid, as it is a set of sections: it resembles those prehistoric-animal models that are put together from cross-sections, and it can be interpreted best when the sections are slightly offset rather than one directly behind the other. Saxby (7) gives a report of a presentation by Laurie Wright on the subject (Fig 25.2).

The third area concerns the development of high-quality prostheses, eg in hip replacement joints. This is a straightforward use of double-exposure interferometry. Interferograms are made showing the stress patterns on normal hip-bones and their artificial replacements, and the latter can then be modified so as to bear the closest possible resemblance to the behavior of a healthy joint. The subject is under active development. It is also possible to make series of X-ray photographs into holographic stereograms, using standard techniques (see Chapter 17), and to use these as complementary to CAT-scan holograms.

Acoustic and Microwave Holography

These techniques do not really come within the scope of this book, but the picture would be incomplete without a mention of them. Certainly the results are true holograms even if the techniques are alien. *Acoustic (ultrasonic) holography* has had a somewhat checkered history, having been subject to patents which were not exploited by their owners, with the result that development was slow between 1968 and 1978. Acoustic holography employs coherent ultrasonic beams with a frequency around 1 MHz for *insonation* of the subject matter and for the reference beam. The technique has the advantage over optical holography that ultrasound can penetrate tissue. Separate transducers can be used for object and reference beams (and the system can even manage without a reference beam as such), and images can readily be produced in real time.

Initially, acoustic holograms were made by techniques closely analogous to those of optical holography, which served to show up its very real drawbacks. One of these was the difficulty of finding a suitable recording medium. One of the more successful suggestions was to use dilute photographic fixer as the ultrasonic transmitting medium, so that the emulsion was fixed in the pattern of the interference fringes; it could then be exposed to light and the fringes developed. A better idea was to form standing waves at the surface of the liquid, which could then be side-lit and photographed. The images were replayed by laser light. The wavelength discrepancy (of the order of 100 times) placed the image far away and much elongated, so that it had to be examined through a telescope. Things were improved by using acoustic lenses to form a real image at the surface (the equivalent of a focused-image optical hologram), and some success was achieved (Metherell (8)).

The next improvement was the detection of the object beam by an array of pressure-sensitive detectors; as these responded by generating electrical signals, a reference beam could be simulated electrically and fed directly into the output. The final output could be decoded by a computer programmed to carry out the required transform operations. The story up to 1978 is given by Greguss (9), and this is updated by Hildebrand and Doctor (10).

Microwave Holography was originally connected with synthetic-aperture radar, and this aspect is outlined by Cutrona *et al* (11). Nowadays, microwave holography is taken to mean holography of a more or less conventional kind, but carried out at centimeter and millimeter wavelengths. Anderson (12) summarizes progress in this field. The most important aspect of this technique is that it can penetrate solid objects to depths of at least 500 mm, and can be used for examination of buried pipes, cables, etc. The large size of the interference pattern makes it possible to reconstruct the image by digital rather than optical means, and thus avoid mismatches in wavelength.

* * *

Conclusion

We are now seeing holography applied through a large part of the electromagnetic spectrum, from radar and microwave imagery at wavelengths of several centimeters through to the ultraviolet – and beyond. Holograms at X-ray wavelengths may well be the next step; these would greatly simplify crystallographic research. But a sufficiently coherent source of X-rays seems a long way off, unless you count the X-ray laser that uses a nuclear explosion as its energy pump. It is beginning to look as though real-time holograms made using ultrasound are going to take their place alongside other types of ultrasonic imaging, and that optical holographic techniques will enhance the value of both conventional X-ray photography and computerized tomography. The holographic portrait studio is already a fait accompli. Holograms appear by the million on credit cards, in cereal packets and in souvenir shops. No prestige exhibition stand is complete without a large hologram. Forecasts are dangerous, though: we may never see the early predictions of a giant ketchup bottle suspended in the air in a supermarket realized (there are still laws of optics – thank goodness). On the other hand, to look at the present moribund state of research into the possibilities of holographic cinema and decide that it is a dead duck, might eventually make one look very silly.

What is so encouraging about holography is that, just as in photography a hundred years ago, and in astronomy today, a large proportion of the exciting discoveries are still being made by amateurs – and it should be borne in mind that the word 'amateur' does not mean 'beginner' but 'enthusiast'.

Now let's make some more holograms.

References

1 Collier R J, Burckhardt C B and Lin L H (1971) *Optical Holography*, Chapter 20. Academic Press.
2 Smith R W and Empson T R (1982) Holography applied to stereomicroscopy. *Proceedings of the SPIE*, **368**, 104–9.
3 Heflinger L O, Stewart G L and Booth C R (1978) Holographic motion pictures of microscopic plankton. *Applied Optics*, **17**, 951–4.

4 Briones R A, Heflinger L O and Wuerker R F (1978) Holographic microscopy. *Applied Optics*, **17**, 944–50.
5 Pollen D A, Lee J R and Taylor J H (1971) How does the striate cortex begin the reconstruction of the visual world? *Science*, **173**, 74–7.
6 Higgins S T (1983) Using Holography in the Medical Photography Department at the Royal Sussex County Hospital, Brighton. *British Journal of Photography*, 21 October 1983, 1111–3.
7 Saxby G (1985) Holography at the Royal Sussex County Hospital. *British Journal of Photography*, 22 March 1985, 327–8.
8 Metherell A F (1969) *Acoustical Holography*. Plenum Press.
9 Greguss P (1980) *Ultrasonic Imaging*, pp 125–55. Focal Press.
10 Hildebrand B P and Doctor S R (1980) Acoustical holography advancements. *Optical Engineering*, **19**, 705–10.
11 Cutrona L J, Leith E N, Porcello L J and Vivian W E (1966) On the application of coherent optical processing techniques to synthetic-aperture radar. *Proceedings of the Institute of Electrical and Electronic Engineers*, **54** 1026–32
12 Anderson A F (1987) In *McGraw-Hill Encyclopedia of Science (6th Edition)*, Vol. 8, 485–92.

APPENDIX

1

Lasers and Safety

'Come back!' the Caterpillar called after her. "I've something important to say!"
LEWIS CARROLL, *Alice in Wonderland*

For the purpose of defining laser safety, there are four general classes of lasers. Class I (inherently safe) comprises very weak lasers not suitable for making holograms. Class II (relatively safe) comprises lasers of powers up to 1 mW, which will not cause retinal damage unless you force yourself to stare down the beam. Class IIIA comprises lasers up to 5 mW which have beams that are sufficiently expanded for the power entering the unaided eye not to exceed 1 mW, and these are therefore grouped with Class II for safety purposes, with the proviso that they must not be looked at with light-gathering devices such as telescopes or theodolites. Class IIIB comprises lasers up to 500 mW (0.5 W), which go up to the limit of safe viewing by diffuse reflection. Finally, Class IV comprises all lasers of higher power, where viewing by diffuse reflection can cause eye damage, and the direct beam can cause skin damage.

The British Standards Institute (BSI) recommendations for safety (1) were updated in 1982. The United States has Governmental Regulations (2), and the American National Standards Institute (ANSI) gives further guidelines (3). There is also a comprehensive book by Sliney and Wolbarsht (4). All these documents make tedious reading, not least because they deal with all kinds of industrial situations and with lasers that have no relevance to holography.

As a holographer you will be working with Class IIIB lasers, which includes all HeNe lasers with powers of more than 1 mW, and possibly with Class IV which includes Ar^+ lasers, some Kr^+ lasers, and ruby pulse lasers. With Class IIIB lasers of powers up to about 15 mW an accidental exposure to the undiverged beam would be unlikely to do any permanent damage to your retina, as the blink reflex is rapid enough to prevent a dangerous amount of energy from entering the eye; above this there could be permanent damage, especially if the beam were to be focused on the fovea (the central region of the retina, where the focus is sharpest). Class IV laser beams can cause severe damage to the retina and possibly to the cornea too. The BSI and ANSI specifications lay down what is a safe exposure to various types of laser radiation, and these can be helpful when installing powerful lasers for large holograms. However, in practice the circumstances are so varied that published figures are not very helpful. The golden rule is to regard anything over 5 mW as

FIG A1.1 *International warning notice (black on yellow) for laboratories containing lasers.*

potentially dangerous and anything over 0.5 W as actually dangerous. Also, it is not enough to be aware of this yourself; you have a duty to others too.

Warning Notices

The first thing is to place a permanent sign on the laboratory door. There are international symbols indicating laser radiation (Fig A1.1), and the appropriate notices should be on the door and on any laser within the lab. If you are working in an establishment where there are visitors, you should have a notice on the outer door that the lab must be kept locked at all times when not in use. You should also have an illuminated sign in parallel with the laser power supply, saying 'Laser on: Do not enter'. But don't lock yourself in when you are working; if you have an accident nobody will be able to get in to help you.

Avoiding Accidents

As a rule, it is not the direct beam that carries the real risk. People are usually careful to keep out of its way; in any case, if you really want to look down the bore of a laser you will not find it easy, as both the laser beam and the pupil of your eye are pretty small in diameter. It is the slightly-diverging specular reflections from lenses, metal clips, screws, etc, that are dangerous. Many workers have received an eye-watering flash from a piece of unblacked metal equipment, and have been more careful afterwards. This is very important when there are two of you working on a set-up: the one who is not actually manipulating the beam should wear laser goggles (see below). As you set up the table, card off all spurious specular reflections from prisms, lenses, etc, as near to their points of origin as possible. This is particularly important for any beams that are reflected at an upward angle. Laser power supplies are probably a greater danger than the beam, as they operate at several thousand volts, so never operate the laser with the cover off either the tube or the power supply.

Ion Lasers

With ion lasers you must be even more careful, as the blink reflex is not fast enough to prevent serious damage to the retina in the event of a direct hit. Set up the table using the lowest possible power, and turn the intensity up only when you are satisfied that all stray beams have been carded off. Aluminized surfaces can be evaporated by powers of over 0.5 W, and can become dangerously hot at lower powers. Check the water-cooling system every time you use the laser; water leaks do not go well with high-voltage power supplies.

Laser Goggles

In the early days of lasers protective goggles were like welders' goggles; when you wore them it was nearly impossible to see anything at all. The latest generation of goggles is completely different; although they have a density of 6 (ie an attenuation of 1 million) to the wavelength to which they are tuned, they are almost totally transparent to other wavelengths. For example, when wearing a pair of goggles tuned to 694 nm, one seems to be wearing only pale blue-tinted glass, yet even standing directly in the expanded beam of a pulse laser one sees only the white light of the flashtube. For making adjustments to laser mirrors such goggles are essential. They are rather less effective as a protection when setting up a table, as the laser beam is completely invisible! These goggles can be obtained in both narrow- and broad-band versions; in general, the narrow-band version is preferable, as it is cheaper, and is transparent to all wavelengths other than that of the laser.

Pulse Lasers

Once you get into the area of pulse lasers you are in a completely different world as far as safety is concerned; to appreciate this you need only fire a pulse laser at a piece of black paper and see what happens. The undiverged beam can cause skin burns, and it is risky even to look at a white surface illuminated by it. The more powerful lasers (~ 10 J) need the beam to be expanded to about 12 cm diameter before being passed through an etched-glass diffuser. No visitor should be permitted to enter a pulse-laser lab without matched laser goggles. A flashing light on the door should indicate when the laser is armed. The makers of these lasers take elaborate precautions for safety as far as the energy-storage capacitor is concerned, and it *should* be impossible to override the safety circuits – but remember Murphy's law, and treat any power supply with respect.

The Laser Itself

It is not only users who are at risk with high-power CW and pulse lasers; if the beam is retroreflected down the tube it can damage the laser. With the more powerful HeNe lasers the result is usually merely an unstable beam; with an Ar^+ laser it can damage the mirrors and cause the plasma tube to overheat. With a pulse laser it can blow the ends off the ruby rods, causing internal havoc. In a pulse laser-laser set-up, therefore, all optical components should be slightly offset to prevent any chance of this happening.

Finally, to keep things in proportion, there is a delightful spoof appendix in Sliney and Wolbarsht (4) on the dangers of using – hammers! Certainly, the risk to eyesight in a holography lab is minuscule compared with those on the average building site, as anyone who has spent a morning's observation in an eye hospital will confirm. Nevertheless, if you intend to employ anyone in your lab, book them for an eye checkup before they first start work. You don't want to find yourself being sued for retinal lesions that may have occurred before your employees began working for you. And perhaps you *will* be that little bit more careful about their welfare too.

References

1. BS 4803 (1982) *Radiation Safety of Laser Products and Systems.*
2. *US Governmental Regulations for the Administration and Enforcement of the Radiation Control for Health and Safety Act, 1968*, Ch 21, CFR Subs. J.
3. ANSI Standard Z-136.1 (1976) *Safe Use of Lasers.*
4. Sliney D and Wolbarsht M (1978) *Safety with Lasers and Other Optical Sources.* Plenum Press.

APPENDIX
2

Mathematical Matters

'Let me see: four times five is twelve, and four times six is thirteen, and four times seven is – oh dear! I shall never get to twenty at that rate!'
LEWIS CARROLL, *Alice in Wonderland*

Formation and Reconstruction of a Hologram

There are several ways of writing down the equation of a traveling wave; some are more rigorous than others, but demand a higher level of mathematical attainment. The treatment below uses cosines only; if you have done a little trigonometry and can remember the relationship

$$\cos X \cos Y = \tfrac{1}{2}\cos(X+Y) + \tfrac{1}{2}\cos(X-Y)$$

that is all you need.

Any *cosinusoidal* waveform or *cosine wave* can be represented by an equation of the form $a = A\cos(Bx)$ where a and x are the variables, A is the amplitude and B is a constant related to the wavelength. In a cosinusoid the origin is at a point where the value of a (ie the instantaneous amplitude) is positive and the curve is symmetrical about the vertical axis (Fig A2.1). However, the wave in question may not have the point of symmetry directly above the origin. If it does not, we have to put a term into the equation to tell us how the waveform relates to a similar wave that *is* centered on the origin. This comparison term is called the phase term, and is denoted by the symbol ϕ (phi). This more general equation is written

$$a = A\cos(Bx + \phi)$$

(Fig A2.2). Now B is related to the wavelength λ by the formula.

$$B = \frac{2\pi}{\lambda}$$

The constant $2\pi/\lambda$ is called the angular frequency and is given the symbol ω (omega). Thus the general equation of a traveling wave with respect to time at a specific point in space is

$$U = A\cos(\omega t + \phi_{[x,y]})$$

where U represents the position of the wavefront at a given time t: ω (omega) is called the angular frequency, and is measured in radians.

MATHEMATICAL MATTERS

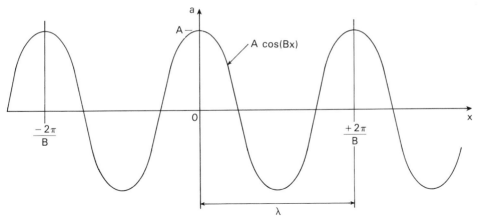

FIG A2.1 *A cosinusoidal wave of amplitude A and wavelength $2\pi/B = \lambda$. The equation of this wave is of the form $a = A\cos(Bx)$.*

If we represent the reference and object wavefronts by U_1 and U_2 respectively, remembering that ϕ_1 and ϕ_2 are for some given point x,y of the recording medium, their intensities I_1 and I_2 are given by $<\tfrac{1}{2}U_1^2>$ and $<\tfrac{1}{2}U_2^2>$, where the angular brackets mean that the values are time-averaged; for convenience these will be omitted from here on. Now, if U_1 and U_2 are mutually incoherent (uncorrelated) their combined intensity is simply $\tfrac{1}{2}(U_1^2 + U_2^2)$. But if the beams *are* correlated, their combined intensity I is $\tfrac{1}{2}(U_1 + U_2)^2$, which is

$$I = \tfrac{1}{2}(U_1^2 + U_2^2 + 2U_1U_2)$$

Writing out the equations in full gives

$$I = \tfrac{1}{2}A_1^2\cos^2(\omega t + \phi_1) + \tfrac{1}{2}A_2^2\cos^2(\omega t + \phi_2) + A_1A_2\cos(\omega t + \phi_1)\cos(\omega t + \phi_2)$$

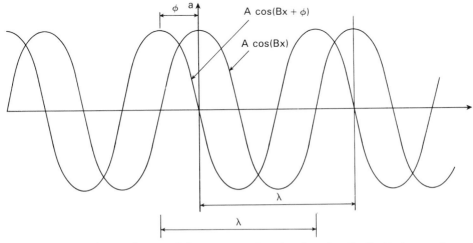

FIG A2.2 *A second wave of the same wavelength referred to the first in terms of its phase difference ϕ.*

Using the identity $\cos X \cos Y \equiv \tfrac{1}{2}\cos(X+Y) + \tfrac{1}{2}\cos(X-Y)$ for the third term gives

$$I = \tfrac{1}{2}A_1^2 \cos^2(\omega t + \phi_1) + \tfrac{1}{2}A_2^2 \cos^2(\omega t + \phi_2) + \tfrac{1}{2}A_1 A_2 \cos(2\omega t + \phi_1 + \phi_2)$$
$$+ \tfrac{1}{2}A_1 A_2 \cos(\phi_1 - \phi_2)$$

The final term, as you can see, contains information on both A_2, the object wave amplitude, and ϕ_2, the object wave phase, in an expression that does not contain t and is therefore not time-dependent.

This expression is now to be time-averaged. The time-average of a $\cos^2$ function of the form $A\cos^2 B$ is simply $\tfrac{1}{2}A^2$, and the time-average of a cos function is zero. The expression thus becomes

$$I = \tfrac{1}{4}A_1^2 + \tfrac{1}{4}A_2^2 + \tfrac{1}{2}A_1 A_2 \cos(\phi_1 - \phi_2)$$

The final term is not affected by time-averaging as it does not contain t. Now, U_1 and U_2 both vary across the plane of the recording medium: U_1 is unmodulated and has a constant value for A_1, and if it is incident obliquely on the recording plane, ϕ_1 will be proportional to the distances x and y; U_2 is the modulated object beam, and both A_2 and ϕ_2 will be complicated functions of x and y. For any point (x, y) the intensity will be given by

$$I_{(x,y)} = \tfrac{1}{4}(A_1^2 + A_2^2 + 2A_1 A_2 \cos[\phi_1 - \phi_2])_{(x,y)}$$

This represents a locally-cosinusoidal fringe pattern which is perturbed by variations in A_2 and ϕ_2 from point to point.

This pattern is recorded on a material which turns intensities into amplitude transmittances, the amplitude transmittance at any point being directly proportional to the intensity of the illuminating wave at that point. This record is the hologram.

To show how the hologram reconstructs the object beam we need to examine what happens when U_1 is incident on the transmission grating we have produced. At any point the emergent wave U_3 can be obtained by multiplying U_1 for that point by the grating transmittance function for that point. Since this is proportional to I, the emergent wave U_3 will be proportional to $U_1 \times I$.

Hence $U_3 = U_1 \times I$ ($\times$ some constant)
$$= \tfrac{1}{4}U_1[A_1^2 + A_2^2 + 2A_1 A_2 \cos(\phi_1 - \phi_2)]$$
$$= \tfrac{1}{4}U_1(A_1^2 + A_2^2) + \tfrac{1}{2}U_1 A_1 A_2 \cos(\phi_1 - \phi_2)$$
$$= \tfrac{1}{4}U_1(A_1^2 + A_2^2) + \tfrac{1}{2}A_1^2 A_2 \cos(\omega t + \phi_1)\cos(\phi_1 - \phi_2)$$

Expanding the last term by the relationship

$$\cos X \cos Y = \tfrac{1}{2}\cos(X+Y) + \tfrac{1}{2}\cos(X-Y)$$

we have

$$U_3 = \tfrac{1}{4}U_1(A_1^2 + A_2^2) + \tfrac{1}{4}A_1^2 A_2[\cos(\omega t + \phi_2) + \cos(\omega t + 2\phi_1 - \phi_2)]$$
$$= \tfrac{1}{4}(A_1^2 + A_2^2)U_1 \qquad (= \text{constant} \times U_1)$$
$$+ \tfrac{1}{4}A_1^2 A_2 \cos(\omega t + \phi_2) \qquad (= \text{constant} \times U_2)$$
$$+ \tfrac{1}{4}A_1^2 A_2 \cos(\omega t + 2\phi_1 - \phi_2) \, (= \text{constant} \times \text{modified } U_2)$$

The first term is identical with the reference beam/replay beam, attenuated. The second term is identical with the object beam, also attenuated. The third term is an oddity: it is certainly in the correct form for a traveling wave, but emerges on the other side of the reference beam (turned through an angle 2ϕ), and reversed in phase (ie pseudoscopic).

Readers who are familiar with complex algebra may like to try the same exercise using the exponential form for the wave equation

$$U = \exp(i\omega t + \phi)$$

where $i = \sqrt{(-1)}$, which reaches the same result less laboriously; and perhaps a little more rigorously, as it turns up a fourth term representing noise. This is the method adopted in most textbooks dealing with the theory of holography; but in the author's experience many people who can cope fairly happily with cosines become uneasy when confronted with complex variables.

Traveling and standing waves The previous section dealt with wavefronts at specific points in space, and so a simplified wave equation in which the instantaneous amplitudes fluctuated only in terms of time has been used up to this point. In describing an entire wave we must also think of the way in which the amplitude fluctuates with distance at a specific point in time. In the first (variation in time) we would have an equation of the form

$$a = A \cos\left(\frac{2\pi}{\lambda} t\right) \quad (\lambda = \text{wavelength})$$

and in the second case (variation in space) an equation of the form

$$a = A \cos\left(\frac{2\pi}{T} x\right) \quad (T = \text{period of wave})$$

We can combine these into a single equation by writing

$$a = A \cos(\omega t - kx) \quad \text{where} \quad \omega = \frac{2\pi}{\lambda} \quad \text{and} \quad k = \frac{2\pi}{T}$$

(ω and k are both measured in radians) and this is the trigonometrical form for describing a *traveling wave* (or *running wave*), in this case traveling from left to right. For a wave traveling from right to left the sign of kx is changed.

If two traveling waves moving in opposite directions occupy the same space their instantaneous amplitudes add algebraically, and we have

$$a \text{ (resultant)} = A\,[\underset{L \to R}{\cos(\omega t - kx)} + \underset{R \to L}{\cos(\omega t + kx)}]$$

Using the identity $2 \cos X \cos Y = \cos(X + Y) + \cos(X - Y)$ gives

$$a = 2A \cos(\omega t)\cos(kx)$$

Distance and time are now interlinked, and A and a vary with distance (Fig A2.3). The result is a waveform with a cosinusoidal profile that has an amplitude varying from $+2A$ to $-2A$ with time. The nodes, or regions of no excursion in amplitude, are one half-wavelength apart, as are the antinodes, or regions of maximum

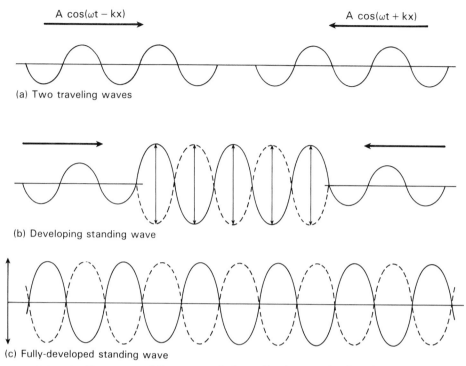

FIG A2.3 *Two waves of the same period traveling in opposite directions combine to produce a standing wave.*

excursion, and their positions are fixed in space. Hence this wave is described as a *standing* (or *stationary*) *wave*.

NOTE: Cosine equations have been used throughout for consistency. Textbooks commonly use the form $a = A \sin(kx - \omega t)$, and in this case the final equation becomes $a = A \sin(kx)\cos(\omega t)$; this puts a node at the origin, which is more convenient for the analysis of the vibration of plucked strings, organ pipes, etc, an important application of standing-wave theory.

Zone Plates

It has been pointed out more than once in the text that a hologram behaves like a lens: it has a focal length, an *f*-number, image conjugates, and so on. To see why this should be so, let us look at the geometry of a *Fresnel zone plate*. This is a transparent plate on which are drawn concentric opaque zones. These are alternated with clear zones, and the radius of successive zone edges in the same sense (eg clear to opaque) is in the proportion of the square roots of successive integers, starting at 1. The result is that at a certain distance along the perpendicular to the plate which passes through its center, all the optical paths of the diffracted beams differ by an integral number of wavelengths and thus interfere constructively. Figure A2.4 shows how this happens.

MATHEMATICAL MATTERS

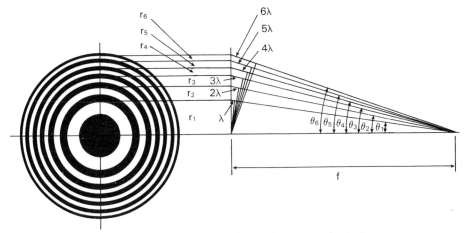

FIG A2.4 *A Fresnel zone plate. The radii of the concentric circles are proportional to* $\sqrt{(1)}$, $\sqrt{(2)}$, $\sqrt{(3)}$ *and so on, and alternate zones are opaque and clear. The focal length of the zone plate is* r^2/λ *where* r *is the radius of the innermost ring.*

By similar triangles, provided θ is small so that $\sin\theta \cong \tan\theta$,

$$\frac{f}{r} = \frac{r}{n\lambda} \quad (n = 1, 2, 3, \text{etc})$$

Hence

$$r^2 = fn\lambda \quad \text{or} \quad r = \text{constant} \times \sqrt{(n)}.$$

For each point across the innermost transparent radius there is a point in each of the other transparent zones for which constructive interference occurs, and so the zone plate acts like a lens of focal length r^2/λ, where r is the outside diameter of the innermost clear area (it does not matter whether the innermost disk is clear or opaque).

Because of its square-wave transmittance profile, a Fresnel zone plate also has several other focal lengths corresponding to higher-order diffractions, but in practice these are weak or non-existent. However, when a hologram is made using a point object and a collimated reference beam, the hologram is a zone plate with a transmittance profile that is cosinusoidal, with a spatial frequency that increases in the same ratio as that of a Fresnel zone plate. This hologram has been given the name of a Gabor zone plate, and it has only one focus.

Because a hologram of a point object is a Gabor zone plate, it is possible to consider a hologram of an extended three-dimensional object as consisting of a very large number of Gabor zone plates each corresponding to a point on the object. This is a sound model, and can be helpful in giving an intuitive grasp of the principles underlying holography to those who find the Fourier approach difficult; on the other hand, a rigorous analysis of holographic imagery is much more difficult with this model than with the Huyghens/Fourier approach. However, it is particularly useful when considering holographic optical elements (HOEs). Holographically-made lenses (see pp. 231–2) are off-axis Gabor zone plates. They are affected by chromatic aberration, the focal length for blue light being greater than that for red light in inverse proportion to the wavelength; they also exhibit the other lens aberrations such as coma, but they obey the lens laws in the same way as refracting lens elements.

The Lens Laws

The lens laws were first formulated by Sir Isaac Newton, and they rule the geometry of both photographs and holograms. They can readily be derived by geometry, and you can find the derivations in any textbook of geometrical optics.

The first lens law, and the most basic, is derived from the laws of refraction, using the approximation $\sin \theta \cong \theta$ for small angles, where θ is measured in radians. In geometrical optics the conventional symbol for the focal length of a lens is f, that for the distance of the object from the lens is u, for the image distance v, and for the refractive index of the lens with respect to air μ (mu) (with respect to empty space the symbol n is used). The sign convention most often used is known as the 'real is positive' convention; all distances involving real focal points and real images, and the radii of curvature of convex surfaces, are given positive signs, while distances to virtual foci and images, and radii of curvature of concave surfaces, are regarded as negative. Using this convention, the focal length of a lens with radii of curvature r_1 and r_2 is given by the formula

$$\frac{1}{f} = (\mu - 1)\left(\frac{1}{r_1} + \frac{1}{r_2}\right)$$

This is the formula used for calculating the focal lengths of liquid-filled lenses and one-dimensional beam expanders.

The best known of the lens laws is the law of reciprocal distances. It is derived from the geometry of similar triangles in a ray diagram, and relates the object and image distances and the focal length of a lens or optical mirror:

$$\frac{1}{f} = \frac{1}{u} + \frac{1}{v}$$

By the same geometry it can be shown that the image scale m is given by:

$$m = \frac{v}{u}$$

These relationships are important in the making of HOEs. If the object and reference sources are chosen appropriately, the HOE can be made with any desired focal length. In general, the u and v distances are not in a straight line in a HOE.

In Newton's original papers the object and image distances are measured, not from the lens, but from the two principal foci, thus avoiding the question of the finite thickness of the lens (which the lens equation ignores). If we call these distances x and x', such that $x = u + f$ and $x' = v + f$, then

$$xx' = f^2$$

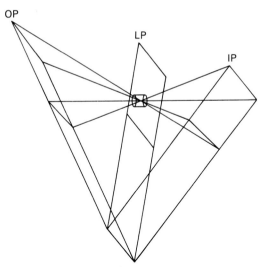

FIG A2.5 *Scheimpflug's condition. The Newtonian condition* $xx' = f^2$ *applies to every conjugate on the object/image planes.* OP = *object plane;* IP = *image plane;* LP = *lens plane. For sharp focus all over* IP *it can be shown that* OP, IP *and* LP *meet in a single line.*

This is known as the Newtonian condition. It is not often used in photographic optics, but it plays an important part in the geometrical proof of what is known as the *Scheimpflug condition*. This is a difficult and long-winded piece of geometrical logic, but if you are curious to investigate it you can find it in manuals of aerial photogrammetry (1). What finally emerges is that when the plane of the object is tilted with respect to the axis of the lens, in order to obtain overall sharp focus the plane of the film also has to be tilted, in such a way that the planes of the object, the lens and the film all intersect in a common straight line; all the object–image distances measured through the optical center of the lens satisfy the Newtonian condition (Fig A2.5).

This requirement, which is part of the bread-and-butter considerations of studio photographers working with large-format cameras, is equally important in any form of transfer holography where the planes are not strictly parallel, as in achromatic transfers.

Axial Magnification

In photographic optics the term 'magnification' is often used instead of 'scale', even when (as is usual) it is less than unity. The relationship between lateral and axial magnification in an optical image is not at all well known, even though it is responsible for the disappointing perspective of stereograms. It is totally ignored by photographic textbooks (not altogether surprisingly, as it is irrelevant to two-dimensional imagery), so a full derivation is given here.

Let us consider the lens law

$$\frac{1}{u} + \frac{1}{v} = \frac{1}{f}$$

with a small increment Δu (Delta-u) added to u, which results in a small decrement $-\Delta v$ added to v (Fig A2.6).

Then
$$\frac{1}{u + \Delta u} + \frac{1}{v - \Delta v} = \frac{1}{f}$$

But
$$\frac{1}{u} + \frac{1}{v} = \frac{1}{f}$$

Subtracting gives

$$\frac{1}{u} - \frac{1}{u + \Delta u} + \frac{1}{v} - \frac{1}{v - \Delta v} = 0$$

or
$$\frac{u + \Delta u - u}{u^2 + u \Delta u} + \frac{v - \Delta v - v}{v^2 - v\Delta v} = 0$$

i.e
$$\frac{\Delta u}{u\Delta u + u^2} + \frac{\Delta v}{v\Delta v - v^2} = 0$$

Expanding out,

$$\Delta u(v\,\Delta v - v^2) + \Delta v(u\,\Delta u + u^2) = 0$$

or
$$v\,\Delta u\,\Delta v - v^2\,\Delta u + u\Delta u\Delta v + u^2\,\Delta v = 0$$

Regrouping gives

$$\Delta u\,\Delta v(u + v) + u^2\,\Delta v = v^2\,\Delta u$$

Now if we allow Δu (and thus Δv) to become very small, $\Delta u\,\Delta v$ will become very small indeed, and in the limit can be ignored, leaving

$$u^2\,\Delta v = v^2\,\Delta u$$

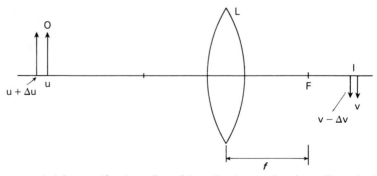

FIG A2.6 *Axial magnification.* O = *object;* I = *image;* L = *lens;* F = *principal focus;* f = *focal length of lens;* u, u + Δu = *object distances;* v, v − Δv = *image distances.*

or

$$\frac{v^2}{u^2} = \frac{\Delta v}{\Delta u} = m^2$$

That is, axial magnification = (lateral magnification)². The implications of this are discussed in various places throughout the text; they are particularly important in slit transfer and focused-image holography. The lens laws also apply wherever there is a change in beam divergence angles; and whenever the wavelength changes the effect is the same as if the refractive index of the 'lens' changed in proportion.

f-number and Parallax Angle

The f/no of a lens is its focal length divided by its diameter, and a holographic 'lens' also has an f/no. The concept can be extended to any kind of hologram, with the distance of the image from the plane of the hologram constituting the focal length (for a virtual image the focal length is negative). This assumes that the reference and reconstruction beam are both collimated. The angle of parallax is determined by the points (in a horizontal plane) at which the image of the object is cut off by the edges of the hologram, or of the edges of the image of the master hologram, in the case of a transfer hologram (Fig A2.7).

If D, the parallax distance, is less than the separation of the eyes it will not be possible to see the image stereoscopically, though there will still be parallax, and by closing one eye and moving around, you can confirm that the image is nevertheless three-dimensional.

In a focused-image hologram the parallax angle can be determined from the (effective) f/no of the lens (Fig A2.8). In a transfer hologram the aperture is the real image of the master hologram. It is between the viewer and the hologram as a rule,

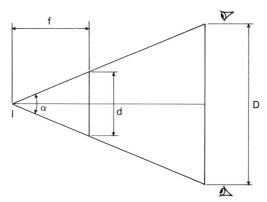

FIG A2.7 *Parallax angle of a hologram.* I = *position of image;* d = *diameter of hologram;* D = *parallax distance;* α = *parallax angle, given by* tan(α/2) = d/2f *or* α = 2 tan⁻¹(d/2f). *If* α *is measured in radians, parallax angle* ≅ d/f.

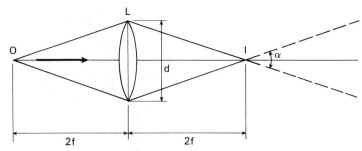

FIG A2.8 *Parallax angle in a focused-image hologram. O = position of object; I = position of optical image of image; L = lens; d = diameter of lens; object distance = image distance = 2f. α = parallax angle, given by tan(α/2) = d/4f or α = 2 tan⁻¹(d/4f). If α is measured in radians, parallax angle ≅ d/2f = ½(f/no of lens).*

and unless all the beams used to make and display it have been collimated (or at least one has been converging), it will be larger and farther from the hologram plane than it was in the making of the hologram. The more divergent the reference and/or replay beams, the larger and farther away will it be. As the distance increases faster than the size (by the axial-magnification rule above), the use of divergent beams actually restricts the parallax angle (Fig A2.9).

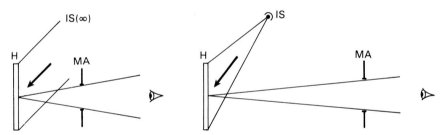

FIG A2.9 *When a transfer hologram made with a collimated beam is reconstructed using a similar beam, the real image of the master aperture appears in its geometrically-correct position. If the hologram is reconstructed using a diverging beam the master aperture is imaged considerably farther away from the hologram, and the parallax angle is reduced. IS = illumination source; H = hologram; MA = real image of master aperture.*

This applies to both vertical and horizontal parallax. However, in a rainbow hologram with a narrow slit, the vertical parallax angle is close to zero. With a focused slit the hologram behaves like a full-aperture transfer hologram in the horizontal plane, but like a restricted-aperture focused-image hologram in the vertical plane.

Effects of Shrinkage During Processing

The fringe planes within the emulsion bisect the angle between the object and reference beams*, the minor angle in the case of a transmission hologram and the major angle in the case of a reflection hologram. If material is lost from the emulsion during processing the angle of the fringes will change. If we suppose the object beam to be perpendicular to the emulsion and the reference beam to have an angle of incidence θ_1, then the angle the fringes make with the normal to the emulsion will be $\theta_1/2$. If we represent the total displacement from top to bottom of the emulsion by d, and the initial thickness of the emulsion is t_1, then $d = t_1 \tan \theta_1/2$ (Fig A2.10).

If, after processing, the new thickness is t_2, the new value of θ, θ_2, will be given by

$$d = t_2 \tan \frac{\theta_2}{2}$$

As d remains constant,

$$t_1 \tan \frac{\theta_1}{2} = t_2 \tan \frac{\theta_2}{2}$$

ie the shrinkage factor

$$\frac{t_2}{t_1} = \frac{\tan \theta_1/2}{\tan \theta_2/2} \cong \frac{\theta_1}{\theta_2}$$

Thus the proportionate change in the replay beam angle needed to satisfy the Bragg condition is equal to the proportionate change in the thickness of the emulsion (the grating condition depends on d and is unchanged).

The proportionate change in the fringe spacing in a reflection hologram is roughly equal to t_2/t_1 and this also gives the proportionate change in the wavelength reflected, if the replay beam angle remains the same as the reference beam angle. In order to reconstruct an image at the same wavelength as was used to make the hologram, the replay beam will have to be moved nearer to the normal by an angle α where $\cos \alpha = \lambda/\lambda_1 \cong t_2/t_1$.

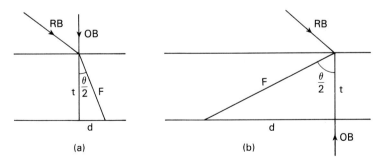

FIG A2.10 (a) *Transmission*, (b) *reflection geometry for the formation of fringes in the emulsion. F indicates the fringe plane in each case.*

---

* The refractive index of the emulsion causes a change in the fringe angle owing to refraction of the beams, but as the effect is cancelled out on emergence it is ignored here.

Modulation and Contrast

As used by the layman (and, in a different way, by the photographer) 'contrast' is an ambiguous word. To the photographer, contrast refers variously to the subject matter, to the photographic image, and to the relationship between the two. For measurement purposes a logarithmic scale is invariably used: for the subject matter the contrast is the common logarithm of the ratio of the maximum to the minimum luminance. For the photographic image, the contrast is the difference between the maximum and minimum densities present in it. Photographic density is defined as the (negative) logarithm of the transmittance. The inherent contrast of the photographic emulsion is the ratio of the contrast of the image to the contrast of the subject and is equal to the slope of the H & D curve of density vs log-exposure (Fig A2.11).

When the useful part of the H & D curve is slightly S-shaped, as in most amateur films, the average gradient is obtained by joining the maximum and minimum values by a straight line. In some specialized emulsions, of which holographic emulsions are an example, the useful portion is substantially straight, and its slope is called gamma (γ), usually spelt out. However, these methods of specifying contrast are satisfactory only for coarse imagery; they break down when fine detail is being considered and are useless for measuring the contrast of either the fringes (subject contrast), the photographic record of the fringes (contrast of photographic image) or their ratio (relative contrast). There are a number of reasons for this:

1   Although we are concerned with time-averaged intensity in the subject contrast,

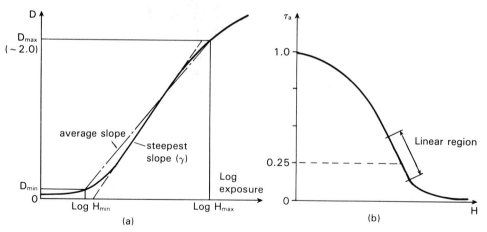

FIG A2.11   (a) H & D *curve for a typical amateur film emulsion. The inherent contrast of the film is indicated by the slope of the line joining the maximum and minimum useful densities. This is somewhat less than the steepest slope of the curve (the gamma). Both scales are logarithmic. The H & D curve provides a good prediction of the response of the emulsion in a photographic situation, but is not helpful in holography.* (b) *shows a curve of amplitude transmittance τ_a versus exposure H. The linear region in this curve is short, and in the diagram is centered on $\tau_a = 0.25$, corresponding to a density of about 0.6.*

we are concerned instead with amplitude transmittance in the image contrast, and it is the relationship between these two quantities that has to be linear.

2 The possible range of densities goes up to infinity, and the range of log-luminances goes down to minus infinity. This is plainly inappropriate, as a density of only 3 looks totally black. It would seem sensible to have a range which in the case of both density and luminance went from 0 to 1, especially if such a system could depict contrast much as we actually perceive it.

3 With all photographic emulsions the relative contrast (gamma) decreases as detail in the image becomes finer. To say that a holographic emulsion has a gamma of 5 or 6 may be true, but the measurement of gamma is carried out by the comparison of the densities produced by different exposures on comparatively large areas of emulsion, typically about 4×10 mm. This tells us nothing about the relative contrast of a set of fringes with a spatial frequency of more than 1000 cycles/mm.

The first difficulty can be overcome by measuring the intensity transmittances corresponding to a range of exposures and plotting the square roots of the values obtained against exposure. The linear region will remain linear for all spatial frequencies, regardless of the drop in relative contrast.

The second difficulty can be taken care of by measuring relative contrast in terms of modulation. This is a quantity that is already in use in communications technology. Its value is given by the expression

$$M_\tau = \frac{\tau_{max} - \tau_{min}}{\tau_{max} + \tau_{min}}$$

where M is the modulation and τ is the transmittance.

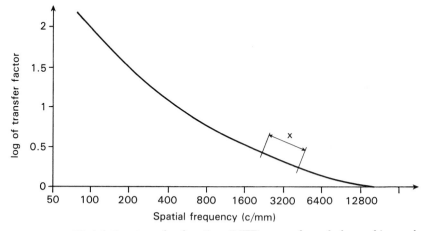

FIG A2.12 *Modulation transfer function (MTF) curve for a holographic emulsion. The modulation transfer factor represents the contrast of the developed fringes relative to that of the interference pattern projected on the emulsion during exposure. Although the emulsion has a very high gamma (4 or more) the contrast at the range of spatial frequencies actually being recorded (x in the diagram) is much lower. The graph corresponds very roughly to a plot of gamma against fineness of detail.*

Modulation can take any value between 0 and 1.

We are now in a position to overcome the third difficulty. Using modulation as the definition of contrast, the relative contrast is simply M_r/M_I, where M_I is the image (ie the fringe) modulation. If we plot the value of this term, which is called the modulation transfer factor, against the spatial frequency of the fringes, we shall have a modulation transfer function, which will give us the relative contrast for any spatial frequency, no matter how high (Fig A2.12).

Designing a Liquid-Filled Lens

Given a refractive index of about 1.5, a plano-convex cylindrical lens will have a focal length that is approximately equal to twice its radius of curvature. In Fig A2.13, $f = 2r$, c is the diameter of the lens and a is the distance across the diameter along the curved surface. Then if the diameter of the lens subtends an angle θ at its center of curvature, (Fig A2.13).

$$\frac{c}{2} = r \sin \frac{\theta}{2} \tag{1}$$

and

$$\frac{a}{2} = r \frac{\theta}{2} \tag{2}$$

From (1),

$$\frac{\theta}{2} = \sin^{-1}\left(\frac{c}{2r}\right) = \sin^{-1}\left(\frac{c}{f}\right)$$

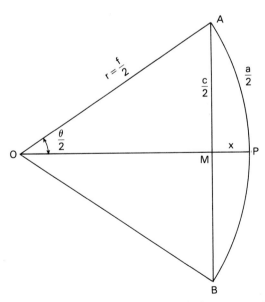

FIG A2.13 *Parameters for a liquid-filled cylindrical plano-convex lens (see text).*

Substituting into (2) gives

$$\frac{a}{2} = r \sin^{-1}\left(\frac{c}{f}\right)$$

ie,

$$a = f \sin^{-1}\left(\frac{c}{f}\right)$$

To find the width needed for the end pieces we need to find the quantity x in Fig A2.13, and add the thickness of the acrylic sheets and the depth of protective rebate required.

The distance from the chord AMB to the center of curvature O is $r - x = r \cos \theta/2$, so that

$$x = r - r \cos \frac{\theta}{2}$$

or

$$x = \frac{f}{2}\left(1 - \cos \frac{\theta}{2}\right)$$

We have already shown that $\theta/2 = \sin^{-1}(c/f)$,

so we have

$$x = f/2\{1 - \cos[\sin^{-1}(c/f)]\}$$

To this we must add the thickness of the acrylic sheets ($= 3$ mm $+ 1.5$ mm) and 2×1 mm for the overhang, so that the final width w for a plano-convex cylindrical lens will be

$$w = \frac{f}{2}\left\{1 - \cos\left[\sin^{-1}\left(\frac{c}{f}\right)\right]\right\} + 6.5 \text{ (mm)}$$

For a bicylindrical convex lens the value of x is doubled and the allowance for the thickness of the acrylic sheet is 2×1.2 mm $= 3$ mm, so the width is given by

$$w = f\left\{1 - \cos\left[\sin^{-1}\left(\frac{c}{f}\right)\right]\right\} + 5 \text{ (mm)}$$

To find the required volume of liquid:

Area of sector $\qquad OAPB = r^2 \dfrac{\theta}{2}$

Area of triangle $\qquad OAB = r^2 \sin \dfrac{\theta}{2} \cos \dfrac{\theta}{2} = \dfrac{1}{2} r^2 \sin^2 \dfrac{\theta}{2}$

Hence area of segment $\qquad APB = r^2 \dfrac{\theta}{2} - \dfrac{1}{2} r^2 \sin^2 \dfrac{\theta}{2}$

$$= r^2 \left(\frac{\theta}{2} - \frac{1}{2} \sin^2 \frac{\theta}{2}\right)$$

Substituting $\frac{\theta}{2} = \sin^{-1}\left(\frac{c}{f}\right)$ and $r = \frac{f}{2}$ gives:

Area of segment $\quad APB = \frac{f^2}{4}\left[\sin^{-1}\left(\frac{c}{f}\right) - \frac{1}{2}\left(\frac{c}{f}\right)^2\right]$ mm^2

If the depth d is in millimeters, the volume v is given by

$$v = \frac{df^2}{4}\left[\sin^{-1}\left(\frac{c}{f}\right) - \frac{1}{2}\left(\frac{c}{f}\right)^2\right] \times 10^{-6} \text{ liters}$$

for a plano-convex cylindrical lens. For a bicylindrical lens the quantity is doubled, and the depth $= c$.

References

1 Graham R W and Read R E (1986) *Manual of Aerial Photography*, pp 261–2. Focal Press.

APPENDIX
3
─────────────── ✳ ───────────────

The Fourier Approach to Image Formation

'I think I should understand that better,' Alice said very politely, 'If I had it written down; but I can't quite follow it as you say it.'
LEWIS CARROLL, *Alice in Wonderland*

Until comparatively recently, the discussion of image formation has relied on one of three models for the behavior of light. The so-called ray model uses a geometrical approach, and is satisfactory for describing the basic geometry of the formation of an image by an optical device, but it does not describe the nature of the fine structure of an optical image, nor does it predict the phenomenon of diffraction. The Huyghens wave model does describe diffraction fairly well, but is difficult to use with gratings that are other than square in transmittance profile. The Maxwell electromagnetic model is capable of describing all optical phenomena except photochemical and photoelectronic effects, but its descriptions demand a level of mathematical ability well beyond the reach of most people.

A fourth approach to the formation of optical images is much younger than the others. It is known as the Fourier model; it first appeared in the 1940s, but was not at that time taken seriously, and no book on Fourier optics was published before 1960. Unlike the other models, Fourier optics has never been a *theory* of light. The great advantage of the Fourier model over the other three models is not so much that it subsumes them (it does not altogether do so), but that it predicts *all* the phenomena of image formation, including the existence of an optical Fourier transform in the rear focal plane of a lens (a phenomenon that could not have been observed before the advent of the laser), and the interference pattern the recording of which results in a hologram. And though Fourier optics has a rigorous mathematical background, this need not be beyond the grasp of anyone with 'A' level mathematics or the equivalent. Anyone trained in electronics will already have mastered it. But the model can be used at the intuitive level, with hardly any mathematics at all, once the underlying concepts have been grasped.

Let us begin with the simplest possible object, and see how and why it diffracts light. Now, you might guess that the simplest possible object would be a circular aperture, but you would be wrong: its diffraction pattern is in fact quite difficult to analyze. Let us try again, this time with a one-dimensional object (ie some sort of pattern that varies in one direction only, and is uniform in the other). A grating of equally-spaced apertures (usually called a square grating) is one such object, and if you guessed this you would be nearly right. There is nevertheless an even simpler object, a grating in which the transmittance varies cosinusoidally with distance. We say 'cosinusoidal' rather than 'sinusoidal' because a cosine function is symmetrical about the vertical axis ($\cos \theta \equiv -\cos \theta$) (Fig A3.1)

How does a cosine grating diffract light? We can reach the answer by finding an assemblage of light beams that would produce the same pattern of amplitudes at the plane of the grating. The first thing to notice is that transmittance has to be positive (it can never be less than zero), and therefore the function shown in Fig A3.1 must contain a component that is uniform and positive. It is simply a constant value of 0.5, which raises the cosinusoid so that it sits on the horizontal axis. This leaves a second component that is positive and negative in a cosinusoidal manner, but fixed in space. Now, it is a fundamental property of all models of the propagation of light that if the direction of travel of the light energy is reversed the result will be geometrically the same. So if we could find two traveling waves that would interfere with one another to form our pattern we would have solved the problem; and we can. Two waves traveling in opposite directions create a stationary or standing wave (see Appendix 2, pp. 361–2) which has troughs and peaks of energy (nodes and antinodes) at intervals of exactly one half-wavelength. However, if the grating period (ie the spacing of the maxima) is greater than this, the two traveling waves that interfere to produce that precise pattern will be at some angle to one another that is less than 180°. We can calculate this angle fairly easily. In order to produce constructive interference (ie bright fringes) the path difference between the waves must be 0, λ, 2λ, 3λ, 4λ, etc. In order to produce destructive interference (dark fringes) the path difference must be $\lambda/2$, $3\lambda/2$, $5\lambda/2$, etc. If the grating period is d, the angle θ made by the direction of the beams with the normal is given by the formula

$$d \sin \theta = \lambda$$

This formula turns up repeatedly in diffraction theory; it is known as the *grating condition* (see Appendix 5).

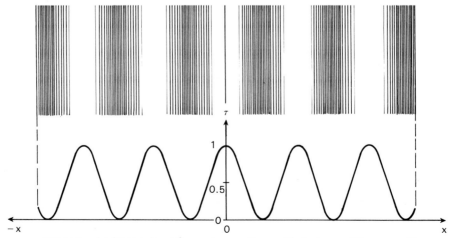

FIG A3.1 *An impression of a cosine grating, with its transmittance profile.*

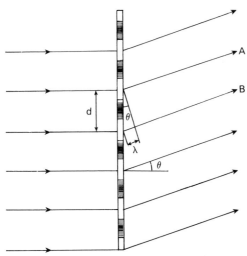

FIG A3.2 *Derivation of the grating condition. For constructive interference the optical path difference between wavefronts emerging from successive spatial periods of the grating must be one wavelength (λ). If the grating period is d, the angle of diffraction θ is given by the equation $d \sin \theta = \lambda$, or $\theta \cong \lambda/d$, if θ is small and measured in radians.*

From this formula we can calculate θ, as we already know λ and d. As noted above, it is a fundamental property of light that its direction can be reversed without altering the geometry of the system. If, therefore, a beam of light is incident normally on a cosine grating, half of the total amplitude is transmitted unaltered in direction (the zero-order beam) and the remaining amplitude will be divided into two equal beams in directions $\pm \theta$, exactly as it was for the beams that would produce such a pattern. The value of θ can be calculated from the formula

$$\theta = \frac{\lambda}{d}$$

where $\theta \cong \sin \theta$. (This is true for angles up to about 20° provided θ is measured in radians (2π radians = 360°).) We usually describe a cosine grating in terms of its spatial frequency q, which is related to the period d by the formula

$$\theta = q\lambda = \frac{\lambda}{d}$$

As half the total amplitude is present in the undiffracted beam, the two diffracted beams must each possess $\frac{1}{4}$ of the total amplitude of the incident beam (Fig A3.3).

As the time-averaged intensity $I = <\frac{1}{2}A^2>$, where '$<>$' means 'time-averaged', it will be seen that the maximum diffraction efficiency for either diffracted beam is only $(\frac{1}{4})^2$ or 6.25%. In practice it is usually less than this, as the modulation of the grating is in general less than unity, ie it is nowhere fully transparent or opaque, so that the undiffracted beam carries more than half the total amplitude.

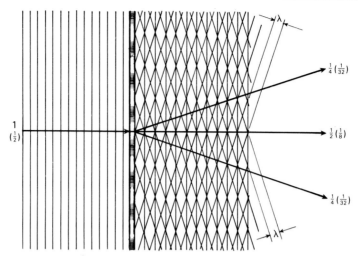

FIG A3.3   FIG A3.2 *redrawn to show wavefronts rather than rays. It will be seen that the interference pattern of the two diffracted waves matches the period of the grating. It can be shown that for a fully-modulated cosine grating (ie opaque-to-clear) the emerging beams will have amplitudes in the proportions $\frac{1}{2}:\frac{1}{4}:\frac{1}{4}$ for an incident amplitude of 1. The intensities are in proportion to the squares of the amplitudes, and are shown in parentheses.*

Fourier Series

To return to the main line of the argument: let us suppose that the value of q is increased, ie the spatial period (grating spacing) is reduced, then from the formula it is clear that θ is increased; the two beams are thus emitted at a greater angle of emergence. So what happens if we superpose two gratings of different spatial frequencies, say with one having three times the spatial frequency and one-third the amplitude? (Fig A3.4(a)). Well, the resultant amplitude transmittance is a more complicated transmittance pattern (Fig A3.4(b)) which is no longer cosinusoidal, though it does have the same period.

What happens when we pass a beam of light through this grating is predicted by the Fourier model: the two gratings produce their sets of plane waves independently, each as if the other did not exist (Fig A3.5).

We can add further gratings and produce a more complicated waveform, but the light is still diffracted as if the component gratings were acting completely independently. If we choose the frequencies and amplitude transmittances of the gratings carefully they can be made to add up to some familiar profiles. For example, if we choose gratings with spatial frequencies of $q, 3q, 5q, 7q$, etc, with amplitudes respectively of $A, \frac{1}{3}A, \frac{1}{5}A, \frac{1}{7}A$, etc in the correct phase relationships (see Fig A3.9 below), we shall finish with a square wave. We can choose other combinations of gratings that produce triangular, sawtooth and many other transmittance profiles. In fact, *any* periodic function can be shown to be made up of nothing but sinusoidal and consinusoidal components.

We have now accounted for two diffraction phenomena. The first is that when

THE FOURIER APPROACH TO IMAGE FORMATION

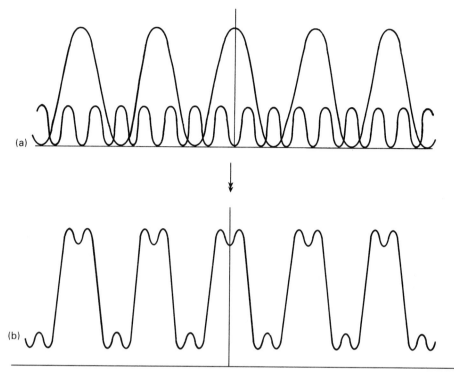

FIG A3.4 *If two cosine gratings of differing spatial frequencies are superposed, a more complicated pattern results; but if the frequencies are simply related, the result is still a periodic waveform.*

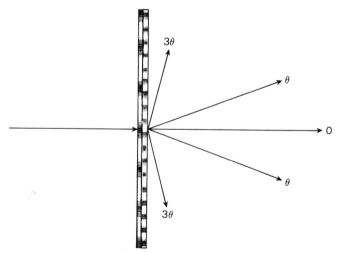

FIG A3.5 *Each of the cosine gratings from* FIG A3.4 *produces its own diffracted beam independently.*

we illuminate a cosine grating with a laser beam we get just two diffracted spots (Fig A3.6). We can insert a lens to bring the plane waves into focus, and provided θ is fairly small and measured in radians, we can use the approximation $\sin \theta \cong \tan \theta \cong D/f$, where D is the distance of the spots from the optical center.

$$D = f\theta = fq\lambda$$

Thus D is proportional to q, the spatial frequency of any component of the grating.

The second is the converse of the synthesis of gratings. We can say with confidence that any repeated waveform, of any shape, behaves as though it is the sum of a large (usually infinite) number of sinusoidal and cosinusoidal components of frequencies that are integral multiples of the fundamental frequency (Fig A3.7).

All the waveforms of Fig A3.7 – indeed, any repetitive waveform – can be synthesized in a similar manner to our square wave, and if gratings are made with these transmittance profiles the positions and intensities of the diffracted spots will be as predicted. Notice that the 'sawtooth' profile is not symmetrical, and is in fact synthesized only from sine terms, which are antisymmetrical ($\sin(-\theta) = -\sin\theta$): the spots which add up on one side of the zero-order beam cancel on the other side. The so-called *blazed grating* which gives rise to this profile has a high diffraction efficiency for the beams on one side of the zero-order beam and a low diffraction efficiency on the other side.

The examples of Fig A3.7 should give you some idea of the principles of Fourier series. The general relationships which have emerged (and which are capable of mathematical proof) are:

1 Any regularly-repeated one-dimensional function such as a grating transmittance

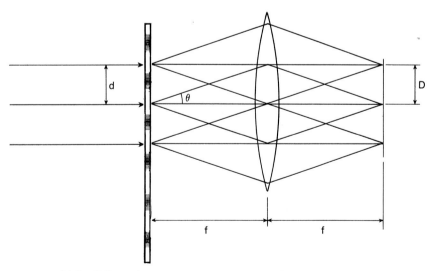

FIG A3.6 *If the diffracted beams are focused by a lens of focal length* f, *a cosine grating will produce three spots separated by a distance* D *such that* D = f sin θ = fq *where* q *is the spatial frequency of the grating (=1/spatial period). In general,* D *will be much larger than* d.

THE FOURIER APPROACH TO IMAGE FORMATION

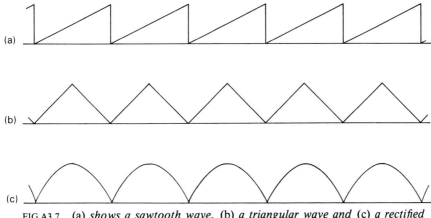

FIG A3.7 (a) *shows a sawtooth wave,* (b) *a triangular wave and* (c) *a rectified cosine wave. All these waveforms (and, indeed, any other periodic function) can be analyzed using Fourier methods into pure sinusoidal and cosinusoidal components.*

function will behave as though it were the sum of a number of sinusoidal functions; if the pattern is symmetrical about any point these will all be cosinusoidal functions. The spacing of any pair of spots in the diffraction pattern is proportional to the spatial frequency of the grating component producing them.

2 The intensity of the spots is directly related to the modulation of the spatial-frequency component causing them to appear.

3 The orientation of the pairs of spots is at right angles to the orientation of the (cosine) bars of the grating which was responsible for them.

Fourier Transform

The Fourier spectrum of a regularly-repeated waveform can be plotted on a graph in terms of amplitude and (spatial) frequency. The Fourier spectrum of a plain cosine grating is the simplest (Fig A3.8(a)).

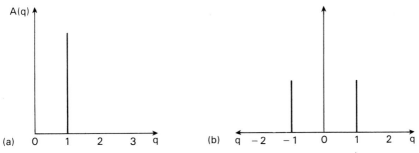

FIG A3.8 (a) *The Fourier spectrum of a cosine wave. It contains only one frequency.* (b) *shows a way of making the spectrum symmetrical, by remembering that* $\cos\theta$ *is the same as* $\cos(-\theta)$, *and putting the energy on either side of zero.*

In practice, for reasons connected with the mathematical analysis of the Fourier spectrum, we remember that $\cos\theta \equiv -\cos\theta$ and split the non-zero spatial frequencies into two, one half positive and the other half negative. This also has the merit of providing a better model for the world of diffraction spots (which always come in pairs) and their amplitudes. Thus the plot of Fig A3.8(a) becomes as in Fig A3.8(b).

The relative intensities of the spots are obtained by squaring these values. A square grating is made up of a zero-order term equal to $2/\pi$ of the incident amplitude, and an infinite series of odd multiples of the basic spatial frequency with their amplitudes decreasing in proportion to their order. One way of representing the sum of these components is as in Fig A3.9.

A second, neater way is shown in Fig A3.10. This is the Fourier spectrum of the square grating, and corresponds to the spatial frequencies and amplitudes of its components. Each component spatial frequency q produces a pair of spots at a distance $d = fq\lambda$ from the axis, and the intensities of the spots are proportional to the squares of their amplitudes. Note, however, that some of the waves are in antiphase with respect to others; if the spots of the diffraction pattern are recorded photographically, the information about phase is irretrievably lost, as if you take the square root of the intensities you get back to the magnitude of the amplitude but not to whether it is positive or negative. If you get the sign wrong for any of the components you will not get a square wave but one of a quite different profile. For example, if you guessed that *all* the components had a positive amplitude together, you would be describing something like a triangular wave, not a square wave (Fig A3.11). This is one of the problems that FT holography helps us to solve.

It is clear from the formula above that the higher the spatial frequency in the

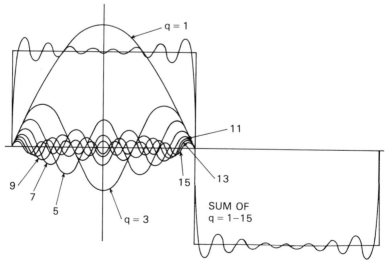

FIG A3.9 *The cosinusoidal components of a square wave up to the fifteenth harmonic. Only odd harmonics are present, with amplitudes in inverse proportion to their frequency. Alternate harmonics are in opposite phases.*

THE FOURIER APPROACH TO IMAGE FORMATION 383

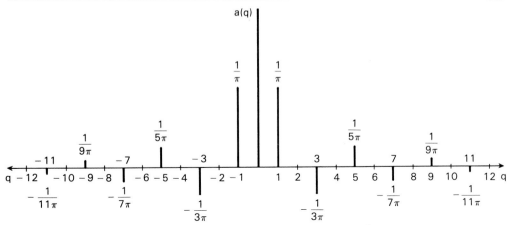

FIG A3.10 *An alternative way of depicting the components of a square wave as a Fourier (frequency) spectrum. Alternative harmonics are indicated as positive and negative.*

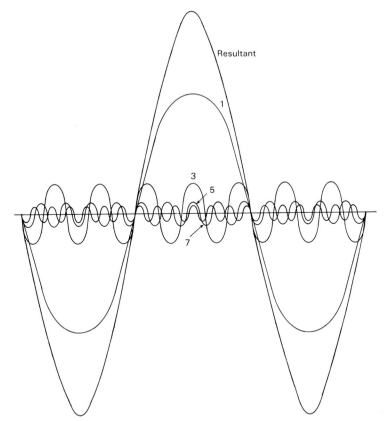

FIG A3.11 *If all the frequency components that go to make up a square wave were positive-going, the shape of the final composite periodic function would look quite different.*

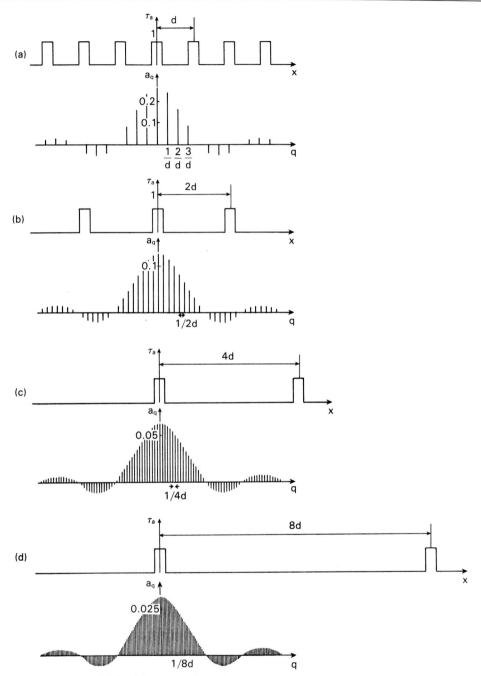

FIG A3.12 (a) *If the rectangular pulses that make up a square wave have their spacing doubled, the number of frequency components doubles too. As the spacing betwen the pulses increases towards infinity, the frequencies become ever closer together, until they merge into a continuous spectrum.*

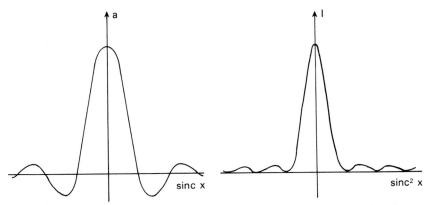

FIG A3.13 *The Fourier spectrum of a single rectangular pulse (a top hat function) is a* sinc *function. In the Fourier-transform plane of a lens the diffraction pattern of a single slit (the optical equivalent of a top hat function) appears as a* sinc^2 *function, since intensity is proportional to the square of amplitude. The width of the central lobe of the diffraction pattern is inversely proportional to the width of the slit.*

original grating, the larger the separation of the spots in the rear focal plane of the lens – which we may now call the Fourier-transform plane. There is a reciprocal relationship between the period of the grating and the spacing of the spots. Let us now consider a 'square' grating in which the squares are gradually moved farther apart. As the fundamental spatial frequency decreases, the separation of the spots also decreases (Fig A3.12). As the adjacent square pulses move away from each other towards infinity, the number of spatial frequencies present in the Fourier transform increases; in the limit they completely fill the space of the envelope which has contained them, so that there is a continuous spectrum of spatial frequencies, and this will give rise to a diffraction pattern of which the intensity is proportional to the square of the envelope of the curve. The associated mathematics shows the amplitude function to be of the form $\sin x/x$, an expression which is called a *sinc function*, and is written 'sinc x'; the intensity pattern is of the form $\text{sinc}^2 x$ (Fig A3.13).

The same reasoning can be used to develop any repeated function into an isolated pulse of the basic shape. Mathematically, what happens is that a 'Σ' (Sigma or repeated addition operation) turns into a $\int$ (integration operation). The situation becomes slightly more complicated when the single pulse is asymmetrical, but in the vast majority of cases which concern holography it is symmetrical.

Reciprocal Relationship of x-space and Fourier Space

The width of the sinc function that is the FT of the square pulse (or single-slit function) is inversely proportional to the width of the slit, as you might expect. As the slit becomes narrower and narrower, its FT becomes broader and flatter until in the limit, as the slit shrinks to infinitesimal width, its FT becomes a uniform straight line. You can actually see this happen if you put a variable slit in front of a laser output port. You can see the sinc² pattern gradually broadening as you make the slit narrower, until just before it vanishes it becomes very wide indeed. It also becomes very faint, as the total amount of light energy is being steadily reduced as the slit is made narrower. In order to keep the energy constant we would have to increase the amplitude of the pulse to match, until we reached a pulse of infinitesimal width, when it would have to be of infinite amplitude. This would be mathematically meaningless. However, an ingenious dodge devised by Paul Dirac and called the *delta function* takes care of the difficulty, and this is the name given to such spikes of infinitesimal duration but unit energy (Fig A3.14). A pulse of infinitesimal width thus contains all frequencies in equal amount.

What happens if we go the other way, and make one single square pulse indefinitely wider? The width of the sinc function becomes progressively less and less, until in the limit, as the square pulse becomes of infinite width, its FT shrinks to a delta function (Fig A3.15).

This is an example that shows the reciprocal relationship between *x*-space and frequency space. The FT of a delta function is a constant: the FT of a constant is a delta function. Another example is that of a cosine function, which, of course, has a single (spatial) frequency (Fig A3.16).

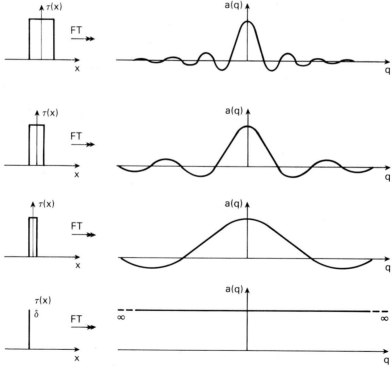

FIG A3.14 *If a rectangular pulse is made narrower, its FT becomes broader, until, in the limit, when the pulse becomes an infinitely-narrow delta function, its FT is a uniform straight line.*

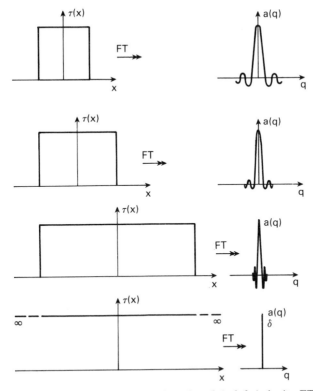

FIG A3.15 *Conversely, if the pulse is broadened indefinitely its FT becomes narrower; in the limit, as the pulse becomes indefinitely wide, its FT shrinks in width to a delta function.*

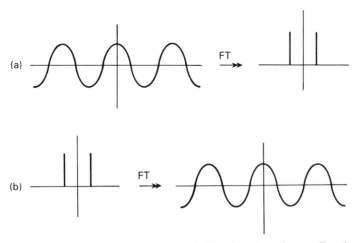

FIG A3.16 *A cosine function and a pair of delta functions form a Fourier pair: each is the FT of the other. This can be shown optically by (a) the pair of spots produced by a cosine grating, and (b) the cosine pattern produced by two slits.*

Figure A3.16(a) is the example of the cosine grating producing two spots (the center spot is not diffracted and is not concerned in the argument), and Fig A3.16(b) is the well-known example of Young's fringes produced by a pair of narrow slits. These are Fourier pairs; each transforms to the other, and illustrates a fundamental property of Fourier transforms:

$$FT(FT(f)) \equiv f$$

or, if $FT(f) = F$,
then $FT(F) = f$,

give or take the odd mathematical constant. (There are several variations of the formula for mathematically deriving a Fourier transform, each involving a constant. Reluctant mathematicians (including the author) are constantly irritated by finding an unwanted $1/2\pi$, or something equally unattractive, hanging about at the end of the calculations and destroying their symmetry. Such encumbrances are easier to deal with if you use the complex-number format.)

All of this is beautifully illustrated in practice. If you pass a laser beam through a cosine grating you will get the two spots, and if you send it through a pair of narrow slits you will get a cosine pattern ($\cos^2$ in fact, because intensity varies as the square of the amplitude). If you pass the beam through a variable-width slit you will see the diffraction pattern become broader and broader as you narrow the slit. By the way, you don't really need the lens. You see, if you use a throw of several meters you need a lens with a focal length of several meters — but this is very little different from having no lens at all; so if your throw is more than about a meter or so you can forget the lens, and use the distance of the screen from the object in the equation instead of the focal length.

The Fourier Convolution Theorem

So far we have considered only what happens when we add two functions in *x-space*: in frequency space the FTs are also added. However, when two functions are multiplied in *x*-space, something quite different happens in frequency space. The principle of this is fundamental to image processing and to many other techniques, including imagery from satellite and spacecraft photography and radar.

If you set up a regular line of point apertures (ie delta functions) in a laser beam, you would expect to get a similar line of points of light on the screen, with a spacing inversely proportional to that of the original delta functions (Fig A3.17).

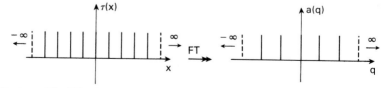

FIG A3.17 *The FT of an infinitely long array of delta functions (a 'comb' function) is a similar array, the spacing in inverse proportion to that of the original function.*

THE FOURIER APPROACH TO IMAGE FORMATION 389

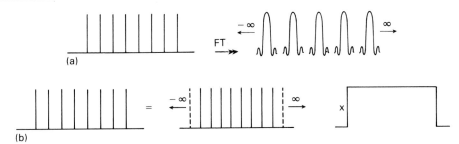

FIG A3.18 *A 'short comb' is an infinite comb multiplied by a top hat. In frequency space the FT of the top hat (sinc function) is dealt out to each spike of the FT of the (infinite) comb function.*

(This has not been proved here, but it is true.) Thus an array of delta functions, known as a *comb function*, produces an FT that is also an array of delta functions, with a spatial frequency that is inversely proportional to that of the generating function. But if we limit the comb to only a small number of spikes the pattern of the FT changes: each spike turns into a narrow sinc function. Now, the sinc function is the FT of a single square pulse (a top hat), and, if you think about it, you will appreciate that a 'short comb' is in fact an infinite comb multiplied by a top hat (Fig A3.18). What is happening in frequency space is that each delta function has been dealt the FT of a top hat, ie a sinc function. We say that the two functions are 'convolved,' and use the symbol ⊛. Convolution follows similar laws to multiplication, in that $A \circledast B \equiv B \circledast A$, and $A \circledast (B \circledast C) \equiv (A \circledast B) \circledast C$. (Read '$A \circledast B$' as '$A$ convolved with B'.) When two functions are multiplied in x-space their FTs are convolved in frequency space, ie

$$FT(f) \circledast FT(g) \equiv FT(f \times g)$$

This is known as the *convolution theorem*.

The converse is also true. If you take a series of narrow (but not infinitely narrow) slits and illuminate this object with a laser beam, you will get a series of narrow lines which fade off in a manner reminiscent of the diffraction pattern of a single slit (Fig A3.20).

This time the convolution is in the object: the comb function is convolved with a narrow top hat function. In frequency space the FTs of the comb and top-hat are multiplied, so that the row of delta functions is limited by a broad sinc function. This is an example of the converse of the convolution theorem:
When two functions are convolved in x-space, their FTs are multiplied in frequency space.

$$FT(f) \times FT(g) \equiv FT(f \circledast g)$$

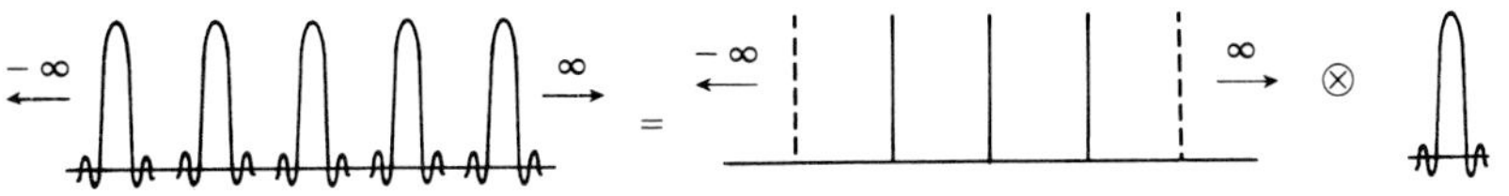

FIG A3.19 *When one function is dealt out to another, they are said to be 'convolved'.*

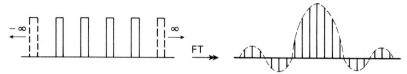

FIG A3.20 *When functions are convolved in x-space, their FTs are multiplied in Fourier space. In this case a comb function is convolved with a top hat, so in Fourier space the FT of the comb function (another comb function) is multiplied by the sinc function.*

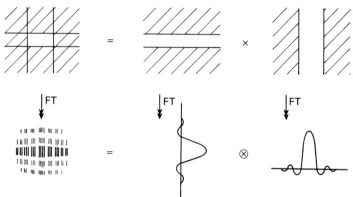

FIG A3.21 *In two dimensions a rectangular aperture is the product of two slits at right angles. The FTs of the two slits are convolved, in two dimensions.*

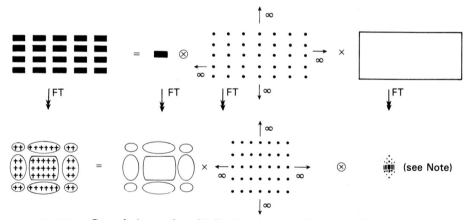

FIG A3.22 *Convolution and multiplication in a single figure. The rectangular 'motif' is convolved with a two dimensional comb, or 'bed-of-nails' function, and this is multiplied by a large rectangle. In Fourier space the FT of the rectangle is convolved with the FT of the bed-of-nails function, and the whole is multiplied by the FT of the motif, which is a large version of* FIG A3.21. *Thus what was the mask in x-space becomes the motif in frequency space, and vice versa. Notice that the largest item has the smallest FT and the smallest item the largest FT.* NOTE: *The final motif at the bottom right has been magnified. It resembles the top left pattern in* FIG A3.23, *but is very much smaller.*

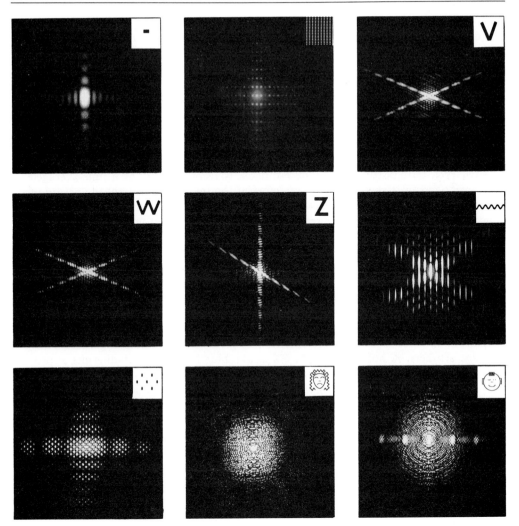

FIG A3.23 *Far-field diffraction patterns. In each case the original object is inset. Objects drawn by Steve Jeffs. The nature of each pattern can be predicted from the nature of the corresponding object by using the rules given on pp. 388–90, and many features of the object can be deduced from an examination of the pattern (eg in the diffraction pattern of the little face at the bottom right, the hairline can be deduced from the strong series of horizontally-orientated spots – though this same set of lines could equally well have been a moustache).*

Two-dimensional Objects

As soon as we begin working in two dimensions the implications of the convolution theorem become important. For example, when two slits are placed at right angles, they produce a rectangle. This is the *product* (ie the multiplication) of the two slits. Their FT is the convolution of two sinc functions at right angles (Fig A3.21).

A motif (such as the rectangle above) dealt out to (convolved with) a two-dimensional array of delta functions (called a *bed-of-nails function*) will appear in frequency space as another bed-of-nails function with its aspect ratio reversed, masked by (ie multiplied by) the FT of the rectangle (Fig A3.22).

We have considered the smallest item (the motif) in an array (the lattice) which is limited by the overall shape (the mask). Many images can be broken down in this way, from microscopic crystals to views of great buildings. In the FT plane the motif becomes the mask, as we have seen: the lattice retains its general pattern but reverses its orientation. The FT of the original mask becomes the motif in frequency space (remember that the largest item in x-space is always the smallest in frequency space). Now, even though the field in the FT plane is recorded only as a time-averaged intensity and relative phases are lost, most of the information about the nature of the original object in x-space can still be retrieved, and by intelligent use of the addition and convolution theorems it is possible to make an informed guess as to the type of object that produced the diffraction pattern. Indeed, this is the basis of the techniques of X-ray crystallography. It is not possible as a rule to be certain of the exact position of any detail, however, as this information is contained in the phase relationships, and the phase information is lost. This information is what FT holography is able to retrieve, and why it is so useful in information processing. For example, in a photograph blurred through movement the blur function is a short straight line. Its FT is a sinc function, in which the negative-going lobes cause spurious reversal of certain fine details; this gives rise to double edges and other image artifacts. A hologram of the FT of the blur function, including the appropriate phase reversals, placed in the FT plane of an imaging lens, can retrieve most of this corrupted information and save the imagery. This technique is discussed on pp. 334–6.

The Fourier model is equally powerful in incoherent optics. A camera lens operating in incoherent light does produce an FT in its rear focal plane, but this FT is convolved (by the rules) with every point on the surface of the light source, and with every wavelength of the spectrum, all diffracted to different points on the plane. Hence the FTs and their associated diffraction patterns are totally smeared out; but that they do indeed exist is indicated by the fact that an inverted image (the double FT) is formed at the image focal plane. It has needed only the coming of the laser to show what was really going on behind the scenes.

PS Just in case you are wondering what the FT of a circular aperture is, it is a *Bessel function* of the first order. Its profile looks something like a sinc function, but the spacing of the zeros is different – and, of course, it is circularly symmetrical. The associated diffraction pattern is our old friend the Airy pattern (Fig A3.23).

APPENDIX
4

The Holodiagram

And the crew were much pleased when they found it to be
A map they could all understand.
      LEWIS CARROLL, *The Hunting of the Snark*

Nils Abramson invented the holodiagram in 1969 to represent the geometrical relationships between reference source, object and hologram. His subsequent book (1) makes extensive use of the holodiagram, as well as the moiré concept, to clarify the reasoning underlying the geometry. The object is considered as small, so that it sends out a spherical wavefront in the same way as the reference source. If we draw a set of concentric circles representing successive wavefronts from both object and reference beams, and fill in alternate rhomboids as in Fig A4.1 to simulate moiré patterns, the moiré fringes form a set of ellipses and a set of hyperbolas, in both cases with the two sources as foci. It must be borne in mind that what has been drawn is actually a cross-section of *ellipsoids* and *hyperboloids* of revolution around the line joining the sources. If the two sources are at A and B, then for any two circles of radius R_A and R_B, $R_A \pm R_B = n\lambda$ where n is an integer.

Figure A4.2 shows a single rhombic cell enlarged. The spacing between adjacent wavefronts is $\lambda/2$, and provided this value is small, the rhombs can be thought of as straight-sided. In Fig A4.2 the angles in the rhombs forming the ellipses have been called 2α, and the length of the diagonal DF is $\lambda/2 \sin \alpha$, which is the separation of the hyperbolas. Similarly, the length of the diagonal CE is $\lambda/2 \cos \alpha$, which is the separation of the ellipses. The loci of equal separation of the ellipses are all arcs of circles; these circles are found to be also the loci of equal separation for the hyperbolas (Fig A4.3).

The hyperbolas represent stationary surfaces which produce interference fringes located at their intersections with a surface placed in their space. The ellipsoids represent stationary fringe-producing surfaces formed when A is a source and B the focal point of a beam of convergent light. The moiré pattern of Fig A4.1 is a geometrical fiction: the real surfaces are continuous. Their real shapes are shown in Fig A4.4. The rhombs are the result of the approximation to square waves of what in fact are cosine waves.

If it were possible to produce a replica of these, every ellipsoid would reflect light from A towards B (this is a fundamental property of ellipses). The optical path from A to B is also constant for any point on the periphery of an ellipsoid (a fact

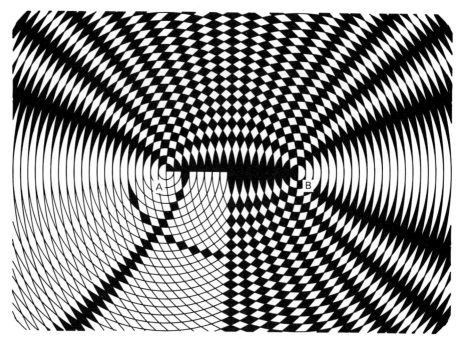

FIG A4.1 *The holodiagram drawn as a moiré pattern. A is the reference source and B the object. The moiré fringes form a set of ellipses and a set of hyperbolas. For clarity only a single ellipse and hyperbola have been masked in the fourth quadrant.*

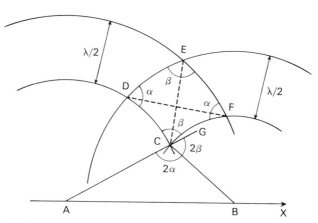

FIG A4.2 *An enlarged rhomb CDEF from* FIG A4.1. *A is the reference source and B the object.* $AC + BC = n\lambda (n = 1, 2, 3 \text{ etc})$. *If λ is very small the sides of the rhombs can be treated as straight lines, and DF (distance between hyperbolas)* $= \lambda/2 \sin \alpha$ *and CE (distance between ellipses)* $= \lambda/2 \cos \alpha$.

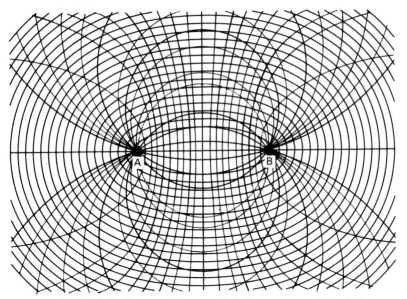

FIG A4.3 *The full holodiagram. The arcs of circles represent the loci of constant separation between ellipses and between hyperbolas.*

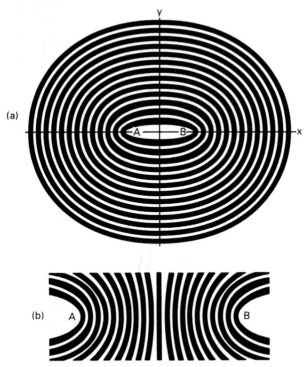

FIG A4.4 *The elliptical* (a) *and hyperbolic* (b) *shells of the holodiagram. Their thicknesses are proportional to* $1/\cos\alpha$ *and* $1/\sin\alpha$ *(from* FIG A4.3*) respectively. Any intersection through the ellipsoids will produce a zone plate which, when illuminated from A, reconstructs B by diffraction.*

which accounts for the ability to construct an ellipse with two drawing pins and a loop of string). In addition, all the path lengths are integral multiples of the wavelength. Hence the Bragg condition is satisfied for every point on the whole family of ellipsoids. A similar result can be proved, somewhat less easily, for the family of hyperboloids: all the light emitted from A is reflected as though it had been emitted from B; thus B is a virtual image of A. If the hyperboloids are intersected by a photographic emulsion, when processed and placed in its original position it will form a Gabor zone plate which diffracts light from A towards B.

Two important conclusions follow immediately from the holodiagram. The first is that the fringe system of any hologram can be found directly from the set of hyperboloids (Fig A4.5). The second is that the sensitivity to in-plane and out-of-plane movement of the holographic plate can be seen from the angle of the fringes within the emulsion. Sensitivity is a maximum at right angles to the fringes, and a minimum parallel to them. (It is never zero, owing to the perturbation of the fringes by irregularities of the object wavefront.)

Abramson's original holodiagram contained only the ellipsoids, separated by a distance equal to the wavelength λ. For a basic single-beam transmission hologram, using a two dimensional diagram, a shift of one fringe follows from the movement of the object from one ellipse to the next. The geometry is optimized if the reference beam also follows a path which touches the same ellipse as the object lies on (this means that the path lengths are closely matched). Two important results follow from this representation. The first concerns interferometric measurements. If the angle subtended at a point P on the object by the reference source and the hologram is 2ψ (psi) (the bisector of this angle being by definition the normal to the

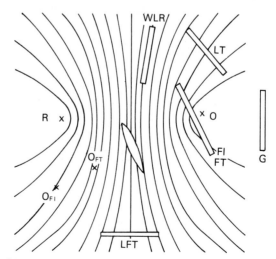

FIG A4.5 *The fringe pattern in the emulsion of any hologram can be deduced from the direction of the hyperbolas of the holodiagram.* R = *Reference source,* O = *Object source,* O_{FT} = *Object (FT hologram),* O_{FI} = *Object (focused-image hologram),* G = *Gabor hologram,* LT = *Laser transmission hologram,* FI = *Focused-image hologram,* FT = *Fourier-transform hologram,* LFT = *Lensless Fourier-transform hologram.* FIGS A4.1–A4.5 *after Abramson* (1).

THE HOLODIAGRAM

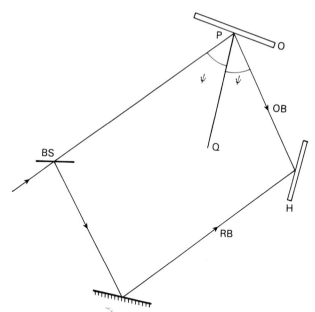

FIG A4.6 *After Hariharan (2). RB and OB are the reference and object beams respectively. By definition, PQ is normal to the ellipsoid on which the object lies, and gives the direction of maximum sensitivity of the displacement fringes.*

ellipse at that point) a shift of one fringe requires a movement of $\lambda/2 \cos \psi$. This means that if the fringe order at P is N, then the component of the displacement normal to the ellipse passing through P is $N\lambda/2 \cos \psi$ (Fig A4.6).

The second result is the ability to plan the geometry of a holographic set-up for recording a large object with a laser of restricted coherence length. The plan is drawn on a holodiagram with ellipses corresponding to increments in optical path difference. The positions of the reference source, object and hologram can be optimized so that (a) the sensitivity of the system to motion of the object in a particular direction is either maximized or minimized, and (b) at the same time the whole of the area to be recorded lies within the bounds of two adjacent ellipses which contain the reference source.

Abramson's book (1) contains much information on the subtler uses of the holodiagram, which are mainly concerned with the analysis of interferometric information.

References

1 Abramson N (1981) *The Making and Evaluation of Holograms*, pp. 28–34, 57–8, 110–5, 144–55, 284–93. Academic Press.
2 Hariharan P (1984) *Optical Holography*, pp. 223–4. Cambridge University Press.

APPENDIX
5

Bragg Diffraction

> She puzzled over this for some time, but at last a bright thought struck her. "Why, it's a looking-glass book, of course!"
>
> LEWIS CARROLL, *Through the Looking Glass*

Bragg diffraction, named after Sir William Bragg and his son Sir Lawrence Bragg, who first described the phenomenon, occurs when light passes into a medium made up of uniformly-spaced layers of partially-reflecting material, or of alternate high and low refractive index, with a spacing that is of the same order as the wavelength of light. Bragg diffraction is utilized in dielectric mirrors and beamsplitters; the principle also underlies the anti-reflection coating of camera lenses, binocular objectives, etc. Bragg diffraction is of immense importance in the reconstruction of a holographic image.

In theoretical texts on holography, transmission holograms are often treated as though they are of infinitesimal thickness; their diffractive behavior is accounted for by invoking the grating condition (see p. 376). If this were so, the two diffracted beams would be of equal intensity, and the real and virtual images would be equally bright. Plainly this is not the case: in practice the real image is very dim. Indeed, in order to be able to see it at all, we usually have to flip the hologram, which reverses the geometry. This effect is due to the Bragg condition. The reason the grating condition leads us to the wrong answer is simply that the emulsion is *not* infinitely thin. As a rule its thickness is at least six wavelengths. The interference fringes that make a transmission hologram are not just on the surface of the emulsion but go right through its thickness like the slats of a venetian blind. The fringe planes lie parallel to a line bisecting the angle between the object and reference beams* (Fig A5.1(a)). If the object beam is perpendicular to the emulsion, and the

---

* In fact, they bisect the angle between the *refracted* object and reference beams. The change in direction of a wavefront crossing the interface of two optical media is given by Snell's Law

$$n_1 \sin \theta_1 = n_2 \sin \theta_2$$

where n_1 and n_2 are the refractive indices of the two media and θ_1 and θ_2 are the corresponding angles made with the normal. On emergence the situation is reversed and the effect is cancelled out, so for our present purpose this complication can be ignored. In fact, its only important consequence is that because the fringe planes are closer together inside the emulsion than they are in air, the resolving power of the emulsion has to be some 50% higher than would appear to be necessary from the formula used here.

BRAGG DIFFRACTION

reference beam is at an angle of incidence θ, then the fringe planes will be at an angle $\theta/2$ to the perpendicular. (They also lie along the bisector of the angle between the two beams when neither of them is perpendicular to the emulsion.) The spacing d of the fringes parallel to the emulsion surface is given by the grating condition as

$$\lambda = d \sin \theta$$

where λ is the wavelength of the laser beam.

At this point the Bragg condition enters. For the intensity of the diffracted beam to be a maximum, the wavefronts diffracted by successive fringes must all be in phase or, to be more precise, any phase difference between the wavefronts must be an integral number of wavelengths. If we think of the fringes as acting like simple plane mirrors we can see the conditions necessary for this to occur (Fig A5.1(b)). If the grating spacing is appropriate to the wavelength of the reconstruction beam, a bright image will result.

The geometry determining the spacing of the fringes (which we will call the Bragg planes) is shown in Fig A5.1(c). If we call the separation of the Bragg planes w, we can find the angle of incidence of the replay beam that gives maximum intensity in the diffracted beam. The Bragg condition requires that the optical path difference for light diffracted at each successive plane is $n\lambda$, where n is an integer (for a cosine grating $n = 1$ only). Now, we already know the grating condition

$$\lambda = d \sin \theta$$

where d is the grating spacing parallel to the emulsion surface and θ is the angle between the beams, but we are more interested in the Bragg condition, and we need to know the spacing w of the Bragg planes. From From Fig A5.1(c) we can see that this is given by

$$w = d \cos \frac{\theta}{2}$$

By using the trigonometric relationship $\sin A = 2 \sin A/2 \cos A/2$, we can rewrite the grating equation

$$\lambda = 2d\sin\frac{\theta}{2}\cos\frac{\theta}{2}$$

so that

$$\frac{\lambda}{2} = w \sin \frac{\theta}{2}$$

and this is the Bragg condition for reconstruction.

The real-image beam satisfies the grating condition, but does not satisfy the Bragg condition, as it is at an angle of 2θ, and the mismatch increases as θ increases. The zero-order beam also becomes increasingly mismatched to the Bragg condition (Fig A5.1(d)). As θ approaches $90°$ the angle of the Bragg planes approaches $45°$, and their length increases to 1.4 times the thickness of the emulsion. Now, it is clear from the Bragg equation that if θ is fixed then λ is also fixed. The hologram is thus wavelength-sensitive; and the more extended the Bragg planes are, and the more of them there are within the thickness of the emulsion, the more wavelength-sensitive is the hologram. For a grazing incidence, such as is typical of a cylindrical hologram

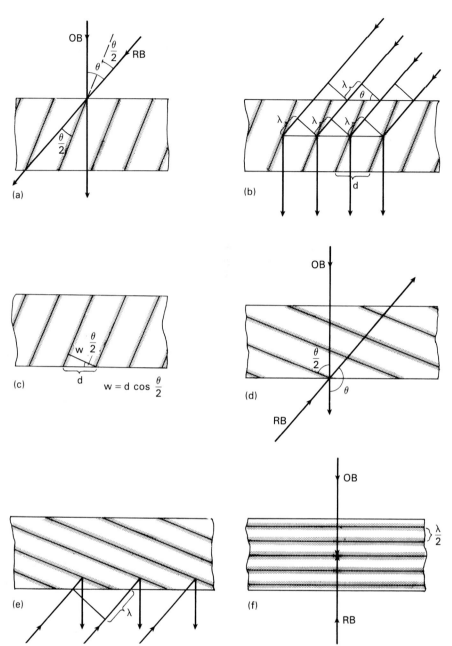

FIG A5.1 (a) *In a transmission hologram the fringes are like a venetian blind, and their planes are parallel to the bisector of the reference and object beams. (b) On reconstruction rays emerging in the direction of the original object beam differ in optical path by an amount equal to exactly one wavelength. (c) The relationship between grating spacing and fringe plane spacing. (d) When the reference and object beams are incident from opposite sides of the emulsion, the fringe planes are more nearly parallel to the emulsion surface. (e) Reconstruction in a reflection hologram. (f) When $\theta = 180°$ the fringe spacing is one half-wavelength.* NOTE: *for simplicity the effect of the refractive index of the emulsion has been ignored.*

(see pp. 128–30), such a hologram will reconstruct a passable image when illuminated with white (or preferably amber) light. When θ exceeds $90°$, so that the reference and object beams are incident on the emulsion from opposite sides, the lengths of the Bragg planes and the number of planes through the thickness of the emulsion are increased so that they resemble the pages of a book rather than a venetian blind (Fig A5.1(d)). This, of course, is the geometry for making a reflection hologram (Fig A5.1(e)). When the two beams are exactly opposed, $\theta = 180°$ and $w = \lambda/2$ (Fig A5.1(f)). There are up to twelve Bragg planes within the thickness of the emulsion, and the zero-order and real-image beams are totally suppressed, as are all inappropriate wavelengths. Of course, for practical reasons we make reflection holograms with $\theta \simeq 135°$ rather than $180°$, but we can still achieve very high diffraction efficiencies and narrow bandwidths.

If the angle of incidence of the (white) replay beam is increased to a more glancing angle, the image hue shifts towards blue. This is because the optical path difference for a given Bragg plane spacing becomes less as the angle of incidence increases. Conversely, if the angle of incidence is decreased the image becomes redder. This property is sometimes used to tune reflection master holograms to give a bright reconstruction by laser light. Unfortunately, the change in angle also causes some image distortion.

If material is lost from the emulsion during processing, the Bragg planes will in general become closer together. This means that a master hologram cannot be set up using the original geometry and give a bright reconstruction for transfer purposes. With transmission master holograms the fringe planes may become S-shaped as a result of shrinkage, resulting in a lowered diffraction efficiency and loss of image sharpness. In a white-light transmission hologram the hue will be changed towards shorter wavelengths.

If a reflection hologram is processed in a tanning developer, the cross-linking effect of the developer products will to some extent preserve the spacing of the Bragg planes. However, under certain processing conditions (depending to some extent on the emulsion batch) the tanning effect may not be constant throughout the thickness of the emulsion, with the result that the spacing of the Bragg planes may vary from the outer surface to the inner surface of the emulsion. This condition, sometimes referred to as chirped fringes, results in a broad-band reconstruction which may produce an almost achromatic image. The trade-off is a loss of sharpness that is progressive out of the plane of the hologram. Nevertheless, for shallow objects this can result in exceedingly bright images, and the phenomenon is exploited in commercial DCG holograms.

APPENDIX
6

The Reproduction of Color

> The Gryphon lifted up its paws in surprise. 'Never heard of uglifying!' it exclaimed. 'You know what to beautify is, I suppose?'
> LEWIS CARROLL, *Alice in Wonderland*

The Young-Helmholtz Model for Color Perception

In 1802 Thomas Young suggested that there were three types of color receptor in the human eye, each sensitive to approximately one-third of the visible spectrum, with the sensitivity ranges to some extent overlapping. The peak sensitivities of the three types of receptor accounted for the ability to perceive the separate *primary hues* red, green and blue. The *secondary hues* yellow, magenta (purple-red) and cyan (blue-green) resulted from the stimulation of pairs of color receptors, giving hues intermediate between the primaries. Thus the sensation of yellow was evoked whenever light stimulated both the red and green receptors together; the sensation of cyan came from simultaneous stimulation of the green and blue receptors; and magenta was perceived when both red and blue receptors were stimulated. The sensation of magenta could not be evoked by pure spectral light, as magenta is not in the spectrum. Other non-spectral colors such as browns, pinks, greys and white resulted from the stimulation of all three types of receptor in varying proportions. Hermann Helmholtz developed the theory further, and showed that *any* color could be synthesized from a suitable mixture of lights of the three primary hues. He went on to quantify the mixtures, and suggested modifications to the hypothetical sensitivity curves of the receptor cells of the eye to take account of some known anomalies in color perception.

In 1862 James Clerk Maxwell made a partially-successful attempt to show that if a photographic record was made of a colored object on three negatives through red, green and blue filters respectively, then if the projected images of black-and-white transparencies made from these negatives were superposed using the same filters, it would be possible to produce an image on the screen that would appear in true natural colors. Maxwell was hampered both by his poor grasp of practical photographic techniques and by the limitations of contemporary photographic materials. Nevertheless, the demonstration was sufficiently convincing to lead to the eventual design and production of color photographic materials. The principle of superposition of primaries is known as *additive color synthesis*, and it is the basis of color television, which employs a fine matrix of dots of red, green and blue phosphors, and of the Polaroid process for instant color transparencies, which uses a raster of fine red, green and blue lines. With the exception of this process, present-day photography uses the principle of *subtractive color synthesis*, developed by Louis duCos du Hauron from Helmholtz's work. It differs from the additive principle in employing dyes of the secondary hues cyan, magenta and yellow to remove differing amounts of red, green and blue from white light. As these dyes have a wide passband (approximately two-thirds of the visible spectrum) the subtractive principle is not appropriate to color holography.

The CIE Chromaticity Diagram

There are a number of ways in which a color can be specified quantitatively. Usually, a color patch appearing on a screen has its color specified in terms of three variables, called respectively *hue*, *saturation* and *lightness*. Hue, the 'name' of a color, is described by the dominant wavelength, ie the wavelength that has a hue exactly matching that of the patch. Purples and magentas, which are hues not found in the spectrum, are defined in terms of the dominant wavelength that has a hue complementary to that of the patch, and would generate neutral grey if mixed with it additively; this wavelength is given a negative sign. Saturation, also known as *chroma*, is a measurement of the vividness of the color. Its two extremes are spectrally pure color and neutral grey; the dominant wavelength remains unchanged throughout the scale. An example is the range of blues from royal blue (high saturation) to slate grey (low saturation). Lightness is specified in terms of luminous intensity: if a patch of color is produced by a projector with a filter over the lens, and the intensity is changed without changing anything else, it is the lightness that changes. Both *luminosity* and *value* are acceptable alternative terms for lightness. The term 'brightness' is ambiguous in this context and should be avoided. (It is used in this book in a colloquial sense, without reference to color, where to talk of 'luminous intensity' would be pedantic.)

As lightness is independent of color quality, diagrams which quantify color usually depict only hue and saturation. The CIE (Commission International de l'Éclairage) chromaticity diagram is the most widely-used of these diagrams. Red content is plotted along the horizontal axis and green content along the vertical axis. The blue content is given by the relationship Blue = 1 − (Red + Green). The horseshoe-shaped curve represents the pure spectral colors. (If the eye were a perfect receptor this curve would be a 45° right-angled triangle.) The hues are specified by wavelengths along the curve. The range of hues between blue and red, which cannot be matched by a single wavelength, are represented by a straight line joining the extreme values for red and blue (380 and 700 nm). The saturations at the various hues lie along straight lines radial from a neutral center, the position of which is determined by the nature of the illuminating light (daylight, tungsten filament, etc) (Fig A6.1). A full description of the theory underlying this diagram is to be found in Hunt (1).

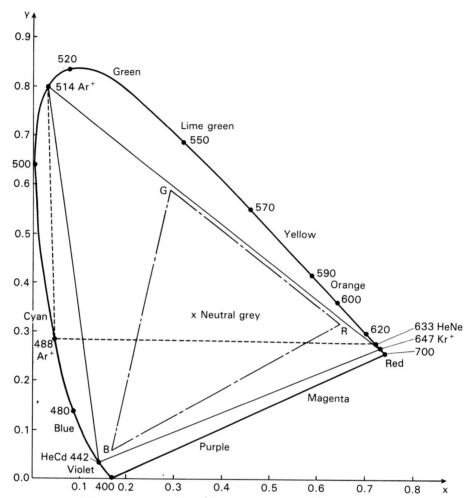

FIG A6.1 *CIE chromaticity digram. The large triangle shows the range of colors that can be produced using a* Kr^+ *laser at* 647 nm, *an* Ar^+ *laser at* 514 nm *and a HeCd laser at* 422 nm. *The interrupted line indicates the possible range of colors using the* 488 nm Ar^+ *laser line. A typical RGB color television display is included for comparison. Neutral grey is at a point corresponding to* 0.33 R, 0.33 G, 0.33 B. *The numbers round the curve are wavelengths in nm, and approximate hues are indicated round the edge (after Hariharan (2)).*

The importance of this diagram is that it can tell us how good any set of colors is at synthesizing the gamut of visible colors. If a color hologram is made using light at 633 nm from a HeNe laser and at 514 and 488 nm from an Ar^+ laser, the color range is somewhat limited. By using the 477 nm, line of the Ar^+ laser, at some sacrifice of power, the rendering can be significantly improved. The optimum combination is a Kr^+ laser at 647 nm with an Ar^+ laser at 514 nm, and a HeCd laser at 442 nm. This combination can give a range of colors greater than is possible in either color photography or TV. However, as explained in the text (p. 64) this does not necessarily mean that any color recorded holographically will appear in its true color when replayed.

References

1. Hunt R W G (1973) *The Reproduction of Colour*. 3rd edn. Fountain Press.
2. Hariharan P (1983) Colour holography. *Progress in Optics*, **20**, 265–324.

APPENDIX
7

Geometries for Creative Holography

> Humpty Dumpty took the book, and looked at it carefully. 'That seems to be done right —' he began.
> 'You're holding it upside down!' Alice interrupted.
> 'To be sure I was!' Humpty Dumpty said gaily, as she turned it round for him.
> 'I thought it looked a little queer.'
> LEWIS CARROLL, *Through the Looking Glass*

Recording Geometries for Multicolor Holograms

This method was originally worked out by Steve McGrew (1) and appears here in a rearranged form, with the author's permission. It covers cases where precise registration of multicolor images is important, and it is valid for both reflection and transmission holograms. While not as rigorous as Benton's mathematical approach (2)*, its graphical approach generates a scale diagram of the table geometry required to make the final transfer hologram once the desired display conditions have been established. This approach is based on two principles: the chromatic properties of diffraction gratings and the existence in all holographic geometries of a *hinge point*.

If the mathematics of the laws of diffraction are applied to produce a diagram (Fig A7.1) in which the angle of incidence of the incident ray is arbitrary, it is not difficult to show that if the emergent rays (ie the zero- and first-order diffracted rays) are intersected by a circle centered at the grating, the sum of their y-intercepts (D in Fig A7.1) is always the same (the *angle* between the rays is not the same). It depends only on the spatial frequency of the grating and the wavelength of the diffracted light.

For a transmission hologram the hinge point is the point at which a straight line passing through the object and the reference source intersects the plane of the hologram. For a reflection hologram the mirror image of the reference source reflected in the hologram plane is used instead (Fig A7.2)

The hinge point (Fig A7.2) corresponds to a point where, if the recording plate were large enough to include it, the spatial frequency of the fringes perpendicular to the plate would be zero. This happens when the reference beam and the object

---

* In particular, it does not give accurate registration of the color images if the real-image projection beam is not a perfect conjugate of the master hologram reference beam.

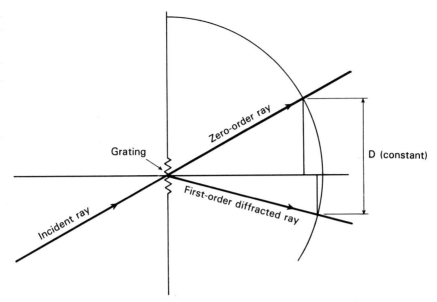

FIG A7.1 *Geometry of diffraction.*

beam are aligned: in a reflection hologram at this point the fringes are parallel to the plane of the plate, whereas for a transmission hologram at the same point there are no fringes at all.

Considered from a horizontal viewpoint a rainbow hologram behaves like a holographic lens, with the replay source and the real image of the slit as conjugate foci. However, the position of the slit image varies with the wavelength of the replay beam; if a white-light point source is used, the image of the slit is spread into a spec-

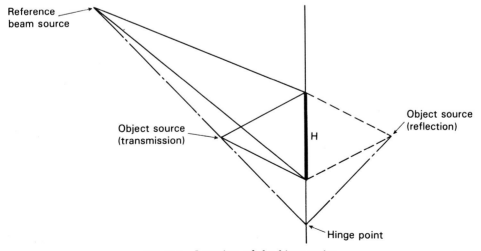

FIG A7.2 *Location of the hinge point.*

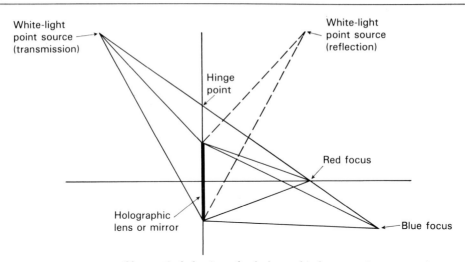

FIG A7.3 *Chromatic behavior of a holographic lens or mirror.*

trum lying along a line passing through the source. For *all* wavelengths the hinge point, which is the point where this line cuts the plane of the hologram, is the point at which no fringes would exist in the hologram (Fig A7.3).

For a reflection hologram the equivalent rainbow configuration can be used to locate the hinge point. A virtual light source, the reflection in the hologram plane of the real light source, is used, the geometrical construction being otherwise the same. In the reflection configuration the ray from the hinge point to the spectrum corresponds uniquely to a ray from the source reflected specularly; for this to occur the fringes at the hinge point would have to lie parallel to the emulsion surface. In a reflection hologram, all but a narrow band of wavelengths fail to satisfy the Bragg requirement and are suppressed; nevertheless, the rule is obeyed in principle.

Locating the Hinge Point and Illumination Axis

First decide the position of the replay source and the optimum viewing distance. Draw a diagram to scale on squared paper, with lines passing through the replay source and center of the display hologram H_2, and through the replay beam and the optimum viewing point. The plane of the hologram is the y-axis and the center of the hologram is the origin. The hinge point is located at the intersection of the second of these lines with the y-axis, and the first of these lines, the illumination axis, is also the axis on which the reference beam source will lie when the hologram is made (Fig A7.4).

Rainbow Hologram: Multiple-Strip Set-up

Choose a suitable distance for the 'red' master strip S_1 on the viewing axis (this will be determined by whether you want the final image to be virtual, real or hologram-

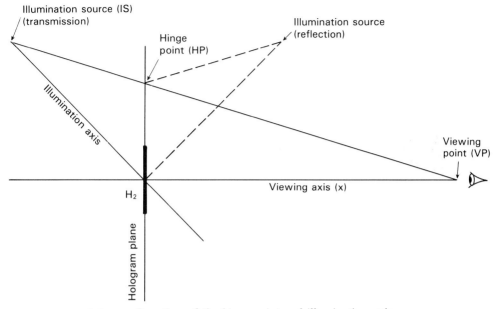

FIG A7.4 *Location of the hinge point and illumination axis.*

plane). Draw a line from the hinge point through S_1 (the recording axis) and continue the line until it intersects the illumination axis. The point of intersection is the correct position for the reference beam source (Fig A7.5).

To determine the positions that S_2 (green) and S_3 (blue) will occupy, draw a circle centered at the origin and passing through R, the reference beam source. Measure the *y*-coordinate of R by counting squares from R vertically to the *x*-axis. To find the position of S_2 there are three steps:

1 Multiply the value of the *y*-coordinate by λ_1/λ_2. λ_1 is 633 nm for a HeNe laser. λ_2 can be estimated from the following table:

Hue	Wavelength range (nm)
red	700–620
orange	620–580
yellow	580–570
green	570–480
blue/violet	480–400

2 Count upwards from R in squares until you have reached a point $y \times \lambda_1/\lambda_2$ above it and draw a horizontal line from this point until it intersects the circle.
3 Join this point of intersection to the origin. S_2 should be located at the intersection of this line and the recording axis (Fig A7.6).

Use the same method to identify the position of S_3 (blue).

410 GEOMETRIES FOR CREATIVE HOLOGRAPHY

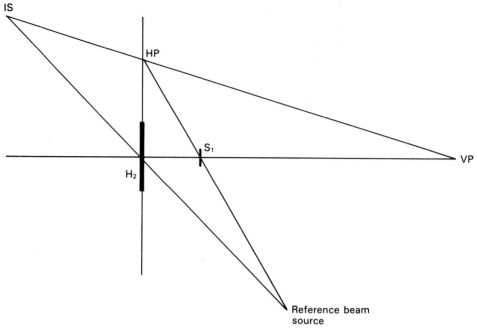

FIG A7.5 *Position of the reference beam source for a rainbow hologram (multiple slit).*

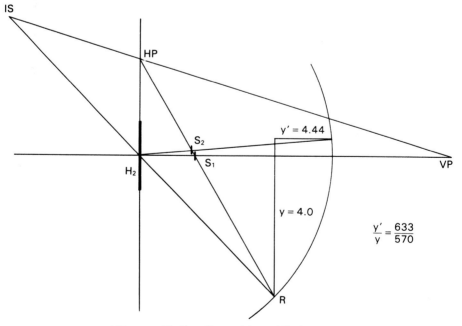

FIG A7.6 *Finding the position of S_2 (green).*

Your table geometry is now complete. Notice that unlike the simplified set-up with all the strips in the same plane, you cannot expose all the strips simultaneously as you have the master reconstruction beam coming from the same direction as the reference beam; the three master slits overlap when seen from this direction. You will need to remember this when setting up the table for the master slits.

Rainbow Hologram: Multiple Reference-beam Set-up

The multiple-slit set-up is space-saving and comparatively quick, but has the disadvantages of requiring three plate-holders (which must not overlap) and of making it difficult to adjust exposures between the three strips if a first test shows differences in brightness between the three images. The multiple reference beam configuration avoids these difficulties, though the plate receives three times as much reference beam energy in toto, and the beam ratio may have to be altered to take care of this.

The initial diagram (Fig A7.7) is similar to that for the multiple-strip configuration. To determine the position of R_2 (green) and R_3 (blue) you use a method similar to the preceding one, but this time y' is *outside* the circle and below the x-axis. As before, begin by drawing an arc of a circle centered at the origin and passing through the 'red' reference source R_1, as in Fig A7.8. Again there are three steps:

1 Measure the y-intercept of R_1 by counting squares, and multiply it by λ_1/λ_2.
2 Count downwards from the x-axis in squares to a point $-y \times \lambda_1/\lambda_2$ below it and draw a horizontal line from this point until it intersects the circle.
3 Join the origin to this point of intersection and continue the line until it intersects

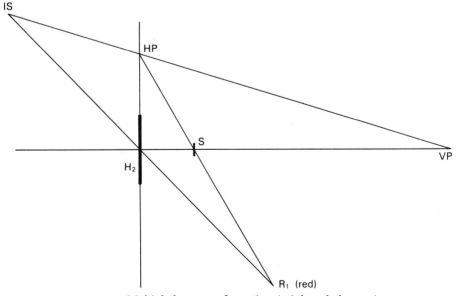

FIG A7.7 *Multiple-beam configuration (rainbow hologram).*

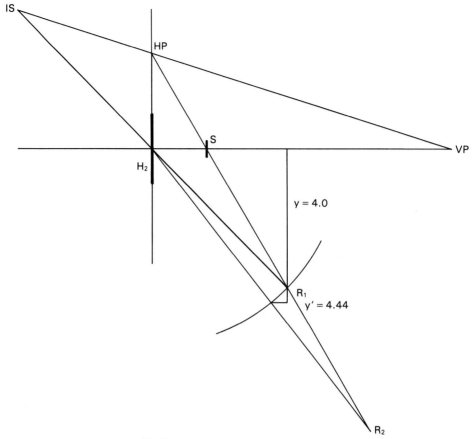

FIG A7.8 *Finding the position of R_2 (rainbow hologram).*

the recording axis. This new point of intersection R_2 is the correct location for the source of the 'green' reference beam.

Use the same method to find the correct location for R_3 (blue).

It will be apparent from Fig A7.8 that the multiple-reference-beam configuration takes up more room than the multiple-slit configuration. If you have limited space you will need to adjust the geometry so that the angle between the illumination and recording axes is as large as possible within the constraints set by the requirements of imaging and display. If you do not do this you may well find that R_3 is off your table. However, it is possible to cheat a little and bring R_3 nearer without the loss of accurate register becoming obvious.

Reflection Hologram: Multiple Reference-Beam Set-up

It might appear that a multicolor reflection hologram would not need a change in reference beam source position between exposures, as the image hue is derived from

GEOMETRIES FOR CREATIVE HOLOGRAPHY   413

differential swelling of the gelatin, but this is not so. In processing the swelling agent is removed, and as a result each set of fringes has its own preferred distance and direction of illumination source. As there can be only one illumination source for reconstruction, and as the Bragg requirement makes it difficult to have the master holograms in different positions and at the same time get equally bright reconstructions for all the images, it is necessary to use the multiple-reference-beam set-up.

This is a mirror image of the previous set-up. The first task is to locate the hinge point, as in Fig A7.4. The table diagram is a reflection of the layout of Fig A7.8. In order to find the recording axis you need to construct a virtual image or reflection of H_1 (Fig A7.9). Begin by drawing an arc of a circle centered at the origin and passing through the 'red' reference source R_1, as in Fig A7.8, then:

1 Measure the y-intercept of R_1 by counting squares up to the x-axis, and multiply the figure by λ_1/λ_2.
2 Count downwards from the x-axis in squares to a point $-y \times \lambda_1/\lambda_2$ below it and draw a horizontal from this point until it intersects the recording axis. This new point of intersection R_2 is the correct location for the source of the 'green' reference beam.
3 Use the same method for finding the correct location for R_3 (blue).

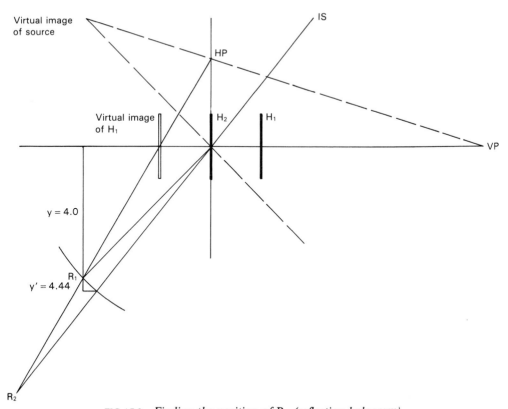

FIG A7.9 *Finding the position of R_2 (reflection hologram).*

It is easier to fit the set-up on the table if the distance between S_1 (or H_1) and H_2 is fairly short, and if the distance and angle of incidence of the illumination source are large.

There are many possible variations of these set-ups, all of which can be analyzed in terms of a hinge point and of the angles found by the graphical procedures described above. Conversely, an examination of an existing hologram along these lines in terms of its display characteristics can by deduction furnish the details of its set-up geometry.

A Worksheet for a Multicolor White-light Transmission Hologram

Suzanne St Cyr (3) has provided a comprehensive and detailed worksheet for calculating the geometry of a multicolor white-light transmission hologram. The worksheet is very long, and by making a few approximations she has been able to produce an abridged worksheet which is much more manageable. It is reproduced here with permission. All wavelengths are in micrometers (μm) and all distances in centimeters (cm). Although somewhat cumbersome, St Cyr's symbols and suffixes have been retained so as to remain consistent with those used in her papers (3) and (4).

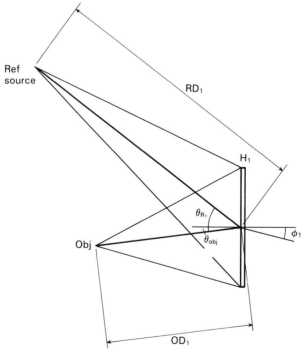

(a) *Making the master*. SH_1 *Shrinkage factor for master*, RD_1 *Reference source distance (−ve or ∞)*, OD_1 *Object distance (−ve)*, θ_{R_1} *Angle of incidence of reference beam*, θ_{obj} *Angle of incidence of object beam*, ϕ_1 *Angle between fringes and normal*, λ_L *Laser wavelength*.

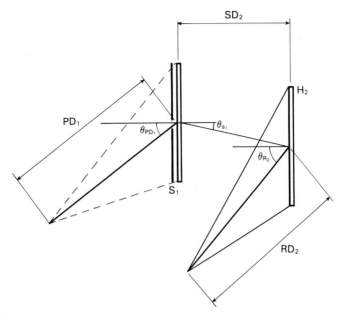

(b) *Making the transfer.* PD_1 *Slit illumination distance,* θ_{PD_i} *Angle of incidence of slit illumination beam,* θ_{s_1} *Angle of emergence of image beam,* θ_{R_2} *Angle of incidence of transfer illumination beam,* SD_2 *Separation of slit* S_1 *and* H_2, RD_2 *Reference source distance* $(-ve)$.

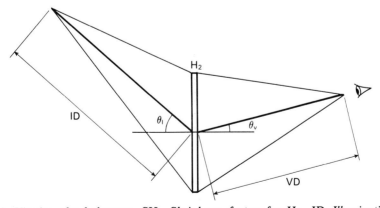

(c) *Viewing the hologram.* SH_2 *Shrinkage factor for* H_2, *ID Illumination distance,* θ_I *Angle of incidence of illuminating beam,* VD *Viewing distance,* θ_v *Angle of emergence of viewing beam,* λ_V *Viewing wavelength.*

FIG A7.10 *Parameters for multicolor white-light transmission hologram worksheet.*

The sign convention is that the direction of light travel and distance measurement is from left to right; distances measured along diverging beams are negative and distances along converging beams are positive (eg hologram to real image). The convention for angles is that angles of illumination are measured with respect to the normal to the hologram. If the light source is above the hologram the angle is positive; if it is below, the angle is negative. Figure A7.10 shows the various parameters.

NOTE: To find the shrinkage factor SH, illuminate the hologram with a laser beam and adjust the angle for maximum brightness. Then $SH = \sin \theta \text{ ref} / \sin \theta \text{ ill}$

References

1. McGrew S (1982) A graphical method for calculating pseudocolor hologram recording geometries. *Proceedings of the International Symposium on Display Holography* **1**, 171–83.
2. Benton S A (1982) The mathematical optics of white-light transmission holograms. *Proceedings of the International Symposium on Display Holography*, **1**, 5–14.
3. St Cyr S (1984) A holographic worksheet for the Benton math. *Holosphere*, **12**, 8, 4–16.
4. St Cyr S (1985) One-step pseudocolor WLT camera for artists. *Proceedings of the International Symposium on Display Holography*, **2**, 191–229.

GEOMETRIES FOR CREATIVE HOLOGRAPHY

Worksheet for Multicolor WLT Holograms

*Project title*_____ *Date*_____

Enter display parameters

$\lambda V =$ ___μm; $\theta V\ (=0°)$; $\theta I =$ ___°; $VD =$ ___cm: $ID =$ ___cm.

Enter transfer parameters

$\lambda L =$ ___μm; $SH_2 =$ ___cm; $RD_2 =$ ___cm.

Transfer geometry

1. $\theta R_2 = \sin^{-1}\left(\dfrac{\lambda L}{\lambda V}[SH_2 \sin \theta I]\right)$ = ___°

2. $SD_2 = \dfrac{1}{\dfrac{1}{RD_2} - \left(\dfrac{\lambda L}{\lambda V}\left[\dfrac{1}{VD} - \dfrac{1}{ID}\right]\right)}$ = ___cm

Note: This value will be negative.

Enter master parameters

$SH_1 =$ ___cm; $RD =$ ___cm; $PD =$ ___cm; $\theta R_1 =$ ___°; $\Theta obj =$ ___°

Master geometry

3. $OD_1 = 1\dfrac{1}{\dfrac{1}{RD_1} + \dfrac{1}{PD_1} + \dfrac{1}{SD_2}}$ = ___cm

4. $\phi_1 = \dfrac{\theta R_1 - \theta Obj}{2}$ = ___cm

5. $\theta PD_1 = \sin^{-1}\left(\dfrac{\sin \theta R_1 + \sin \theta Obj}{2SH_1 \cos\phi_1}\right)$ = ___cm

6. $\theta S_1 = \sin^{-1}\left(\dfrac{\sin \theta R_1 + \sin \theta Obj}{2SH_1 \cos\phi_1}\right)$ = ___°

NOTE: If $\theta R_1 = \theta Obj$, then $\theta PD_1 = \sin^{-1}\left(\dfrac{\sin \theta R_1}{SH_1}\right)$

APPENDIX

8

Monomode Optical Fibers in Holography

'I beg your pardon', said Alice very humbly: 'You had got to the fifth bend, I think?'

LEWIS CARROLL, *Alice in Wonderland*

Optical fibers are used routinely where it is necessary to convey light from one place to another and where straight paths, with or without mirrors, are out of the question; a familiar example is the decorative use of optical fibers in lamps, producing points of light at their ends. But fiber optics finds more serious uses: in car dashboard illumination, in the lighting of microscopic specimens, in endoscopy, in communications, and now, increasingly, in holography.

Multimode Fibers

For most of the above purposes the light is kept within the fiber by total internal reflection. This phenomenon occurs at the interface of an optically-dense medium with one less dense; a ray incident at an angle greater than the critical angle cannot escape through the interface, and is reflected. Optical fibers for taking illumination into awkward spots can be quite thick, their diameter being restricted chiefly by the lack of flexibility of the material. When the overall diameter of the source is required to be large, a bundle of fibers is used. In general, these bundles are manufactured without any particular correlation between the positions of the fibers in the bundle at the beginning and at the end. The result is called an 'incoherent fiber-optics light guide'. However, under certain circumstances the bundle may be required to convey an image from one plane to another, in which case the fibers have to retain their spatial relationships to one another at both ends of the light guide; such a guide is said to be 'coherent'. Note that this term has nothing to do with coherent light: such bundles of fibers are used for white light, and the resolution of the image is limited by the diameter of the individual fibers.

It is, of course, possible to send laser light down one of these light guides. Temporal coherence is retained, as nothing occurs to affect the bandwidth; but the varying delays caused by the different paths taken by parts of the beam mean that spatial coherence is destroyed. Thus, although fiber bundles can be used as object illuminating beams for holograms, they cannot be used to convey the reference beam.

Single-mode Propagation

If the diameter of the fiber is made very small, small enough to be comparable with the wavelength of the light, the light behaves like a single ray, bending with the bends in the fiber and never touching the sides. Of course, the ray model does not provide an account of the way this happens; the electromagnetic wave model, on the other hand, does. It tells us that in a monomode fiber there is constructive interference at the center of the fiber and destructive interference at the walls; the energy is thus confined within the fiber, and is propagated as a coherent traveling wave. The propagation has exactly the same nature as the propagation of microwaves in a waveguide. The critical diameter at which monomode propagation occurs is in theory not much more than a wavelength, though in practice it occurs when the fiber diameter is up to seven or eight times the wavelength. For HeNe light the upper limit is about 6 μm. If the fiber is truly circular the polarization can be destroyed, and fibers of elliptical and D-shaped cross-section have been designed to avoid this; but in practice small errors in the cross-sectional geometry preserve the polarization, the only effect being a small amount of rotation of the plane of polarization, which can be controlled by rotating the end of the fiber.

The Use of Monomode Fibers in Holography

The importance of using monomode optical fibers in holography is that for most of the optical pathway the beams are fully protected from air currents. The optical paths will remain matched even if the reference-beam fiber has a knot in it. Once the light leaves the fibers, however, the usual stability controls are necessary.

The light that leaves the fiber is fully spatially filtered and has a divergence which is typically about 12°, roughly the same as would be obtained using a ×10 microscope objective. The angle of divergence depends on the numerical aperture of the system; the larger the difference in refractive index between the fiber and its cladding, the larger the divergence. The beam can be given a greater divergence by placing a small concave lens near the output end of the fiber; alternatively, a convex end can be formed on the fiber itself by merely holding a match close to it to soften the end, while the beam is monitored. Another method, suitable for more powerful lasers, is to reflect the beam back onto the end of the fiber with a mirror.

Launching the Beam

Launching a laser beam into an optical fiber 6 μm in diameter presents the same sort of problem as feeding it through a 6 μm pinhole. The problem is approached in the same way. The holder resembles a spatial filter, but in place of the pinhole is a nipple holding the optical fiber, and in addition to the usual x, y, z movements it has x and y tilts. This is because cleaving the fiber usually leaves an error of about 1° or so at the end face, and this is sufficient to produce a significant effect on the amount of energy that gets into the fiber. After lining up the fiber in the x, y, z directions by the usual method for a spatial filter, you need to align the angle. The adjustment is not as difficult as it might seem, as the diameter of the fiber and cladding together is quite large, and to begin with most of the light passes down the cladding. It is fairly easy to distinguish this diffuse beam from the concentrated patch produced when the beam is confined to the fiber itself.

The usual method of cleaving a fiber is by the scratch-and-pull method. There is a diamond tool made especially for the purpose, though a razor blade or a scalpel will do if you are prepared to put up with a high proportion of failures. First, remove the protective plastic coat with your thumbnail. Then stretch the fiber over a curved surface about 25–50 mm in diameter; a piece of plastic pipe will do. Scratch the fiber with the knife – how much pressure to exert is a matter of trial and error – then take the fiber and pull it apart, firmly but not roughly. If you have access to a microscope, examine the end of the fiber; if it is completely clean, you can use it. If not, try again. Another method, which you can use if your launcher requires the fiber to be mounted in a metal holder or nipple, is simply to cut off the fiber with scissors, mount it in the nipple or ferrule, then polish the end down flat, using lapping paste and finally jeweler's rouge. You can use the equipment made for preparing metallurgical specimens if it is available.

Making a Fixed-ratio Fiber Coupler

In fiber-optics technology a beamsplitter is called a fiber coupler. This operates on the principle of evanescent-wave coupling. An evanescent wave exists just outside an interface between two media at which total internal reflection is taking place. If two optical fibers are sufficiently close together, this evanescent wave can be coupled from the first one into the second one: anything from 1% to nearly 100% can be transferred. Sheen and Giallorenzi (1) used a technique which etched away most of the cladding, keeping the fibers under tension; however, this technique is not easy to reproduce. A simpler method of producing fixed-ratio couplers appears in a paper by Kawasaki *et al* (2), and this does work well. The paper is somewhat coy about the details of the method used, and the author is indebted to David Jackson of the University of Kent for supplying working instructions, which have been found to give good results.

The first step is to launch a laser beam into a single-mode fiber, and to set up the other end so that the expanded beam is projected onto a screen. Set up the second fiber alongside it. Remove about 4 cm of the plastic sheathing from each of the fibers and clean the exposed cladding with alcohol. Twist the fibers round each other one whole turn and, holding them under slight tension, warm them gently with a Bunsen burner or similar heat source. As the fibers soften and begin to stretch and fuse together, light will begin to appear from the end of the second fiber. Kawasaki *et al* quote a waist about 15 μm in diameter and several mm long, the actual coupling being a function of the length and diameter of the waist of the tapered region. The joint is fragile, and is best protected by potting. The simplest method of doing this is to fit the system inside an old ballpoint pen case and seal the ends with Blu-tack. Kawasaki *et al* suggest coating the tapered part with a silicone grease of low refractive index (Dow Corning C-20057 compound) and potting with epoxy resin into a plastic tube. By allowing the beams to fall on the cells of two light meters, the power levels in the two beams can be compared during the fusion process. There is no need to remove the stub of the second fiber; indeed, it can be made use of in a fringe-stabilizing device (see Appendix 9). Polarization is preserved in this type of coupler, and the beamsplitting ratio is independent of the direction of polarization. It does depend on the wavelength, however; less of the beam is transferred at shorter wavelengths.

Variable Couplers

Variable couplers are now becoming available off the shelf. If you have accurate machining facilities you can make your own. The principle is described by Bergh *et al* (3). Two quartz blocks are slotted to take single-mode fibers from which the plastic jacket has been removed, using a wire saw to make the slots with a slight downward curvature at the ends (radius of curvature about 20 cm). The fibers are then bonded into the slots and the quartz blocks carefully polished down until the cladding is reached. At this point the blocks can be put together and the fibers tested for coupling (Fig A8.1).

Polishing is continued until coupling is at least 50% and optical flatness in the neighborhood of the fiber is no more than one or two fringes. Index-matching fluid is run into the interspace by capillary action. The blocks are then mounted so that one block can be moved relative to the other by micrometer adjustments, the coarse control being the x-control (across the fibers) and the fine control being the z-control (along the fibers).

Practical Use in Holography

The most important potential application of monomode fiber optics in holography is in the production of transfer holograms from master holograms, as this requires

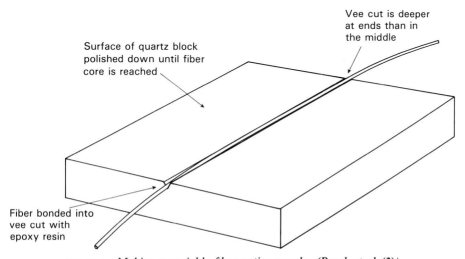

FIG A8.1 *Making a variable fiber-optics coupler (Bergh et al (3)).*

only two beams. There is a comparatively easy method for fringe stabilization (see Appendix 9). The limitation to two fibers, at least when fringe stabilization methods are used, makes this method less attractive for making master holograms; however, there is no limit to the number of couplers that can be inserted, and losses are negligible.

Stray light in the cladding can be troublesome, and this can be cured by stripping off the plastic coating near the ends of the fibers and passing them through a box of silicone grease (petroleum jelly also works well). The effect is to leak out all the unwanted light in the cladding, leaving the wanted beam confined within the fiber.

References

1 Sheen S K and Giallorenzi T G (1979) Single-mode fiber-optical power divider: encapsulated etching technique. *Optics Letters*, **4**, 29–31.
2 Kawasaki B S, Hill K O and Lamont R G (1981) Biconical-taper single-mode fiber coupler. *Optics Letters*, **6**, 327–8.
3 Bergh R A, Kotler G and Shaw H J (1980) Single mode fiber optic directional coupler. *Electronics Letters*, **16**, 260–1.

APPENDIX
9

Fringe Stabilization

'How puzzling all these changes are! I'm never sure what I'm going to be from one minute to another! However, I've got back to my right size.'
LEWIS CARROLL, *Alice in Wonderland*

The most common cause of failure of a hologram is movement of the primary fringes during exposure. The problem of fringe stability is particularly serious when one is working with low-power lasers or with non-silver materials, where exposures can be tens of minutes. The main causes of instability are listed on p. 147, pp. 150–2 and pp. 160–2. It is not always possible to control these; random air movement and small changes in ambient temperature can seriously degrade the quality of an image. It is not generally appreciated that a single brief disturbance of the air in the vicinity of a hologram does not settle down for as long as 15 min. During this time the primary fringes will be jittering by as much as half a wavelength, enough to destroy the fringes during the time the jitters occur – which may be for more than half the total exposure.

The answer is an active fringe stabilizer. This works on the principle of *servomechanisms*, usually called *servo controls*. The basic elements of a servo system are an error detector, an amplifier and an output which corrects the error. In the case of the fringe stabilizer the error detector detects movement of the fringes, sends an 'error' signal to the amplifier, and the output of the amplifier is fed to a device which moves the fringes back into their original position (Fig A9.1). There are many methods of dealing with this sort of situation and the one that is most appropriate depends on the nature of the error. In the case of holographic fringes, the main difficulty is their small dimensions. They are typically only about 1 μm apart, a far smaller distance than the size of any detector, and they are usually of low intensity. Let us look at the general principle first, then see what special techniques we need to provide the stability of the fringes that is essential for good holograms. Figure A9.1 shows the basic elements of a servo control system. In our particular case the 'device' is the optical system which generates the fringes. The 'error detector' is a light-sensitive device which detects any movement of the fringes, coupled to a differential amplifier. The output of the error detector is amplified and a correction signal fed back to the device, altering the length of one of the optical paths so as to move the fringes back to their original position.

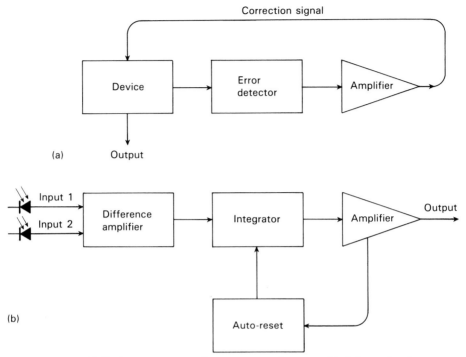

FIG A9.1 (a) *Basic components of a servo control system.* (b) *Schematic of error detector component.*

Error Detector

Let us consider the error detector first. This will contain some kind of light sensor, and its output has to be compared with a reference standard. The most obvious problem is the very small scale of the fringes, much smaller than the diameter of any photodiode; we shall have to consider methods of expanding the fringes. But for the moment let us suppose they are large enough (2–3 mm) to span a pair of P-I-N diodes. We can place one diode in the fringe pattern and compare its output with a fixed reference voltage, but this will require careful pre-setting and constant adjustment for fluctuations in the laser output. It is simpler to use a pair of P-I-N diodes set a few millimeters apart, and to arrange them so that they are on opposite slopes of a fringe; any fringe movement will then increase the output of one and decrease the output of the other (Fig A9.2(a)).

The fringe pattern will not always be made up of straight bars. Some optical configurations give circular fringes, or contour fringes which are very close together. In Fig A9.2(b) detector 1 covers only one fringe, but detector 2 covers two, and its variation in output will be less than that of detector 1. In Fig A9.2(c) the output from detector 2 will be constant, as it is covered by many fringes, and the sensitivity of the system will be halved.

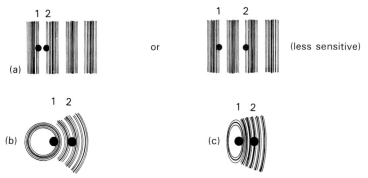

FIG A9.2 *Possible positions for detector elements.*

Expanding the Fringes

There are several methods of expanding the fringes. In a transmission layout the pattern can be sampled directly by placing a ×60 microscope objective directly behind the hologram and placing the detectors in its focal plane (Fig A9.3).

However, the light level obtained by this method is so low that very sensitive detectors, such as photomultipliers, are needed. In practice, it can be used only when making diffraction gratings of low spatial frequency, where the magnification need not be large. Nick Phillips (1) refers to a method using moiré or secondary fringes. Before the actual hologram H_2 is made, another hologram H_1 is made in the space immediately behind it. This is developed and bleached, and placed back in position with just sufficient lateral displacement to generate large secondary fringes with the live interference pattern. The detectors are placed behind the H_1 hologram, and they control the position of the primary fringes falling on the actual hologram H_2 (Fig A9.4).

Phillips points out that the hologram H_1 is specific to the subject matter and cannot be used with any other subject matter, as the fringe pattern would then be

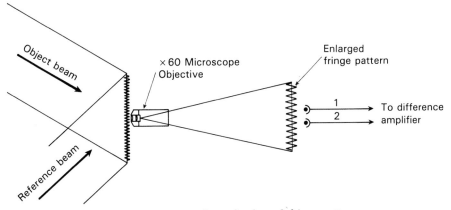

FIG A9.3 *Direct sampling of enlarged fringe pattern.*

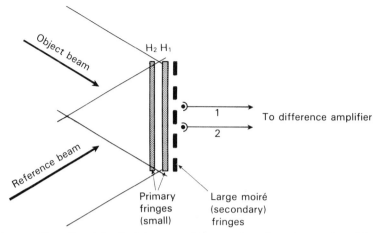

FIG A9.4 *Use of moiré principle to produce greatly enlarged secondary fringes.*

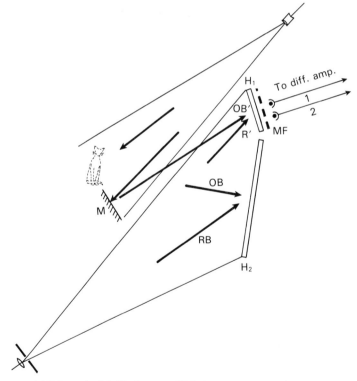

FIG A9.5 *Addition of trivial hologram (H_1) for generation of secondary fringes. H_1 derives its object beam OB' via a mirror M, and its reference beam R' directly. It is exposed, processed and repositioned before H_2 is made. Live moiré fringes are generated in the space behind it and detected by the light sensors 1 and 2.*

FRINGE STABILIZATION

entirely different. However, if H_1 is a trivial hologram, ie a plain diffraction grating, secondary fringes will be generated for any object, though they will be somewhat weaker. Richard Rallison (2) describes an improved version which sites H_1 out of the pattern of H_2 and in its own space (Fig A9.5).

As H_1 is a trivial hologram it does not have to be positioned accurately, though it should not be rotated in plane with respect to the fringes. The mirror does not throw any light on the hologram H_2 and should therefore be invisible in the final image. Rallison suggests that reflection holograms should have part of one of the beams diverted by means of a beamsplitter or mirror to produce large live fringes (Fig A9.6). For fringe detection in a set-up for making a reflection hologram from a transmission master hologram, Rallison suggests the configuration of Fig A9.7. These two systems produce fringes of the type illustrated in Fig A9.2(b). The fringes

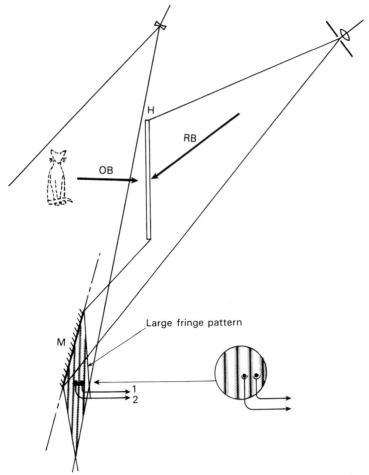

FIG A9.6 *Fringe detection for reflection holograms. Part of the object illuminating beam is deflected by a mirror M to overlap the reference beam at a small angle, generating broad live fringes in the region of overlap.*

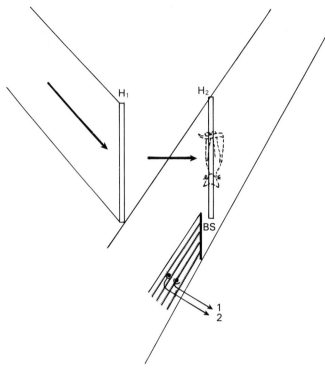

FIG A9.7 *Fringe detection while making reflection transfer hologram from a transmission master hologram, using a beamsplitter BS. It is immaterial whether the light has bypassed or gone through the plates.*

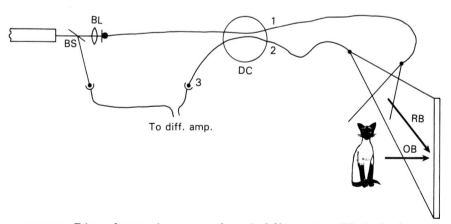

FIG A9.8 *Fringe detector in monomode optical-fiber system. BL is the beam launcher; DC is a directional coupler (fiber-optics beamsplitter); light output 3 receives the beams reflected back from the ends of fibers 1 and 2, acting as a Michelson interferometer. The output is compared with part of the original laser beam divided off by the beamsplitter BS.*

of Fig A9.2(c) are produced when the beamsplitter technique is used for a rainbow transfer; but the grating method of Fig A9.5 is better.

Corke *et al* (3) describe a method of monitoring path lengths when using monomode optical fibers. As mentioned in Appendix 8, when a fiber-optic coupler is fabricated, there is a fourth stub, which is not used in illumination. At the ends of the 'object' and 'reference' fibers, about 3% of the light is reflected back along the fiber, forming a Michelson interferometer together with the fourth fiber. The output from this fiber will be either bright or dark depending on whether the object and reference wavefronts are in or out of phase. If this output is monitored by one photodetector and the output at the laser itself is monitored by another detector, a difference signal will be obtained (Fig A9.8). This system monitors only the light within the optical fibers. If the light path in air is long, it will require a separate stabilization system.

Comparator and Amplifier

The comparator is a differential amplifier based on the principle of the long-tailed pair. This is a balanced-input amplifier which gives an output (positive or negative) that depends on the voltage difference between the inputs. It can be built round a cheap operational amplifier chip such as the 741 (Fig A9.9).

The circuit needs to be stabilized against any form of drift from variations in rail voltage, thermal effects etc.

Transducer

A *transducer* is a device which converts electrical energy into mechanical displacement. The best-known example of a transducer is an ordinary loudspeaker, which

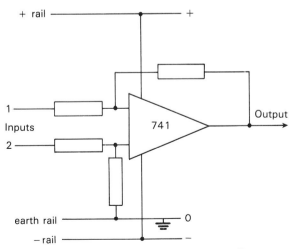

FIG A9.9 *A simple operational amplifier used as a difference amplifier.*

converts an electrical input into vibrations which produce sound waves. If a DC signal is applied to a speaker, the voice coil and the cone attached to it are displaced by a distance proportional to the current passing through the coil. If the speaker has a mirror attached to the cone, and the mirror is set up as one arm of a Michelson interferometer, then any current passing through the voice coil will displace the mirror and thus shift the fringe pattern (Fig A9.10).

If a negative feedback path is established between the fringe detectors and the speaker, any fringe movement will set up a movement of the voice coil which will move the mirror M to such a position as to restore the fringes to their original position. This is the system described by Steve McGrew (4) and marketed by Rallison's company (see Appendix 12). Ordinary cone speakers are not entirely satisfactory as transducers, as there is some freedom to rotate out of plane; and as they are current-operated they are not thermally stable. Phillips suggests the use of a Motorola piezoelectric horn loudspeaker diaphragm as the transducer element. This does not consume any quantity of current, as it is voltage-operated; it does not suffer from out-of-plane rotation; it is relatively unaffected by extraneous soundwaves (microphoning); and it is cheap. A piece of front-surface mirror about 12.5 mm in diameter is cemented to it, and the diaphragm is mounted in a plastic housing (Fig A9.11).

An alternative transducer, reported by Phillips (1) as having been designed by James Cowan of the Polaroid Corporation, uses a glass block in the optical path. This is mounted on a galvo-scanner assembly, and turns when activated, lengthening or shortening the optical path. This has the advantage of being immune from microphone effects, and it does not need the beam to be folded, but, as with the

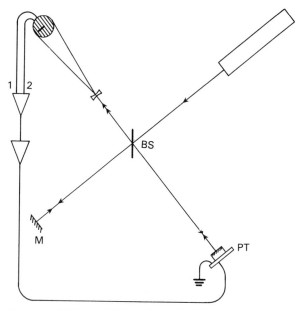

FIG A9.10 *Principle of fringe-shifting piezo transducer* (PT). *For simplicity the transducer is shown operating in one arm of a Michelson interferometer.*

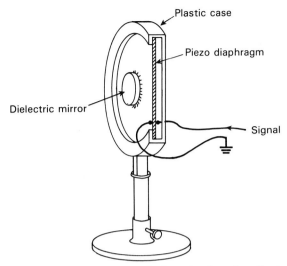

FIG A9.11 *Piezo transducer made from loudspeaker diaphragm.*

moving-coil speaker, it is current-actuated, and requires some damping to prevent overrun.

Fringe locking devices should be placed in the reference beam, preferably at a point where there would be a mirror in any case; otherwise they will need to be one leg of an optical sheepshank. They provide some measure of stability even for systems with more than one object illumination beam.

David Jackson and his team at the University of Kent have designed a very efficient transducer system for use with their fiber-optics configuration. It employs a piezo element in the form of a cylinder round which the reference-beam optical fiber is wrapped. It is described by Jones *et al* (5). The paper mentions six turns, and a voltage of up to 250 V; but with a typical amplifier output of only a few volts, about 100 turns are required.

Fringe stabilization is a comparatively recent development in holography, but its implications for the home holographer as well as for the research physicist who needs to undertake the odd holographic experiment are considerable. The use of fringe locking makes lab-bench and even kitchen-table holography possible. However, it must be stressed that it is a panacea; it is not a substitute for built-in stability. Phillips (1) says '...Fringe lockers do not provide a complete answer to sloppy housekeeping.' He adds that the best use of a fringe locker is as diagnostic device; from the results, the engineering and environmental aspects can be tidied up until the locker is effectively no longer needed.

References

1 Phillips N J (1985) Fringe Locking devices for the stabilization of holographic interference. *International Symposium on Display Holography*, **2**, 111–30.

2 Rallison R A (1984) *Instruction Manual for the DK-10 Fringe locker*. International Dikrotek, 12755 South 200 West, Richmond, UT 84333.
3 Corke M, Jones J D C, Kersey A D and Jackson D A (1984) All single-mode fiber optic holographic system with active fringe stabilisation. *Journal of Physics E*, **18**, 185–6.
4 McGrew S P (1982) An inexpensive fringe stabilizer for long exposure holography. *International Symposium on Display Holography*, **1**, 189–93.
5 Jones J D C, Corke M, Kersey A D and Jackson D A (1983) Single-mode fiber-optic holography. *Journal of Physics E*, **17**, 271–3.

APPENDIX
10

Processing Formulae

'Beautiful soup, so rich and green,
Waiting in a hot tureen!
Who for such dainties would not stoop?
Soup of the evening, beautiful soup.'
　　LEWIS CARROLL, *Alice in Wonderland*

Transmission Holograms

The traditional developer for transmission holograms used to be Kodak D-19, and there are still many professional holographers who will not use anything else for making master holograms for transfer. D-19 was originally designed for X-ray and infrared films, both of which are notorious for low contrast and high fog level. It is based on metol (p-methylaminophenol sulfate) and hydroquinone (1,4-dihydroxybenzene), and to cope with the awkward properties of the emulsions for which it was designed it contains large quantities of sulfite and bromide. The developer thus has a strong complexing action, and with the extremely small silver-halide crystal size, and a development time of up to 5 min, some silver halide is inevitably lost. Metol is not very soluble, and its action is inhibited by the presence of large amounts of bromide; it has been replaced by phenidone (1-phenyl-3-pyrazolidone) in all modern high-energy developers. Kodak DX-80, and, more recently, LX-24, are phenidone–hydroquinone developers designed as replacements for D-19 and similar metol–hydroquinone developers, and, unlike these, can be returned to the bottle after use and re-used virtually without limit as the action of phenidone is not affected by the presence of bromide (there is a replenisher which should be used to top up the bottles)*. For developing holograms it should be diluted 1 : 1. Development time with all these developers is 4–5 min at $20°C$, to an average density of 0.7 for master (H_1) holograms that are not going to be bleached; master holograms that are to be bleached should be given extra exposure so that the final density is about 1.5.

When development is complete, rinse the hologram in running water and immerse it in one of the proprietary rapid-fix solutions diluted according to the manufacturer's instructions for photographic negatives. Do not leave the hologram

---

* It should be mentioned that there is a strong body of opinion against using *any* developing solution more than once.

in the fixing bath longer than about 2–3 min, because after that time the solution will begin to attack the silver. Wash the film in running water for 8 min, emptying the wash tray several times during this period. The final rinse should be in de-ionized water to which you have added a drop of photographic wetting agent (not Photo-Flo) or washing-up fluid. Lay the film on a clean sheet of glass (or the plate on absorbent paper) and remove all surface water with a blade squeegee (the type sold for cleaning windows). Make sure you do not leave any streaks. Wipe surplus drops of water from the back with absorbent tissue or a cellulose sponge, and hang up to dry in a dust-free environment.

Alternatively, immerse the hologram after washing in a 70% methanol (CH_3OH)/water mixture for 2 min, followed by immersion in a 100% methanol bath for a further 2 min, then dry with a current of warm air from a hair dryer or fan heater. You can use industrial-grade methylated spirit instead of methanol, but *not* the blue-dyed type, which contains oily substances that leave a sticky deposit (surgical spirit does the same). Another alternative is isopropanol (propan-2-ol, C_3H_7OH, or isopropyl alcohol). This method is not recommended for final masters on film.

Both of these methods of drying can leave drying marks unless you are very careful to avoid leaving isolated drops on either the emulsion or the back of the film. A completely safe and successful method originating from Stephen Benton's laboratory is to omit the methanol bath, and to stand the processed hologram on a drying rack (if it is a plate) or suspend it from a frame (if it is a film) over a dish of ordinary tap water, covered with a belljar or cloche, until all the surface water on the emulsion has gone (this will take about two hours), after which the hologram can be brought into the open air and will dry without any marks in a few minutes.

If development is carried out to finality the undeveloped silver halide particles that are left are almost totally insensitive to light, and some workers prefer to leave the emulsion unfixed. To avoid possible printout, a final rinse in 1% acetic acid (see p. 89) is recommended.

Unbleached master holograms have a high signal-to-noise ratio, and make excellent transfers – as long as your laser is sufficiently powerful. You can obtain a much brighter image by turning the fringes back into silver halide in a bleach solution. Transfer holograms and focused-image holograms should always be bleach-processed. There are several different routes for bleaching (see pp. 72–4) and different methods are suitable for different types of hologram. The most thorough account of the principles involved is due to Nick Phillips (2), which describes progress in processing since an earlier, equally important paper (Phillips *et al*, 1).

Rehalogenating bleaches for transmission holograms After development in a developer of the D-19 or LX-24 type followed by fixation, a rehalogenating bleach is necessary. A formula that has proved successful is given below. It is intended for holograms that have been developed to a density of approximately 1.5 over a period of 2–4 min then fixed and washed.

De-ionized water	700 ml
Ferric nitrate crystals ($Fe(NO_3)_3 \cdot 9\,H_2O$)	150 g
Potassium bromide (KBr)	30 g
De-ionized water to make	1000 ml

Bleach time is about 2 min. Leave the film in the bleach for about 1 min (but not longer) after bleaching appears complete. Wash for 8 min and dry. This method of processing was originally described by Phillips (1). Fixing is no longer recommended for the best quality master holograms, and a bleach bath for use with holograms that have been developed but not fixed, originally described in Phillips (1) is no longer recommended owing to the extremely unpleasant qualities of p-benzoquinone (PBQ). New formulae have been evolved (eg Phillips, 7), and these are given on pp. 438–9.

Graube (3) suggested the use of bromine vapor for bleaching, as a method which gives considerable protection against printout. However, this was shown to result in breakdown of the gelatin after a few months or years. A better method is to use bromine *water*, preferably (but not essentially) after fixation (Benton, 4). The lack of stain in the bleached hologram gives an impression of cloudiness, but the signal-to-noise ratio is high. Bromine water is made by shaking up water and liquid bromine. Enough of the bromine is dissolved to produce a bleaching action. The solution is poured off the top of the liquid bromine for use, and is poured back into the container afterwards.

NOTE: Bromine water generates irritant fumes, and you would be extremely ill-advised to use it in an unventilated lab. Make sure your extraction equipment complies with regulations, as an accident could cause permanent injury to your lungs. Keep a 10% solution of sodium sulfite (Na_2SO_3) on hand in case of spills.

Reflection Holograms

Holograms intended as reflection masters for transfer have to reconstruct a bright image with laser light. This is possible even when material is lost from the emulsion during processing, provided the original reference beam was at a large angle of incidence and was more or less collimated, though the change in the reconstruction angle may cause some distortion of the image.

Tanning development Alternatively, a tanning development process can be used. This is a process in which the oxidation products of development produce cross-linking effects in the gelatin, making it rigid, so that on drying it does not shrink. The most important developing agent of this type is pyrogallol or 'pyro' (1, 2, 3-trihydroxybenzene). Pyrogallol was almost the first developing agent used in photography, and its tanning and staining properties were exploited for many years in aerial reconnaissance photography, where it was used in association with metol (*p*-methylaminophenolsulphate) to obtain useful imagery from severely under-exposed air films. The system was first adopted for holography by Ruud van Renesse. It is described in a paper by Walter Spierings (5), who dubbed it the 'Pyrochrome' process. It is a particularly good-natured process which produces bright reflection holograms under a range of fairly unpropitious conditions from moderate under-exposure to gross over-exposure and with development times varying from a few seconds to over 5 min at temperatures between $16°$ and $50°C$ and with chemicals measured out with a teaspoon rather than a laboratory balance. Michael Burridge (6) has optimized the exposure and processing conditions, and the author has modified the formula to provide a high and stable pH and give the highest possible emulsion speed.

Solution A
- De-ionized water — 700 ml
- Pyrogallol — 15 g
- Metol — 5 g
- De-ionized water to make — 1000 ml

Solution B
- De-ionized water — 700 ml
- Sodium carbonate (anhydrous) (Na_2CO_3) — 60 g
- Potassium hydroxide (KOH) — 20 g
- De-ionized water to make — 1000 ml

Solution C
- De-ionized water — 700 ml
- Sodium sulfite (anhydrous) (Na_2SO_3) — 100 g
- De-ionized water to make — 1000 ml

NOTE: *Solution A* should be kept in a concertina bottle with all air excluded. Should this not be possible, a pinch of sodium or potassium metabisulfite will discourage atmospheric oxidation; this should not, however, be done for solutions intended for reflection master holograms as it may produce a yellow-orange result.

To make holograms that are the same color as the laser light, take equal parts of *Solutions A* and *B*. Mix immediately before use. The best results seem to be obtained at elevated temperatures, eg 40°C, with a development time not exceeding 45 s, though excellent results can be obtained at 20–25°C with development of 60–75 s. With such short development times, a pre-soak of 1 min in plain water to which a drop or two of wetting agent has been added is strongly advised. The exposure should be adjusted to give a density of about 2.0, though higher densities can be tolerated. The inclusion of metol in *Solution A* is not essential, but it gives an approximately threefold increase in effective emulsion speed. The potassium hydroxide gives a substantially higher pH^* (and fringe contrast) than sodium carbonate alone; the carbonate acts as a *buffer* (pH stabilizer).

Solution C is used for changing the color of the final hologram. In the proportions $A:B:C = 1:1:0.2$ it gives a yellow image. At proportions of $1:1:0.5$ it gives a green image. The actual hue depends to some extent on the emulsion batch and to some extent on the developed density, a higher density tending to give a greener result. Pre-swelling techniques (see below) seem to give brighter results at shorter wavelengths – even into the blue – and do not need the addition of sulfite.

Reversal bleach Reflection holograms developed in pyro can be bleached in a

---

* Too high a pH, however, results in only the surface layer of the emulsion being developed.

reversal bleach, ie a bleach which removes the silver and leaves the unexposed silver bromide. The reversal bleach bath is:

De-ionized water	700 ml
Potassium dichromate ($K_2Cr_2O_7$)	5 g
Sodium hydrogen sulfate crystals ($NaHSO_4 \cdot H_2O$)	80 g
De-ionized water to make	1000 ml

The original pyrochrome bleach used 4 ml sulfuric acid, but sodium hydrogen sulfate, (also known as sodium bisulfate) has been adopted on a suggestion from Jeff Blyth (7), as it is much safer to handle and provides buffer action, so the bleach can be used repeatedly. After development, rinse the emulsion first under the tap, then in de-ionized water (any chlorides in the tap water will have a deleterious effect on brightness), and bleach until the emulsion has been apparently clear for about 1 min. Rinse in de-ionized water after bleaching, then wash for a minimum of 8 min in several changes of running water, and finish with a final rinse in de-ionized water with a few drops of acetic acid as descibed above. In the early stages of the final wash brush the surface of the emulsion with your fingers or give it a wipe with a cellulose sponge, as a certain amount of silver dichromate precipitates onto the surface during bleaching. This does no harm, and sufficiently prolonged washing will remove it anyway; but it is best to be certain.

Non-complexing developers A more recent approach by Blyth (7), which has proved to give outstandingly bright results, is to look for processing methods that do not remove any material from the emulsion, and therefore cause the minimum of disturbance to the fringes. These result in an image which reconstructs using light from the original laser, so that it is eminently suitable for reflection master holograms[*]. If a change in hue is required for a transfer hologram, this can be achieved by pre-swelling. For this process it is necessary to employ a developer that is free from both complexing and tanning action, and a rehalogenating bleach that deposits the silver bromide on the existing (unexposed) crystals, rather like a physical developer in reverse – Blyth refers to it as a 'physical bleach'. Phillips (2) calls it a 'diffusion transfer process', a term which is perhaps more accurate. Most bleaches show this effect to some extent, but earlier workers mistakenly tried to suppress it rather than exploit it.

Almost all developing agents are partially-substituted benzenes: for example, hydroquinone (*p*-dihydroxybenzene) has two hydroxyl groups substituted for hydrogen at opposite ends of the benzene ring; pyro has three adjacent ones. Developing agents containing amino (NH_2) and similar substitutions (for example 1,4-diaminophenol, H_2N—⟨ ⟩—NH_2) have a strong tendency to form soluble complexes with silver halides. One developing agent which does not do this is ascorbic acid ($C_6H_8O_6$). A typical developer of this type, tailored for holographic

---

[*] This does not apply to Ilford emulsions, which have a built-in 10% shrink factor (see p. 439).

emulsions, is:

De-ionized water	700 ml
Ascorbic acid	40 g
Sodium carbonate, anhydrous	100 g
Potassium hydroxide	20 g
Phenidone	0.2 g
De-ionized water to make	1000 ml

The phenidone increases the effective emulsion speed, but is otherwise optional. The chemicals must be dissolved in the order given. Development time is about 2–3 min at 25°C, to a high density, greater than 2.0. The bleach is a completely new formula:

Solution A

De-ionized water	700 ml
Copper (II) bromide (cupric bromide) ($CuBr_2$)	100 g
Lactic acid ($C_2H_5OH.COOH$)	100 ml

Solution B

Phenosafranine	0.5 g
Boiling water	100 ml

The phenosafranine may not all dissolve, but allow it to settle, and add the liquid to *Solution A*; then make up to 1000 ml with de-ionized water.

To give consistent results the emulsion must be completely dry at the start of the exposure. It should be placed in a dehumidifier or left on a radiator sill overnight before use. If you fail to do this you are liable to get a variation of fringe spacing from the center to the edges. In white-light reconstruction this results in variation of hue; in a master hologram it means that the brightness of the image will vary from the center to the edges.

To produce different hues chemically you can add potassium bromide up to a maximum of 60 g/l for a bright yellow. Blyth, working with the author, has recently found that by changing the copper bromide bleach formula somewhat it is possible to obtain a range of hues from bright yellow to green and even blue, depending on the length of immersion (up to 35 min). The formula is as follows:

De-ionized water	700 ml
Copper (II) bromide	130 g
Acetic acid, glacial	75 ml
De-ionized water to make	1000 ml

A reliable method of producing a variety of colors is to pre-swell the emulsion with triethanolamine (TEA) : 5% for yellow, 10% for green and 15% for blue, squeegeeing and drying before use as described on p. 265. Blyth (7) states that the brightest 'pyrochrome'-processed holograms are those which have been pre-swollen in 6% TEA. A master hologram that has shrunk too much can be permanently re-swollen by immersion overnight in 0.2% PBQ ($C_6H_4O_2$) (Burridge (6)). (Use a fume cupboard!) It is also possible to produce temporary swelling with TEA (5–15%) or sorbitol (10–30%) for making laser copies of reflection holograms that were originally yellow or green, though both substances tend to encourage print-out. Aluminum sulfate ($Al_2(SO_4)_3.16H_2O$) is a temporary hardening agent which will hold the emulsion in a partially-swollen state when dried: it should be used at a concentration of 5–15%, depending on the amount of color shift required. It can be readily washed out. A hologram that is too dark a red can be corrected by desiccation in an oven set to a low temperature, for about 10 min (plates only!).

Phillips (9) has summarized all the problems of bleach-processing, and has evolved a non-toxic bleach which keeps well and can be partially regenerated by leaving it in an open dish overnight. The formula is as follows:

De-ionized water	700 ml
Ferric sulfate ($Fe_2(SO_4)_3$)	30 g
EDTA disodium salt	30 g
Potassium bromide	30 g
Sulfuric acid (H_2SO_4), concentrated	10 ml
De-ionized water to make	1000 ml

(EDTA is ethylenediamine tetraacetic acid, commercially known as Sequestrene.) As mentioned earlier, sulfuric acid in its concentrated form is a dangerous substance: not only is it corrosive, but it generates so much heat when mixed with water that careless mixing can result in spitting of the acid to a considerable distance. Always add sulfuric acid to water, never water to acid. Measure the quantity with a pipette equipped with a proper bulb – you can get these from suppliers of chemical glassware. Phillips, in a private communication, has suggested the substitution of 50 ml glacial acetic acid for sulfuric acid; it is less corrosive and does not present mixing difficulties. However, in testing this formula the author has found that the substitution of 30 g of sodium hydrogen sulfate crystals for the sulfuric acid (as in the Pyrochrome reversal bleach) gives equally good results.

Ilford Ltd publish technical information on their holographic film, with separate booklets for red- and green-sensitive material. When processed in a non-tanning developer and bleached without fixation in a rehalogenating bleach, both these emulsions shrink in thickness by approximately 10%. This gives a golden-yellow image to reflection holograms made on red-sensitive material SP673 with a HeNe laser, and produces the right change in reconstruction wavelength to allow pulse-laser reflection master holograms to be transferred by a HeNe or Kr^+ laser. To produce reflection holograms of the same hue as the laser light that made them, both SP673 and the green-sensitive SP672 should be developed in a pyro developer and bleached in a rehalogenating bleach. Ilford produce a proprietary-formula non-tanning developer designated SP678C, which is diluted 1:4 for use. Development time is 2 min at 30°C (86°F). It is emphasized that for best results the exposure should be such that the correct density is obtained with the recommended time of development, ie development time should not be altered to compensate for errors in exposure, and exposures should not exceed 1–2 s.

Ilford also produce a proprietary reversal bleach, designated SP679C, which produces a single hue shift (red to yellow and green to blue-green) when used with a pyro developer, and a double hue shift (red to green or blue-green and green to blue) when used with a non-tanning developer. Ilford do not make a proprietary tanning developer, but recommend the following formula:

Solution A
    De-ionized water   700 ml
    Pyrogallol   12 g
    Ascorbic acid   12 g
    De-ionized water to make   1000 ml

Solution B
    De-ionized water   700 ml
    Sodium carbonate, anhydrous   60 g
    De-ionized water to make   1000 ml

For use take equal quantities of *Solutions A* and *B* and mix immediately before use. Average development time is 3 min at 20°C (68°F), but exposure and development times may be adjusted to allow control of the image hue. Unless a floating lid is employed the solution deteriorates quickly and must be discarded after use.

Ilford do not make a proprietary rehalogenating bleach; they recommend a ferric EDTA bleach (see p. 439). They also suggest a final 2 min bath of 2.5% potassium iodide (tap-water solution) as a safeguard against printout, though this does increase noise to some extent (it is emphatically not recommended with Agfa-Gevaert materials). Ilford produce a wetting agent, *Ilfotol*, which has none of the alleged shortcomings of Photo-flo, and this should be added to a final rinse in de-ionized water at the rate of a few drops per liter. Films should be squeegeed before being dried in a drying cabinet at not more than 40°C (140°F). Holograms treated with potassium iodide should be squeegeed and dried without any final rinse.

Stain Removal

Pyro stain is a positive help in suppressing noise, and with inherently noisy emulsions it can improve image contrast considerably. There is evidence that it contributes to the fringe structure, and it is possible to remove all the silver from the emulsion after pyro development and still obtain an image (of sorts) from the stain alone. However, when more than one hologram is used to produce a multiple-image piece, the stain in the outermost hologram may make the other image(s) unacceptably dim. With virtually noiseless emulsions such as Ilford SP673 stain removal will actually leave the image brighter than before.

Solution A
    Potassium permanganate (KM_nO_4)   10 g
    De-ionized water to make   1000 ml

Solution B
    Sodium metabisulfite ($Na_2S_2O_5$)   10 g
    De-ionized water to make   1000 ml

After washing the bleached film immerse it in *Solution A* for about 1 min, then clear it in *Solution B*. Wash for about 8 min and dry in the recommended way. Some workers add 4 ml concentrated sulfuric acid to *Solution A*, but this may not be a good idea for modern holographic emulsions. This process has been discussed in some detail by Ed Wesly (8).

Colloidal-silver Developers

As discussed above, one of the problems associated with bleached holograms is print-out, the darkening of a hologram exposed to light for long periods. Colloidal silver is a substance which is totally unaffected by light. It is dichroic, ie it reflects green light and transmits red light. A process that can produce a photographic image consisting of colloidal silver and shrink the emulsion by the correct amount to tune it to the wavelength the colloidal silver reflects best should be able to produce a very high diffraction efficiency. The developer GP8 from the Soviet Union achieves this when used with Agfa-Gevaert 8E75HD plates (no information is obtainable at present on Kodak or Ilford material). The formula is as follows:

De-ionized water	700 ml
Sodium sulfite, anhydrous	100 g
Hydroquinone	5 g
Potassium hydroxide	10.6 g
Phenidone	0.2 g
Ammonium thiocyanate (NH_4CNS)	25 g
De-ionized water to make	1000 ml

For use dilute 1:7 with de-ionized water. Development time (Agfa-Gevaert emulsions) should not exceed 2 min at 20°C; do not exceed this temperature. The density of the developed emulsion should be only about 0.6 when viewed against the safelight. Fix, wash and dry; do not bleach. The process seems to be less successful with film than with plates. For longer shelf storage, the potassium hydroxide should be dissolved separately in 1000 ml of de-ionized water; for use the two solutions are mixed in equal proportions and the mixture diluted 1:3.

A more recent developer from the Soviet Union is PRG-1. This is a colloidal-silver developer which uses potassium bromide instead of thiocyanate, and is described by Phillips (9). The formula is as follows:

Solution A

De-ionized water	700 ml
Sodium sulfite, anhydrous	38 g
Hydroquinone	30 g
Potassium bromide	22 g
De-ionized water to make	100 ml

Solution B

De-ionized water	700 ml
Potassium hydroxide	240 g
De-ionized water to make	1000 ml

For use take 1 part *Solution A*, 1 part *Solution B* and 13 parts de-ionized water. Development time for Agfa-Gevaert plates is 6–10 min.

Chemical Blackening of Reflection Holograms

Blyth (7) describes a method of blackening the backs of reflection holograms chemically, by immersing them briefly in a solution of sodium borohydride ($NaBH_4$). This is a powerful reducing agent which will convert silver halide to silver whether it has been exposed or not. The formula is as follows:

Methanol (or industrial meths)	650 ml
Water (tap)	350 ml
Sodium borohydride	0.5 g

Immerse the hologram, which must be grease-free and completely dry, in the solution until the back is sufficiently opaque (not black), then immediately wash vigorously; continue washing for several minutes, and dry as usual. The result is superior to that produced by thin films of most brands of black paint. The method is, of course, suitable only for holograms which are to be viewed through the base.

References

NOTE: The material in these references must be considered in the light of the state of knowledge at the time the papers were written. In examining the literature you should also beware of printing errors and omissions. The information given in the above chapter differs in important respects from that given in some of the reference material, and is the best there is at the time of writing.

1 Phillips N J, Ward A A, Cullen R and Porter D (1980) Advances in holographic bleaches. *Photographic Science and Engineering*, **24**, 120–4.
2 Phillips N J (1985) The role of silver halide materials in the formation of holographic images. *Proceedings of the SPIE*, **532**, 29–38.
3 Graube A (1974) Advances in bleaching methods for photographically-recorded holograms. *Applied Optics*, **13**, 2942–6.
4 Benton S A (1979) Photographic materials and their handling. In *Handbook of Optical Holography*, ed H S Caulfield. Academic Press.
5 Spierings W (1981) 'Pyrochrome' processing yields color-controlled results with silver-halide materials. *Holosphere*, **10**, 7/8, 1–2, 7.
6 Burridge, M J (1985) Private communication.
7 Blyth J (1985) Notes on processing holograms with solvent bleach. *Proceedings of the International Symposium on Display Holography*, **2**, 325–31.
8 Wesly E (1984) You'll wonder where... . *Holosphere*, **12**, 7, 21.
9 Phillips N J (1986) Benign bleaching for health holography. *Holosphere*, **14**, 4, 21–2.

APPENDIX
11

Books, Periodicals, Research Publications and Courses

'Now, *here*, you see, it takes all the running *you* can do, to keep in the same place. If you want to get somewhere else, you must run at least twice as fast as that.'
                    LEWIS CARROLL, *Through the Looking Glass*

As far as learning about holography is concerned, we are all running as fast as we can, just to keep up with the tide of information. This is a long book, probably longer than any other on practical holography that has appeared so far; yet there are still many aspects that have been covered only superficially: sometimes because information is scarce; sometimes because the topic has been only marginally appropriate; and, most frequently, because it is covered elsewhere more fully than would have been possible within a single comprehensive work. Here, then, is a list of the more important literature. Some of the books, inevitably, will have gone out of print by the time you read this, but you can still borrow them from your local library via the national library network.

Books

Abramson N. (1981) *The Making and Evaluation of Holograms*. Academic Press.

> Of all the books written on the more academic aspects of holography, this is the most approachable. It is so elegantly written that one would never guess that English was the author's second language. Abramson has been responsible for many original ideas concerning the possibilities of holography in physics and engineering. Alone among the theoretical textbooks, this book gives insight into the practical problems of holography on the machine-shop floor. The mathematics is within the range of 'A' level physics students, and the photographic illustrations are superb.

Caulfield H J (Ed) (1979) *Handbook of Optical Holography*. Academic Press.

> There is much that is excellent about this large book, including as it does among its authors almost every holographic scientist of repute. It is, indeed, full of immensely useful information (if only you could find it: the index is infuriatingly inadequate). The authors write in a variety of styles, and the academic level and the depth of treatment varies widely from chapter to

chapter. There are gaps in the material which the serious student will find frustrating, and some topics are treated at what seems excessive length in view of their limited importance. Nevertheless, this is an extremely valuable book, which contains much information not readily available elsewhere.

Collier R J, Burckhardt C B & Lin L H (1971) *Optical Holography*. Academic Press.

This book covers all the principles of optical holography in depth, and there is a comprehensive mathematical treatment of the relevant optical theory. It is short on practical advice, and much has happened since the last edition; but it remains an important reference book.

Hariharan P (1984) *Optical Holography*. Cambridge University Press.

This is simply the best-ever book on the theory of holography. It is fully up-to-date, and contains probably the most complete reference list (more than 700 references) of any published book on the subject. The book covers the entire field of optical holography, with rigorous mathematical back-up, making no concessions to those whose knowledge of complex numbers or vector theory is shaky; if you need a refresher course on these, it is up to you to take one before you get down to this book. There is no practical instruction.

Jones R and Wykes C (1983) *Holographic and Speckle Interferometry*. Cambridge University Press.

For those who want to become involved in this area of measurement science this is the complete book. It uses the same level of mathematics as Hariharan's book, but makes the concession to the reader of including an appendix which shows how to manipulate complex functions, and another on vectors. A new edition is in preparation.

Kaspar J E and Feller S A (1985) *The Hologram Book*, Prentice Hall International.

An introduction to the principles of holography, using simple geometrical models to explain the basis. It gives an account of the main types of hologram, and there is an overview of specialized techniques and applications of holography, which is particularly well-written. Elsewhere the style is rambling and discursive, with many minor errors.

McNair D (1983) *How to Make Holograms*. Tab Books Inc.

The main part of this book is a set of step-by-step instructions for building a holography lab in a garage, and equipping it with a sand-table and appropriate optical items with home-made holders, and a processing area. This part of the book is very good indeed. The remainder contains holographic layouts which are sound if limited; also an explanation of the principles of holography that is somewhat less than adequate, and a series of short interviews with some well-known American holographers.

Okoshi T (1976) *Three-dimensional Imaging Techniques*. Academic Press.

A wide-ranging book dealing with the principles of all types of stereoscopic, autostereoscopic and full-parallax imagery. It includes discussions of the physiology and psychology of the perception of depth, and discusses all types of photographic and holographic three-dimensional imaging. In spite of its age, it is surprising how many developments thought of as very recent are discussed in it. The mathematics is approachable, and is in terms of simple wavefront theory; geometrical optics is used where suitable. There is an excellent analysis of the possibilities of holographic television.

Saxby G (1980) *Holograms; How to Make and Display Them*. Focal Press; (in French) Masson.

Still in print in France at the time of writing, this was the author's previous book on the subject. The theory section which begins the book is non-mathematical and fairly lightweight. The practical section still works – but we have all learnt a lot since 1980. The final, predictive section was lucky: most of its guesses were right.

Saxby G (1980) *An Introduction to Holography*. Henry Greenwood.

A summary of modern holography, this is a reprint of six articles which would be useful to anyone asked to give lectures on basic holography to a lay audience.

Steward E G (1983) *Fourier Optics: an Introduction*. Ellis Horwood.

An excellent introduction to the subject at undergraduate level. Topics include Fraunhöfer diffraction, Fourier series and periodic structures, optical and crystal diffraction gratings, convolution and correlation, optical imaging and processing, holography (very briefly), interferometry, spectroscopy and astronomical applications. Unlike many student textbooks, the author's delight in the subject shows through clearly.

Taylor C A (1978) *Images*. Wykeham.

A very thorough discussion of the models used in the investigation of optical images, leading to a unified overview of all types of image information in terms of diffraction. The treatment is non-mathematical and the general level is that of the sixth-form student.

Unterseher F, Schlesinger R and Hansen J (1982) *Holography Handbook*. Ross Books.

This is a do-it-yourself manual of practical holography, a cookbook rather than a textbook. The style reflects the authors' exuberance. All the layouts work, though – no doubt in the interests of simplicity, which the book emphasizes – not all of them have been optimized for image quality. Much space has been used up on whimsical illustrations that have little relevance to the text, and a great many pages from catalogs of equipment have been reprinted without comment. The parts of the book that deal with the theory of holography and with philosophies based on concepts associated with holograms are not altogether satisfactory. Nevertheless, for the complete beginner in holography this book is a mine of useful information and guidance, and is warmly recommended.

Periodicals

A number of periodicals devoted to holography have come and gone over the past decade; of those that remain, the most important is *holosphere*, the quarterly journal of the New York Museum of Holography, 11 Mercer Street, New York NY10013. *Real Image*, the newsletter of the Royal Photographic Society's Holography Group, The Octagon, Milsom Street, Bath BA1 1DN, is published several times a year; *L.A.S.E.R. News* is the organ of the Laser Arts Society for Education and Research, PO Box 42083, San Fransisco, CA 94101, and appears quarterly; the Museum voor Holografie of Lovelingstraat 56, 2008 Antwerp, publishes a glossy newsletter under the name of *Holoblad*, with articles in Flemish and English.

A number of other periodicals regularly give space to holographic matters. Chief among these are the *British Journal of Photography*, *Laser Focus*, *Opto and Laser Products* (a trade journal) and *Lasers and Applications*. The arts magazine *Leonardo* also carries occasional articles on holography. Articles and news items related to holography are sometimes to be found in the *New Scientist* and the *Sunday Times (Business News Section)* (UK) and the *Scientific American*.

Research Publications

Several learned societies and institutes with an interest in holography hold conferences and symposiums on holography and related subjects from time to time. The most important of these is the Society of Photo-optical Instrumentation Engineers (SPIE), PO Box 10, Bellingham, Washington WA 98227–0010. Symposiums devoted entirely to holography are held every few years, organized by Tung Jeong at Lake Forest College, Lake Forest, IL60045. The proceedings are subsequently published by the organizations concerned. In the United Kingdom the Royal Photographic Society also holds conferences, papers from which are published in the *Journal of Photographic Science*, which also publishes refereed research material on holography. Other research journals which include papers on holography include *Optics Letters, Optica Acta, Applied Optics, Optics Communications, Journal of the Optical Society of America, Journal of Applied Physics, Electronics Letters, Journal of Physics (E), Progress in Optics* (annual), *Optical Engineering, Optik, Photographic Science and Engineering, Optical and Quantum Electronics, Japanese Journal of Applied Physics, Optics and Laser Technology, Journal of Optics* (Paris) and *Nature*.

Courses in Holography

In the United Kingdom the main training centers for beginners and advanced holographers are Darkroom Eight in Acton Town, and the Holography Workshop in Camberwell, London. The Royal College of Art offers higher degrees in holography to arts graduates. Salisbury College of Art and Liverpool Polytechnic offer tuition in holography to art students, and Wolverhampton Polytechnic offers projects in holography to students in both applied sciences and visual communications, as well as short courses for beginners which are open to the public. Loughborough University of Technology includes holography as an integral part of engineering courses, and encourages research in both applied and display holography. In the United States the most important training center is Lake Forest College in Illinois, which runs workshops for both beginners and advanced workers. The New York Museum of Holography operates an artist-in-residence program, and the New York Holographic Lab is a training center for holographers. There are two schools of holography in San Francisco: Holografix, and The School of Holography. The UCLA extension in Los Angeles also offers classes. In the midwest, apart from Lake Forest, there is the Chicago School of Holography associated with the Fine Arts Research and Holographic Center; and the School of Art Institute and Fermilab also provide instruction. The Media Laboratory at the Massachusetts Institute of Technology is developing a graduate teaching and research program in holographic imaging especially as applied to 3-D computer graphics. For details of these and other courses in the United States, it is suggested that you should contact the New York Museum of Holography.

In the United Kingdom the Open University offers a second-level half-credit course in applied modern optics called *Images and Information*. The home experiment kit includes a small laser and a 35 mm holographic camera. The correspondence units, which together form a complete textbook, are available separately, and TV programs associated with the course are broadcast at fortnightly intervals from February to September.

Computer Link-ups

INSPEC offers literature search facilities. It holds more than 6000 papers on holography; these can be accessed by most home-computers equipped with modems. Printouts of lists are forwarded by post if required, and copies of papers can be obtained via the public library system.

Mile High Media of Denver, Colorado, has set up a computer-conferencing system for holographers, with a world-wide network. The address is given in Appendix 12.

APPENDIX
12

Suppliers of Equipment, Materials and Information

> He offered large discount – he offered a cheque
> (Drawn 'to bearer') for seven-pounds-ten.
> LEWIS CARROLL, *The Hunting of the Snark*

The names and addresses given below represent only a proportion of the available sources. These have been chosen either because they represent the cheapest sources, or because they are established sources which are not likely to disappear before this book gets into print, or because they are the only sources of which information is available at the time of writing. A very full list of addresses, particularly in the USA, can be found in the *Laser Focus Annual Buyers' Guide*, published by the PennWell Publishing Co, 119 Russell Street, Littleton, MA 01460, (617) 486 9501, and in the UK by PennWell House, 39 George Street, Richmond, Surrey TW9 1HY, (01) 948 7866.

Chemicals

Rayco Instruments Ltd, Blackwater Way, Ash Road, Aldershot, Hants. (0252) 22725
Tri-Ess Sciences, Student Science Service Inc, 622 West Colorado Street, Glendale, CA 91204

Collimating mirrors

Optical Works, The Mall, Ealing, London W5 3TJ. (01) 567 5678
Wise Optics, Unit 9, Hollins Business Centre, Stafford. (0785) 3535
Coulter Optical Company, PO Box K, 54121 Pinecrest Road, Idlewild, Ca 92349. (714) 659 2991

Computer Data Information Services

INSPEC Publication Sales Dept (UK), Institute of Electrical Engineers, PO Box 26, Hitchin, Herts SG5 25A. (0462) 53331
INSPEC (US) IEEE Service Counter, 445 Hoeslane, Piscotoway, NJ 08854. (201) 981 0060

Mile High Media, Holography Computer-Conferencing Service, 3542 East 16th Street, Denver, CO 80206. (303) 329 3113

Courses on Holography

Liverpool Polytechnic (Faculty of Art & Design), 70 Mount Pleasant, Liverpool 1. (051) 207 3581
Royal College of Art, Kensington Gore, London SW7 2DF. (01) 584 5020
Salisbury College of Art, Southampton Road, Salisbury, Wilts SP1 2LW. (0722) 23711
The Holography Workshop, The Millard Building, Cormont Road, London SE5 9RG. (01) 733 3716
Wolverhampton Polytechnic, Wulfruna Street, Wolverhampton WV1 1SB. (0902) 313004
Loughborough University of Technology, Loughborough, Leicestershire. (0509) 263171
Darkroom Eight Ltd, Unit 8, Impress House, Vale Grove, Acton, London W3 7QH. (01) 749 2218/5162
Museum of Holography, 11 Mercer Street, New York NY 10013. (212) 925 0581
New York Holographic Lab, 34 West 13th Street, New York NY 10011. (212) 242 9774
Holographic Workshops, Lake Forest College, Lake Forest, IL 60045. (312) 234 3100
Fine Arts Research and Holographic Center, 1134 West Washington Boulevard, Chicago IL 60607. (312) 226 1007
Holography Institute, 1420 45th Street, Studio 35, Emeryville, CA 94608. (415) 658 3200
School of Holography, 550 Shotwell Street, San Francisco, CA. (415) 824 3769

Cross Holograms

Sapan Holographics, 240 East 26th Street, New York NY 10010. (212) 286 9397
Multiplex Custom Holograms, 3221 20th Street, San Francisco, CA 94110. (415) 285 9035

Embossing

Light Impressions (Europe) Ltd, 12 Mole Business Park, Station Road, Leatherhead, Surrey KT22 7AQ. (0372) 386677
(USA) Light Impressions Inc, 149B Josephine Street, Santa Cruz, CA 95060. (408) 458 1991
See-3 (Holograms) Ltd, 4 Macaulay Road, Clapham, London SW4 0QX. (01) 622 7729
Jayco Holographics Ltd, 29/43 Sydney Road, Watford, Herts WD1 7PY
Global Images Inc, 663 5th Avenue, New York NY 10022. (212) 759 8606
Holoplate, 1239 Central Avenue, Hillside, NJ 07205. (201) 352 0913
Holovision, 43 Pall Mall, London SW1. (01) 289 9969

Electronic Components

Maplin Electronic Supplies Plc, PO Box 3, Rayleigh, Essex SS6 8LR. (0702) 554155
Radiospares, Components Ltd, PO Box 99, Corby, Northants NN17 9RS. (0536) 201201
Tandy and Radio Shack outlets. Throughout the UK and USA

Fringe Stabilizers

International Dikrotek, 12755 South 200 West, Richmond, UT 84333. (801) 258 5461

General Optical Equipment for Holography

Optical Works Ltd, 32 The Mall, Ealing, London W5 3TJ. (01) 567 5678
UK Optical Supplies, 84 Wimborne Road West, Wimborne, Dorset BH21 2DP. (0202) 886831
Wise Optics, Unit 9, Hollins Business Centre, Stafford. (0785) 3535
Proops Brothers Ltd, 52 Tottenham Court Road, London W1P 0BA. (01) 636 4420 (callers only)
Edmund Scientific, 101 East Gloucester Pike, Barrington, NJ 08007. (609) 547 3488 or toll-free 1-800-222 1224

Halide Holograms

Applied Holographics, Braxted Park, Witham, Essex CM8 3XB. (0621) 893030 (also embossed holograms)
Third Dimension Ltd, 4 Wellington Park Estate, Waterloo Road, London NW2 7JW. (01) 208 0788
Smith & Cvetkovich Holography, 1000 W Monroe, Chicago, IL 60607. (312) 733 5462

Holographic Cameras

Holofax Ltd, Netherwood Road, Rotherwas, Herefored HR2 6JZ. (0432) 278400
Holomex, 4 Borrowdale Avenue, Harrow, Middlesex HA3 7PZ
Laser Technology Inc, 1055 W, Germantown Park, Norristown PA 19401. (505) 822 1123

Holographic Commissions and Displays

Light Fantastic Ltd, The Trocadero, 13 Coventry Street, London W1V 7FE. (01) 734 4516
Advanced Holographics Ltd, 315 New Kings Road, London SW6 4RF. (01) 731 4091
Holografix, 8827 SW129 Terrace, Miami, FL 33176. (305) 255 3166

Ion Laser Tube Repairs

Cambridge Lasers Ltd, Brookfield Business Centre, Cottenham, Cambridge CB4 4PS. (0954) 50083

Excitek Inc, 277 Coit Street, Irvington, NJ 07111. (201) 372 1669

Lasers (HeCd)

Liconix: (UK) Laser Lines Ltd, Beaumont Close, Banbury, Oxon OX16 7TQ (0295) 67755

(USA) 1390 Borregas Avenue, Sunnyvale, CA 94089. (408) 734 4331

Kimmon: (UK) Edinburgh Instruments, Riccarton, Edinburgh EH14 4AP. (031) 449 5844

Lasers (HeNe)

Spectra-Physics: (UK) 17 Brick Knoll Park, St Albans AL1 5UF. (0727) 80131

(USA) Laser Products Division, 1250 West Middlefield Road, Mountain View, CA 94039-7013. (415) 961 2550

NEC: (UK) Monolight Instruments Ltd, 2–3 Waterside, Hamm Moor Lane, Weybridge, Surrey KT15 2SN. (0932) 58566

(USA) NEC, 252 Humbolt Court, Sunnyvale, CA 94086. (408) 745 6520

Jodon Laser, 62 Enterprise Drive, Ann Arbor, MI 48103. (313) 761 4044

Melles Griot: (UK) Culdrose House, Frederick Street, Aldershot, Hants GU11 1LQ. (0252) 334411

(USA) 1770 Kettering Street, Irvine, CA 92714. (714) 261 5600

Siemens: (UK) Valiant, 20 Lettice Street, Fulham, London SW6 4EH. (01) 736 8115

(USA) Siemens Components Inc, 186 Wood Avenue South, Iselin, NJ 08830. (201) 321 3400

Lasers (ion)

Spectra-Physics (see above).

Coherent: (UK) Ltd, Cambridge Science Park, Milton Road, Cambridge CB4 4BH. (0223) 68501

(USA) 3210 Porter Drive, PO Box 10321, Palo Alto, CA 94303. (415) 493 2111

Lasers (Ruby Pulse)

Lumonics Ltd, Cosford Lane, Swift Valley, Rugby, CV21 1ON. (0788) 70321

Lumonics Inc, 105 Schneider Road, Kanata (Ottawa), Ontario K2K 1Y3. (613) 592 1460

Materials

Agfa-Gevaert: (UK) NDT Sales Dept, Great West Road, Brentford, Middlesex. (01) 560 2131. Supply warehouse: French's Avenue, Dunstable, Herts. (0582) 64101.

(USA) 275 North Street, Teterboro, NJ 07608. (201) 288 4100 Suppliers: Intergraf, Box 586, Lake Forest, IL 60045. (312) 234 3756
Eastman Kodak (USA only) Rochester, NY 14650. (800) 225 5572
Ilford Ltd, Mobberley, Knutsford, Cheshire WA16 7HA. (0565) 50000

Optical Fibers for HeNe Light

York Technology Ltd, York House, School Lane, Chandler's Ford, Hants SO5 3DG. (04125) 60411

PCB Holders

Radiospares, Components Ltd, PO Box 99, Corby, Northants NN17 9RS. (0536) 201201
Panavise, Colbert Industries, 1071 Adella Avenue, South Gate, CA 90280. (213) 569 8108

Pinholes

Graticules Ltd, Morley Road, Tunbridge, Kent TN9 1RN. (0732) 359061
Optimation, PO Box 310, Windham, NH 03087. (603) 484 2346

Research-grade Optical Equipment

Newport: (UK) Newport Ltd, Pembroke House, Thompsons Close, Harpenden, Herts. AL5 4ES. (05827) 69995
  (USA) 18235 Mt Baldy Circle, Fountain Valley, CA 92708. (714) 963 9811
Ealing Electro-optics plc (Corp. for the USA): (UK) Greycaine Road, Watford WD2 4PW. (0923) 242261
  (USA) 22 Pleasant Street, South Natick, MA 01760. (617) 655 7000
Aerotech: (UK) 19a Livingstone Road, Newbury, Berks RG14 7PD. (0635) 31324
  (USA) 101 Zeta Drive, Pittsburgh, PA 15238-2897. (412) 963 7470
Oriel: (UK) 1 Mole Business Park, PO Box 31, Leatherhead, Surrey, KT22 7AU. (0372) 378 822
  (USA) 250 Long Beach Blvd, PO Box 872, Stratford, CT 06497. (203) 377 8282
Melles Griot: (UK) Culdrose House, Frederick Street, Aldershot, Hants, GU11 1LQ. (0252) 334411
  (USA) 1770 Kettering Street, Irvine, CA 92714. (714) 261 5600
Farsound Engineering Ltd, 23 Englefield Road, London N1 4EJ. (01) 534 4214

Safety Goggles

AG Electro-optics Ltd, Tarporley, Cheshire CW6 0HX. (08293) 3305
Fred Reed Optical Co Inc, PO Box 27010, Albuquerque, NM 87125-7010. (505) 265 3531

Scaffolding Clamps

Kee Systems Ltd, Thornsett Works, Thornsett Road, London SW18 4EW. (01) 874 6566

Kee Industrial Products, PO Box 207, Buffalo, NY 14224.

Spatial Filters and Fiber-optics Beam Launchers

Scie-Mechs Ltd, 8a Wheatash Road, Addlestone, Weybridge, Surrey KY15 2ER. (09238) 62514

Jodon Engineering Association, 62 Enterprise Drive, Ann Arbor, MI 48103. (313) 761 4044

Spotlights

Wotan Lamps Ltd, 1 Gresham Way, Durnsford Road, London SW19. (01) 947 1261

See-3 (Holograms) Ltd, 4 Macaulay Road, Clapham, London SW4 0QX. (01) 622 7729

Halo Power-track (Lighting Division), McGraw-Edison Corp, 6 West 20th Street, New York, NY 10011. (212) 645 4580

Stands and Bases

Scie-Mechs Ltd, 8a Wheatash Road, Addlestone, Weybridge, Surrey KY15 2ER (09238) 62514

Climpex Ltd, Hammers Lane, Mill Hill, London NW7 4DY. (01) 959 1060

UK Optical Supplies, 84 Wimborne Road West, Wimborne, Dorset BH21 2DP. (0202) 886831

Optical Works Ltd, 32 The Mall, Ealing, London W5 3TJ. (01) 567 5678

Akron Tool Co, 21/3 Cherry Hill Rise, Buckhurst Hill, Essex IG19 6EB. (01) 505 8135 (magnetic bases)

Martonair Ltd, St Margarets Road, Twickenham TW1 1RJ (air suspensions)

Tables

UK Optical Supplies, 84 Wimborne Road West, Wimborne, Dorset BH21 2DP. (0202) 886831

Spectrolab Ltd, PO Box 25, Newbury, Berks RG13 2AD. (0635) 35733

Photon Control Ltd, Kings Court, Kirkwood Road, Cambridge CB4 2PF. (0223) 323071

All suppliers of research-grade equipment (see above).

Glossary

'Speak English!' said the Eaglet. 'I don't know the meaning of half those long words, and, what's more, I don't believe you do either!'
                        LEWIS CARROLL, *Alice in Wonderland*

'When I use a word,' Humpty Dumpty said, in a rather scornful tone, 'it means just what I choose it to mean – neither more nor less.'
                        LEWIS CARROLL, *Through the Looking Glass*

... Any writer of a book is fully authorised in attaching any meaning he likes to any word or phrase he intends to use. If I find an author saying at the beginning of his book 'Let it be understood that by the word "*black*" I shall always mean "*white*", and that by the word "*white*" I shall always mean "*black*", I meekly accept his ruling, however injudicious I may think it.
                        CHARLES LUTWIDGE DODGSON, *Symbolic Logic*

ABBÉ PRISM   An optical device for derotating an image, similar to a Dove prism (*qv*) but which can be constructed from mirrors for lightness.

ABERRATION   The optical properties of a lens, mirror, prism or hologram that result in its failure to form a point image in the geometrically correct position from a point source. Individual aberrations are listed under their names.

ACHROMATIC   1 Describes an optical system free from dispersion (*qv*).
                    2 When used to describe a holographic image, colorless.

ACOUSTIC HOLOGRAPHY   A form of holography using ultra-high-frequency sonic beams. Also called 'ultrasonic holography'.

ACTIVE SWITCHING   In pulse-laser technology, the operation of a Q-switch (*qv*) by external means (*cf* passive switching).

ADDITIVE COLOR SYNTHESIS   A method of producing a multicolor image by mixing lights of different hues (*cf* subtractive color synthesis). *See also* CIE Chromaticity Diagram.

AIRY DIFFRACTION PATTERN (DISK)   The far-field diffraction pattern (*qv*) of a circular aperture or an opaque disk. The central portion of the pattern is known as the Airy disk.

ALCOVE HOLOGRAM   A concave cylindrical 180° holographic stereogram which projects a real image of a computer-generated object to its center. The image may continue back from the center through the plane of the hologram to infinity.

ALIASING 1 A problem occurring in computer-generated holography (*qv*), where a continuous function is represented by discrete samples. If these samples are incorrectly spaced, the original image cannot be retrieved.
2 The zigzag or staircasing effect produced in a visual display when a sloping or curved line or edge is imaged by rectangular pixels (*qv*).

AMPLIFIER See Oscillator (2).

AMPLITUDE The maximum value of the displacement of a point on a wave from its mean value. Instantaneous amplitude is the actual displacement at a given instant.

AMPLITUDE HOLOGRAM A hologram in which the information is coded in the form of variations in transmittance (*cf* phase hologram).

ANAMORPHIC SYSTEM An optical holographic system which produces a spatially-distorted image.

ANTIPHASE Describes the relative state of two wavefronts that have phases that differ by one-half of a complete cycle.

ARGON-ION (Ar^+) LASER A gas laser using ionized argon gas as the lasing medium (*see* Ion laser). Its most important lines in the visible spectrum are at 514.5 nm and 488 nm.

ARTIFACT 1 A man-made archaeological find.
2 A blemish on a hologram etc that is not part of the true image.

ASTIGMATISM 1 An aberration (*qv*) of a lens or mirror, characterized by the focusing of an off-axis point source in one plane as a line radial from the principal axis, and in another as part of a circle centered on it.
2 An aberration of a hologram or cylindrical lens system in which the vertical and horizontal aspects of the image are focused in different planes.

AUTOSTEREOGRAM, AUTOSTEREOSCOPIC Describes an image that can be seen stereoscopically without the aid of any optical device.

BANDWIDTH The spread of temporal frequencies produced by a light source.

BEAM EXPANDER An optical system which increases the diameter of a laser beam.

BEAM INTENSITY RATIO 1 Strictly, the intensity ratio of the components of the electric vectors of the reference and object beams that are polarized in the same direction, along a line bisecting the angle between them.
2 In practice, what is measured is the ratio of intensities in the direction of the reference and object beam sources at the hologram plane, using a light meter that may or may not be equipped with a polarizing filter.

BEAMSPLITTER An optical device which divides a beam of light into two beams.

BED-OF-NAILS FUNCTION A regular two-dimensional array of delta functions (*qv*).

BENTON HOLOGRAM See Rainbow hologram.

BENTON STEREOGRAM An achromatic holographic stereogram made on the multiplexing principle (*qv*) which uses a transfer geometry, in contrast to a Cross hologram (*qv*), which does not.

BESSEL FUNCTION   The solution of a particular type of differential equation, useful in the solving of two-dimensional problems. The symbol for a Bessel function is J_x, where x is the order of the function.

BICYLINDRICAL LENS   A lens having two orthogonal cylindrical surfaces.

BINARY RECORDING MEDIUM   A recording material in which the response is either zero or total.

BIREFRINGENCE   The property possessed by certain substances of forming two orthogonally-polarized wavefronts from a single incident wavefront. Also known as 'double refraction'.

BLACK HOLE   A colloquial term for the three-dimensional silhouette left when some part of the subject matter has moved during the exposure of a hologram.

BLAZED GRATING   A diffraction grating (qv) in which the transmittance profile is a sawtooth function: most of the diffracted energy is in the beams on one side only of the zero-order beam.

BLEACH   In holographic processing, a chemical solution which renders the photographic image transparent, changing an amplitude hologram into a phase hologram. Bleaches are described as 'rehalogenating' when they convert the photographic image from silver to silver halide, 'reversal' when they remove the photographic image leaving the undeveloped silver halide, and 'total' when they remove both the developed silver and the undeveloped silver halide.

BLOOM STRENGTH   A measure of the resilience of a sample of gelatin.

BLU-TACK   The trade name of a synthetic putty adhesive marketed by Bostik plc and sold in the United States under the name of Superstuff. The material is also sold under various other names.

BOHR MODEL   A model which describes the behavior of an atom in terms of fixed energy states of its electrons and the transitions between these energy states, which may be visualized as spherical, ellipsoidal, etc orbitals.

BRAGG CONDITION   The condition for efficient reflection of a light beam undergoing Bragg diffraction (qv). In order to satisfy the condition the optical path difference between wavefronts reflected from successive layers of material must be an integral multiple of the wavelength (*cf* Grating condition).

BRAGG DIFFRACTION   The principle by which a stack of parallel reflecting surfaces (or zones of alternate high and low refractive index) will reflect a beam of light if and only if the reflected wavefronts are of appropriate wavelength and orientation to produce constructive interference.

BRAGG HOLOGRAM   Any hologram in which the Bragg condition is more important than the grating condition (qv) in forming the image.

BREWSTER ANGLE   The angle of incidence of a light beam on a surface such that the reflected and refracted beams are orthogonal; typically 56–58° for glass. At the Brewster angle the transmitted beam is partly polarized in the plane containing the incident ray and the normal to the surface (p-polarized); the reflected beam is totally polarized at right angles to it (s-polarized).

BREWSTER-ANGLE WINDOW   An optical window, offset so that both the incident and emergent rays at the Brewster angle. Such a window has zero reflectance for a beam that is p-polarized.

BREWSTER PRISM   A dispersing prism designed so that both the incident and emergent rays are at the Brewster angle. It is used in the optical cavities of multi-frequency lasers to select particular lines, and is tuned by tilting (*cf* Littrow prism).

BUFFER A chemical substance added to a solution to stabilize the pH (*qv*).

BURN-OUT A portion of the holographic emulsion that is grossly over-exposed due to focusing of light from the object. Its occurrence is marked by a black patch in the plane of the hologram obscuring a portion of the image.

CAMERA In holographic terminology, a device for making holograms in which the sensitive material is totally enclosed in a lightproof container. Holographic cameras may be modified versions of photographic cameras or (more commonly) lightproof enclosures containing the entire optical system including the subject matter.

CARDING-OFF The blocking-off of stray or unwanted light on a holographic table, usually by pieces of black card.

CARRIER BEAM An alternative name for a reference beam, drawn from the field of radio communications.

CHEMICAL DEVELOPMENT The removal of halide ions from exposed crystals of silver halide in a photographic emulsion.

CHIRPED A term, borrowed from communications technology, used to describe a reflection hologram in which the Bragg planes vary in spacing throughout the thickness of the emulsion.

CHOPPER A disk containing a number of apertures, set up so as to interrupt a laser beam to produce repeated pulses for stroboscopic illumination.

CHROMA The equivalent in the Munsell color system (*qv*) of saturation (*qv*).

CHROMATIC ABERRATION A lens aberration (*qv*) caused by dispersion, which causes blue rays to be refracted more than red rays; the result is that in an optical image formed by a lens the image for blue light is smaller and closer to the plane of the lens than is the image for red light. In holography the image is formed by diffraction, and as blue light is diffracted less than red light, the result is the reverse: the red image is nearer to the hologram and smaller than the blue image.

CIE CHROMATICITY DIAGRAM A diagram which gives a quasi-objective description of the color quality of a visual stimulus in terms of hue and saturation (*qv*). The color quality is defined in terms of coordinates specifying the equivalent content of red, green and blue stimuli that would give an exact visual match.

CIRCULAR POLARIZATION *See* Polarization.

COHERENCE The degree to which the photons in a light beam are in phase. It is usually defined in terms of temporal coherence, which is related inversely to the bandwidth (*qv*), and spatial coherence, which is related to the degree of correlation between the various parts of the beam.

COHERENCE LENGTH The distance over which the phases of the photons making up a light beam remain correlated, ie the greatest optical path difference between two beams derived from the same source such that, when they are recombined, interference fringes will be formed.

COHERENT FIBER-OPTICS BUNDLE A fiber-optics light guide in which individual fibers retain their positions relative to one another at both ends.

COHERENT LIGHT *See* Coherence

COLLIMATED BEAM (COLLIMATION) (The production of) a beam of light that neither converges nor diverges.

COLLOIDS Systems in which one phase (the dispersed phase) is distributed in another (the continuous phase). The dispersed phase is in the form of particles or droplets typically less than 1 μm in diameter. Sols are dispersions of solid particles in a liquid; aerosols are solid or liquid dispersions in a gas; emulsions are systems in which both phases are liquids; and gels are colloids in which both phases have a three-dimensional network throughout the material (gelatin is a common example). Note that a photographic 'emulsion' is not an emulsion at all: as coated it is a sol, and when dry it is a solid suspension.

COLLOIDAL SILVER A dispersion of extremely small particles of silver metal; it is strongly dichroic, transmitting red light and reflecting green.

COLOR TEMPERATURE A description of the spectral energy distribution of a light source in terms of the thermodynamic temperature in kelvins (abbreviated K) at which a perfectly-radiating body would produce the same energy distribution.

COMA A lens aberration in which an off-axis point appears as a comet-shaped patch aligned radially from the optical center. The same aberration is shown by holographic images.

COMB FUNCTION A uniform one-dimensional array of delta functions (qv).

COMPUTER-GENERATED HOLOGRAM (GCH) Originally a hologram drawn by a computer to produce a specific image or for spatial or correlation filtering. The meaning is now generally extended to include holograms (usually holographic stereograms) of objects drawn by a computer.

CONFOCAL SYSTEM A double lens system in which the rear focal plane of one system is the front focal plane of the other.

CONJUGATE BEAM A beam in which the direction of travel of the wavefront is the exact reverse of that of the original beam.

CONSTRUCTIVE INTERFERENCE *See* Interference.

CONTINUOUS SPECTRUM The luminous output of a light source which radiates a continuum of wavelengths (*cf* line spectrum).

CONTINUOUS-WAVE (CW) LASER A laser that emits a beam of light the intensity of which does not vary with time (*cf* pulse laser).

CONVOLUTION (CONVOLVED) A mathematical expression in which one function is dealt out to every point of another function (the two functions are said to be convolved).

CONVOLUTION THEOREM This states that the Fourier transform of the product of two functions is the same as the convolution of their individual Fourier transforms, and vice versa.

CORRELATION FILTERING A form of holographic data processing in which a Fourier-transform hologram (qv), often computer-generated, is used to identify the presence and position of a particular item in a transparency.

COSINE GRATING A one-dimensional grating with a transmittance that varies cosinusoidally with distance.

COSINE WAVE A wave with an instantaneous amplitude that varies cosinusoidally.

COSINUSOIDAL Describes a function which varies according to the relationship $y = A \cos Bx$, where x and y are the variables. The fluctuations are identical with those of the sinusoid $y = A \sin Bx$, but are shifted in phase by one-quarter of a cycle. The cosine function is an even function (qv).

CRITICAL ANGLE The angle of incidence at the interface of an optical medium with another medium of lower refractive index such that the emergent ray travels along the interface (see also Total internal reflection).

CROSS HOLOGRAM A particular type of holographic stereogram (qv) based on holograms of motion-picture frames, invented and patented by Lloyd Cross. Also known as an integral hologram, an integram and a Multiplex hologram.

CROSS LINKING A phenomenon that can occur in substances made up of long chain molecules such as gelatin and certain polymers. Under certain conditions (such as the action of light) the molecules form bonds at various points in their length with adjacent molecules, causing the substance to change from a liquid or a gel to a rigid and insoluble solid.

CROSS-TALK A term borrowed from communications technology, used in holography to mean the production of spurious images in multiplexed (qv) or color holograms.

CURVATURE OF FIELD A lens aberration characterized by a focal field that is saucer-shaped rather than flat.

CYLINDRICAL LENS A lens that is curved in one direction only.

DECONVOLUTION (DECONVOLVED) The removal of unwanted elements of an image (eg halftone dots, movement smear) by modification of its optical Fourier transform by spatial filtering (qv). The elements are said to have been 'deconvolved'.

DE-IONIZED WATER Water that is almost completely free from chemical contamination. Often used as synonymous with 'distilled water', though the process of removing the contaminants is not necessarily distillation.

DELTA FUNCTION Also known as a Dirac delta function, this is the limiting case of a square pulse of unit area which contracts in width indefinitely. It has infinitesimal width, infinite height and unit area.

DENISYUK HOLOGRAM A single-beam (reflection) hologram (qv).

DENSITOMETER A device for measuring photographic density (qv).

DENSITY See Photographic density.

DEPTH OF FIELD In photography, the distance between the nearest and farthest points that give an image of acceptable sharpness when the camera is focused on a given distance and the lens is set to a given f/no. The use of the term in holography is ambiguous and is not recommended.

DEROTATION See Optical derotation.

DESTRUCTIVE INTERFERENCE See Interference.

DEVELOPING AGENT A reducing agent (qv) which converts silver halide crystals into silver grains if and only if they bear a latent image (qv).

DEVELOPMENT The process of producing a photographic image in metallic silver.

DICHROIC FOG A deposition of colloidal silver which appears red by transmitted light and green by reflected light.

DICHROIC MIRROR A type of beamsplitter (qv) which reflects one band of wavelengths and transmits the remainder of the spectrum.

DICHROMATED GELATIN (DCG) A light-sensitive non-silver emulsion consisting of gelatin sensitized by dichromate ions which promote cross-linking (qv) when stimulated by short-wave light energy, and producing on processing a phase hologram (qv) of high diffraction efficiency.

DIELECTRIC MIRROR   A mirror which operates by Bragg diffraction (qv). It consists of a glass plate coated with layers of material of alternately high and low refractive index, the layers being one half-wavelength in thickness.

DIFFRACTION   The change in direction of a wavefront encountering an object.

DIFFRACTION EFFICIENCY   In a hologram, the ratio of image-forming to incident light, in terms of intensity.

DIFFRACTION GRATING   A one-dimensional grid of ruled lines, used for dispersing a beam of white light into a spectrum. It can be produced holographically by the interference of two coherent light beams.

DIFFUSION TRANSFER   The migration of chemical substances within a solid or gel. In holography the term is used to describe the deposition of silver on the developing fringes (physical development) or of silver halide on the unexposed silver halide crystals in a rehalogenating bleach (physical bleach).

DISPERSION   The separation of a polychromatic (qv) beam of light into its components by wavelength.

DISTORTION   A lens aberration characterized by the outward (pincushion) or inward (barrel) displacement of points in the outer image field from their true geometric positions.

DOPPLER BROADENING   The increase in bandwidth (qv) caused by the rapid random movement of energy-emitting molecules in a hot gas.

DOPPLER EFFECT   When a source radiating at a constant frequency is moving relative to a receiver, the frequency at the receiver is higher when the two are approaching and lower when they are receding from one another.

DOUBLE-EXPOSURE HOLOGRAPHIC INTERFEROMETRY   See Holographic interferometry.

DOVE PRISM   A prism with opposed 45° entrance and exit angles. The beam suffers total internal reflection at the hypotenuse face. The prism produces inversion with preservation of right-left orientation.

DYE LASER   By passing a beam from a laser into a cell containing a fluorescent dye, a range of wavelengths may be produced, tunable via a Littrow prism (qv).

ELECTRIC VECTOR   The electric component of electromagnetic radiation. Linearly-polarized light is defined as having a plane of polarization that corresponds to the plane of the electric vector.

ELECTROMAGNETIC MODEL   A model for the propagation of light which uses the field equations of an electromagnetic disturbance as a basis for the prediction of the behavior of light.

ELECTROMAGNETIC RADIATION   The propagation of radiant energy according to the electromagnetic model.

ELECTRONIC SPECKLE-PATTERN INTERFEROMETRY (ESPI)   A method of measuring small movements in an object by the comparison of movements of the speckle pattern of the image of the object illuminated by a laser.

ELLIPSOID   In this text, the term is used to mean the surface obtained by rotating an ellipse about one of its axes (*cf* Hyperboloid). Strictly, it is known as an ellipsoid of revolution.

ELLIPTICAL POLARIZATION   *See* Polarization.

EMBOSSED HOLOGRAM   A hologram produced by a mechanical process similar to that used for producing audiodisks.

ENERGY BAND   In quantum physics, a group of energy levels which contains so many closely-spaced levels that it can be effectively considered a continuum.

ENERGY LEVEL (STATE)   In quantum physics, a single level at which an electron is in a specified quantum state. In the Bohr model it corresponds to a specific orbital.

ENTRANCE PUPIL   The effective diameter of that part of a beam of light entering a lens which is actually transmitted.

EVANESCENT WAVE   Under conditions of total internal reflection (qv), there is still a small amount of energy outside the surface, which decreases exponentially with distance. This is known as the evanescent wave. If a surface is placed close enough, light can be induced to pass out of the first surface into the second; this is the principle of the evanescent-wave beamsplitter.

EVEN FUNCTION   A function which is symmetrical about the y-axis.

EXCIMER LASER   When the atoms of noble gases are excited they form metastable compounds with halogens; such compounds have powerful lasing properties. The term 'excimer' is a contraction of 'excited dimer'.

F-NUMBER, *F*/NO, *F*-STOP   The focal length of a lens divided by the diameter of the entrance pupil. In a first-generation hologram it is the distance of the object from the hologram plane divided by the horizontal dimension of the plate.

FABRY-PÉROT ETALON   An optical cavity formed by two accurately-parallel glass–air surfaces, used in high-power lasers to increase the coherence length (qv). A beam of light entering the etalon undergoes multiple reflections which interfere constructively only for a wavelength that is an integral submultiple of twice the distance between the faces.

FAR-FIELD DIFFRACTION PATTERN   The diffraction pattern produced by an object at a plane sufficiently far removed for the pattern to be regarded as identical (except for size) with that formed at infinity.

FAR-FIELD HOLOGRAM   *See* Fraunhöfer holography.

FILL-IN LIGHT   The light used in photographic, etc illumination for controlling the lighting contrast.

FIXATION   The removal of undeveloped silver halide from a developed photographic emulsion.

FLAT-BED EMBOSSING   A method of producing embossed holograms (qv) using a flat press similar to that used for making test pressings of audiodisks.

FLIPPING   The rotation of a hologram out of its plane about a horizontal axis, through 180° (*cf* spinning, rotating).

FLOAT GLASS   Glass of near-optical quality made by floating molten glass on a bed of molten tin.

FOCUSED-IMAGE HOLOGRAM   An image hologram (qv) made using the optical image produced by a lens or optical mirror as the object.

FOURIER ANALYSIS   The analysis of a function into cosinusiodal and/or sinusoidal components.

FOURIER MODEL, FOURIER OPTICS   A model for the behavior of light with particular reference to image formation, based on the methods of Fourier analysis.

FOURIER SERIES   The series (usually infinite) of sine and cosine functions that make up a given non-cosinusiodal waveform.

FOURIER TRANSFORM (FT)   The frequency spectrum of any function. If the function is repetitive, the FT is a Fourier series; if the function is not repetitive the spectrum is continuous and is defined by the equation of its envelope.

FOURIER-TRANSFORM HOLOGRAM   Strictly speaking, a hologram made in the FT plane of a lens. Also applied to a hologram made with a spherical reference wavefront originating from a point in the plane of the object. Both produce a pair of images, one erect and the other inverted. The second type is known as a lensless Fourier-transform hologram.

FOURIER-TRANSFORM PLANE   In Fourier optics this is the name given to the rear principal focal plane of a lens. In this plane the electric field is the FT of the electric field in the front focal plane.

FOUR-LEVEL LASER   A laser in which the relevant energy transition is between two energy levels, both of which are normally unoccupied.

FRAUNHÖFER DIFFRACTION   A form of diffraction in which the diffracting object is illuminated by collimated light and the diffraction pattern is examined at a large distance from the object (*cf* Fresnel diffraction).

FRAUNHÖFER HOLOGRAPHY   Holography carried out a considerable distance from the object. Applied in particular to in-line holograms (*qv*).

FREQUENCY   Used without qualification, the term is taken to mean temporal frequency (*qv*).

FREQUENCY SPACE   The domain of the Fourier transform (*qv*) as opposed to 'real' or *x*-space.

FRESNEL BIPRISM   A prism in which the angle is very nearly 180°, so that it can be thought of as two very thin prisms joined back to back. When illuminated by laser light it produces an interference pattern large enough to be seen by the unaided eye.

FRESNEL DIFFRACTION   Near-field diffraction: either the light source or the examining screen or both are close to the diffracting object (*cf* Fraunhöfer diffraction).

FRESNEL HOLOGRAM   The usual type of hologram, where the recording material is close to the object.

FRESNEL LENS (PRISM)   A lens (prism) which is cut back in steps in order to reduce its thickness.

FRESNEL ZONE PLATE   *See* Zone plate.

FRINGE STABILIZER   A device which holds the optical path difference of the object and reference beams constant within narrow limits by means of a servo device. Also known as a fringe locker.

FROZEN-FRINGE TECHNIQUE   A form of holographic interferometry involving two recordings of an object, one before and one after deformation (*cf* live-fringe techniques).

FT   An abbreviation for 'Fourier transform' (*qv*).

FULL-APERTURE HOLOGRAM   A transfer hologram (*qv*) in which the entire area of the master hologram is used to form the secondary image. Also known as an open-aperture hologram.

GABOR HOLOGRAM   Another name for an in-line hologram (*qv*).

GABOR ZONE PLATE   *See* Zone plate.

GAMMA   The gradient of the steepest portion of the density-vs-log-exposure curve for a photographic emulsion; the concept is useful only when this portion is of significant length, as it is in a holographic emulsion. It gives a very rough guide to the amplification of the contrast of the primary fringes that occurs when the holographic record of the fringes is processed.

GAS DISCHARGE TUBE   A device for producing light by ionic transitions. It produces a line spectrum (*qv*).

GAS LASER   A laser in which the pumping energy is supplied by an ionized gas.

GRATING CONDITION   The condition for diffraction at a particular angle for a light beam of given wavelength passing through a plane diffraction grating. The optical path difference between wavefronts diffracted by adjacent periods of the grating must be an integral multiple of the wavelength of the light (*cf* Bragg condition).

GROUND STATE   In the Bohr model of the atom (*qv*), the condition in which all the electrons are in the lowest permitted orbitals and the atom is in a stable state.

H_1, H_2, H_3 etc   Symbols used in holographic literature to indicate a master hologram and subsequent generations of transfer holograms. Slit master holograms are designated S_1 etc.

H & D CURVE   The curve showing the graphical relationship between photographic density (*qv*) and the logarithm of the exposure for a photographic emulsion, named after the initials of its inventors Hurter and Driffield.

HALATION   Loss of image quality due to internal reflection within a plate or film base. Holographic materials have been available from time to time with antihalation backing; such materials are not suitable for reflection holography unless the backing is removed before exposure.

HALF-WAVE PLATE   A retardation plate (*qv*) which has the effect of rotating the axis of polarization; this may be set to any desired angle without loss of light.

HALFTONE PROCESS   A process which permits the reproduction of a continuous-tone picture using only one density of printing ink. The process breaks up the image optically into an array of tiny dots of constant reflectance but differing sizes.

HALIDE, HALOGEN   A halide is a compound of a halogen with another element or group of elements. The halogens are the elements fluorine, chlorine, bromine, iodine and astatine.

HEAD-UP DISPLAY (HUD)   A method of displaying instrument readings etc as optical images at infinity before the eyes of the operator, superimposed on the direct view.

HELIUM-CADMIUM (HeCd) LASER   A neutral-gas laser (*qv*) which employs helium as the pumping medium and cadmium vapor as the lasing medium. Its most important line in the visible spectrum is at 442 nm in the violet.

HELIUM-NEON (HeNe) LASER   A neutral-gas laser (qv) which employs helium as the pumping medium and neon as the lasing medium. Its most important line for holography is at 633 nm in the orange-red, but other, weaker lines are available, in particular at 545 nm, in the green region.

HINGE POINT   The point at which a straight line passing through the object and the reference source (for a transmission hologram) or the reflection of the reference source in the hologram plane (for a reflection hologram) intersects the hologram plane.

HOLODIAGRAM   A geometrical construction developed by Nils Abramson for designing holographic geometries. It consists of the loci of points equidistant from the reference and object sources at intervals that are an integral number of wavelengths apart in total optical path length.

HOLOGRAM   A complex diffraction grating which, when illuminated appropriately, generates an image of an object, usually with full parallax. This grating is a record of the interference pattern generated by the wavefront from the object (the object beam) and an unmodulated wavefront (the reference beam), and is usually, though not invariably, a photographic record. The various types of hologram are listed individually.

HOLOGRAMMETRY   Measurement by means of holography; the holographic equivalent of photogrammetry.

HOLOGRAPHIC CONTOURING   A method of producing a contour map of a solid object by holography, using two wavelengths.

HOLOGRAPHIC INTERFEROGRAM   See Interferogram.

HOLOGRAPHIC INTERFEROMETRY   The techniques of producing interferograms (qv) by means of holography. The various methods are listed separately.

HOLOGRAPHIC OPTICAL ELEMENT (HOE)   A hologram of an optical element which can be used instead of the element itself. HOEs can also be computer-drawn.

HOLOGRAPHIC SHEAR INTERFEROMETRY   A type of double-pulse holography recording the motion of an object.

HOLOGRAPHIC STEREOGRAM   A type of parallax stereogram (qv) in which the series of two-dimensional images is recorded as an array of vertical line holograms; up to 360° parallax is possible.

HOLOMETRY   Holography applied to metrology and measurement science in general.

HOLOMICROGRAPHY   The technique of making holograms which can reconstruct a microscope image, usually in three dimensions.

HOT-FOIL BLOCKING   A method of producing small embossed holograms on an exceedingly thin base. The resulting hologram is flush with the surface. Used for security holograms in credit cards etc. Also known as 'hot stamping'.

HUE   One of the three qualities that describe a color, the others being saturation and lightness (or luminosity). Hue is defined by the dominant wavelength of the color, or, in the case of hues that do not lie within the spectrum (such as magenta) by the dominant absent wavelength.

HUYGHENS WAVE MODEL   A model for the propagation of light in which each point on a wavefront is considered as a source of secondary waves, the envelope of which forms the new wavefront.

HYPERBOLOID   Within the context of this book the term means a hyperboloid of revolution. In the geometric approach to holography it represents the saucer-shaped surfaces that are the loci of equal optical path differences from the reference source and the object.

HYPERSENSITIZING   A chemical or physical process that increases the sensitivity to light of a photographic emulsion, usually (though not invariably) before exposure.

ILLUMINANCE   The quantity that is measured by a light meter placed in the plane of the hologram, and often called simply 'intensity' (incorrectly in this context). The unit of illuminance is the lux (lx), 1 lux being 1 lumen per square meter (lm/m^2). Strictly, the term 'illuminance' applies only to white light, the term 'irradiance' being the correct one for monochromatic or narrow-band radiation.

IMAGE   In a hologram, the three-dimensional representation of the subject matter that is produced by the hologram when illuminated by the replay beam. It may be virtual (behind the hologram), real (in front of it), or partly virtual and partly real, ie straddling the plane of the hologram.

IMAGE BEAM   The diffracted beam that forms the image when a hologram is replayed.

IMAGE HOLOGRAM   A hologram made with a real image as object.

IMAGE-PLANE HOLOGRAM   An image hologram (qv) in which the image straddles the plane of the hologram.

IMAGES, 2D/3D   Embossed holograms (qv) which combine flat graphics in diffraction colors with a background image in depth.

INCANDESCENT (LIGHT) SOURCE   A light source such as a filament lamp which emits a continuous spectrum of light as a consequence of its high temperature.

INDEX-MATCHING (FLUID)   A method of avoiding troublesome internal reflections in sandwiches of optical materials by filling the interstices with a fluid of intermediate refractive index (the index-matching fluid).

INFINITY   1   In mathematics, a quantity having a value that is greater than any assignable value.
2   In imaging science, the distance of a (hypothetical) point source of what is observed as collimated light.

IN-LINE HOLOGRAM   Also called a Gabor hologram, this is a single-beam hologram with the subject matter and the reference source collinear.

IN PHASE   Two traveling waves (qv) are said to be in phase when their relative phases are identical, ie crests coincide with crests and troughs with troughs.

IN-PLANE MOVEMENT   Movement in a plane parallel to that of the holographic emulsion.

INSONATION   The acoustic equivalent of illumination. The neologism 'insonification', which is sometimes seen in papers on acoustic holography, is intended to mean the same thing.

INSTANTANEOUS AMPLITUDE   *See* Amplitude.

INTEGRAL HOLOGRAM, INTEGRAM   Alternative names for a Cross hologram (qv).

INTENSITY   1   Luminous power of a light source, measured in a specific direction. The unit of measurement is the candela (cd).
2   Energy of an electromagnetic wave, proportional to the square of its amplitude. Intensity may be peak or instantaneous (*see* Amplitude). It can also be time-averaged.

3 A colloquial term for illuminance (*qv*).

INTERFERENCE When two coherent wavefronts of the same frequency are superposed, their instantaneous amplitudes add algebraically at every point. If the amplitude of the resultant is greater than that of the component waves the interference is said to be constructive; if it is less, the interference is said to be destructive.

INTERFERENCE FRINGES The pattern of light and dark bars that appear on a screen when it is placed in the path of two interfering light beams.

INTERFEROGRAM The record of a secondary fringe pattern. Secondary fringes (*qv*) are recorded by holography, by speckle photography, by electronic speckle-pattern interferometry (ESPI) or by traditional interferometric methods. Used for measuring very small deformations or disturbances.

INTERFEROMETER An optical device used to produce interference fringes for measurement purposes; the earliest, and most important is the Michelson interferometer (*qv*). Others include the Mach-Zehnder, Twyman-Green and Fabry-Pérot interferometers.

ION An atom which has lost or gained one or more electrons. It bears an electric charge the magnitude of which depends on the number of electrons lost or gained.

ION LASER A laser in which the important transitions are between ionized states (*cf* neutral-gas laser).

IONIZATION The state of being or becoming ionized (*see* below).

IONIZED GAS A gas in which the atoms have received enough energy for many of the electrons to have been removed from their outer shells.

ISO INDEXES Numbers which indicate the speed (sensitivity to light) of photographic materials. In the United Kingdom and United States the arithmetic index is preferred (ISO 100 is twice the speed of ISO 50, etc); in Europe and in photographic laboratories using sensitometric control, the logarithmic scale is used (ISO 30° is twice the speed of ISO 27°).

JOULE (J) The SI unit of energy or work, representing 1 kilogram meter per second squared (1 kg m/s^2). It is the unit used to describe the output of a pulse laser. It is equivalent to 1 watt-second.

KELVIN (K) The SI unit of thermodynamic temperature. A temperature interval of 1 K is the same as an interval of 1°C, but 0 K represents absolute zero, the temperature at which all molecular movement ceases ($-273.15°C$). Its significance as a light measurement is that it can be related to the emission spectrum of a hot object (*see* Color temperature).

KERR CELL A glass cell filled with nitrobenzene, a substance that become birefringent in an electric field. In combination with polarizing filters it can be used as a high-speed shutter (*cf* Pockels cell).

KEY LIGHT Also known as a modeling light, this is the light used in photographic etc illumination to produce the main effect of light and shade (*cf* fill-in light).

KICKER A small spotlight used in photographic etc illumination to produce highlights.

KRYPTON-ION (Kr$^+$) LASERS An ion laser (*qv*) which uses ionized krypton gas as the lasing medium. It produces a large number of lines in the visible spectrum, chief of which are at 476, 521 and 647 nm.

LANTHANIDES A series of elements, also called the rare earths, ranging from atomic number 58 (cerium) to 71 (lutetium) in the periodic table of the elements. These elements have very similar chemical properties, and their main use is in optical glasses. When aluminum oxide crystals are doped with certain lanthanides they exhibit non-linear optical properties, and can be used to change the frequencies of laser outputs.

LASER A device for producing a beam of coherent light by stimulated emission of radiation (qv). The various types of laser are listed separately.

LASER SPECKLE An interference effect which causes a laser-illuminated surface to appear granular. The speckle which is produced by the surface itself is called 'objective speckle'. When this is recorded by an optical system its structure is modified by the system; it is then known as 'subjective speckle'.

LASER TRANSMISSION HOLOGRAM Also called a laser-read hologram. A hologram in which the image is viewed only by laser (or some kind of quasi-monochromatic) light. The replay beam falls on the opposite side of the hologram to the viewer.

LASING MEDIUM In a laser, the substance which produces laser action by stimulated emission (cf pumping medium).

LATENT IMAGE Small clusters of silver atoms formed in silver halide crystals that have been exposed to light. They render the crystals developable (see Developing agent).

LENS LAWS The laws of image formation in terms of geometrical optics.

LIGHTNESS (Also luminosity). One of the three variables describing a color, the other two being hue and saturation (qv). If the intensity of a source is altered without any alteration in hue or saturation, it is the lightness that changes. In the Munsell color system (qv), lightness is called 'value'.

LINE SOURCE 1 A luminous source that is extended in one direction, eg a straight-filament lamp or a fluorescent tube.
2 A luminous source that produces a line spectrum.

LINE SPECTRUM A spectrum consisting of narrow discrete bands at specific wavelengths.

LINEAR REGION That part of a graphical relationship which appears as a straight line.

LINEAR POLARIZATION See Polarization.

LINEAR RELATIONSHIP Two variables are related linearly when the graph of their relationship is a straight line.

LIPPMANN HOLOGRAM Another name for a Denisyuk or single-beam reflection hologram (qv).

LIPPMANN PHOTOGRAPHY A type of color photography invented towards the end of the nineteenth century by Gabriel Lippmann. The colors were produced by Bragg diffraction.

LIQUID GATE A liquid-tight glass-sided plate-holder in which the plate is exposed, processed and viewed with the cell filled with liquid throughout.

LITTROW PRISM A dispersing prism used for selecting particular lines in a multiline laser. Essentially it is half of a Brewster-angle prism (qv) with the side away from the incident beam coated to form a dielectric mirror.

LIVE-FRINGE (TECHNIQUES)   Real-time visualization of deformations or fabrication errors by comparison of the hologram of a component with its own holographic image before stressing, or with that of a pattern component. Electronic speckle pattern imaging (ESPI) (qv) is a non-holographic live-fringe technique.

LLOYD'S MIRROR   A technique for producing large interference fringes, employing a mirror at grazing incidence to fold part of a light beam back onto itself.

LOGARITHM   The power to which a number (called the base) has to be raised to give another, chosen number. Thus the logarithm of 32 to the base 2 is 5, because $2^5 \equiv 32$; and the logarithm to the base 10 (called the common logarithm) of 100 is 2, because $10^2 \equiv 100$.

LOGARITHMIC SCALE   A scale in which equal intervals represent equal multiples instead of equal increments, eg 1, 10, 100, 1000, etc rather than 0, 1, 2, 3, etc.

LUMINANCE   The luminous intensity per unit projected area of a light source, measured from a given direction. The SI unit of luminance is the candela per square meter (cd/m^2).

LUMINOSITY   Another name for lightness (qv).

MACH-ZEHNDER INTERFEROMETER   An optical device in which a light beam is divided by a beamsplitter, traverses two separate optical paths (one of which may be perturbed) and is recombined. The resulting fringe pattern contains information concerning the nature of the perturbation.

MAGNETIC VECTOR   The magnetic component of electromagnetic radiation. It is perpendicular to the electric vector (qv) and in unpolarized and linearly-polarized radiation is in phase with it.

MASTER HOLOGRAM   1   A hologram used to produce a real image to act as the object for a transfer hologram (qv).
2   The hologram used for in electroforming the first metal replica in the embossing process (*see* Embossed hologram).

MAYER BAR   A metal rod wound with stainless steel wire, used in holography for coating dichromated-gelatin and photopolymer emulsions.

METASTABLE (STATE)   Describes a chemical or physical condition which remains the same until disturbed by quantum fluctuations or external disturbance.

MICHELSON INTERFEROMETER   A device which compares by interference the optical path lengths of two (usually) orthogonal beams derived from the same source.

MICRODENSITOMETER   A device which measures fluctuations in photographic density (qv) over distances of the order of 50 μm.

MICROPHONING   Vibration of optical components in response to sound waves.

MICROWAVE HOLOGRAPHY   Holography at electromagnetic wavelengths in the millimeter and centimeter region.

MICROWAVES   Electromagnetic waves with wavelengths in the range from a few centimeters to about 1 millimeter.

MIE SCATTER   Scattering of light by particles of diameter typically 1–10 μm. The scatter is roughly in inverse proportion to the wavelength of the light (*cf* Rayleigh scatter).

MODELING LIGHT *See* Key light.

MODULATION A measure of contrast, obtained by dividing the difference between maximum and minimum transmittances (or luminances) by their sum.

MODULATION TRANSFER FUNCTION (MTF) The part of the optical transfer function (*qv*) that deals with the transfer of contrast from the subject matter to the image. It is plotted as a graph of image modulation ÷ subject modulation vs spatial frequency, and is an accurate method of predicting primary fringe contrast (*cf* Gamma).

MOLE The SI unit of amount of substance. 1 mole (mol) of a substance has a mass equal to its relative molecular mass expressed in grams.

MONOCHROMATIC Describes electromagnetic radiation of a single frequency. Often used loosely to describe in comparative terms the bandwidth (*qv*) of a light source.

MONOCHROMATOR A device for removing unwanted frequencies from an incoherent or partially coherent light source.

MONOMODE/MULTIMODE OPTICAL FIBER *See* Optical fiber.

MULTIPLES AND SUBMULTIPLES When basic units of measurements are too large or too small for a given situation they are multiplied or divided in steps of 1000 and given a prefix as follows:

Multiples
$\times 10^3$ = kilo (k)
$\times 10^6$ = mega (M)
$\times 10^9$ = giga (G)
$\times 10^{12}$ = tera (T)
$\times 10^{15}$ = peta (P)
$\times 10^{18}$ = exa (E)

Submultiples
$\times 10^{-3}$ = milli (m)
$\times 10^{-6}$ = micro (μ)
$\times 10^{-9}$ = nano (n)
$\times 10^{-12}$ = pico (p)
$\times 10^{-15}$ = femto (f)
$\times 10^{-18}$ = atto (a)

See also Scientific notation.

MULTIPLEX HOLOGRAM Another name for a Cross hologram (*qv*).

MULTIPLEXING Placing several images in the same hologram occupying the same space in the emulsion, by analogy with a similar process in communications technology.

MUNSELL COLOR SYSTEM A color atlas for artists based on the visual appearance of colors. It bears a superficial resemblance to the CIE chromaticity principle, but is based on criteria of painting rather than visual stimuli.

NEAR-FIELD DIFFRACTION PATTERN The diffraction pattern produced at a given plane by an object that is relatively close to it (*cf* far-field diffraction pattern). *See also* Fresnel diffraction.

NEUTRAL-DENSITY (ND) FILTER A light filter which has the same transmittance at all visible wavelengths, used for attenuating a light beam by a known factor. The number given to a particular filter (eg ND 0.6) represents its photographic density (*qv*).

NEUTRAL-GAS LASER A gas laser in which the lasing element operates between two energy bands in the neutral atom (*cf* ion laser).

NEWTON'S FRINGES Interference fringes formed when two transparent surfaces are in imperfect contact. They are easily visible to the unaided eye. Also known as Newton's rings, though they are circular only when the surfaces in contact are spherical.

NOBLE GASES The gases helium, neon, argon, krypton, xenon and radon. So called because they do not readily form chemical compounds.

NUMERICAL APERTURE (NA) The sine of half the angle subtended by the entrance pupil (qv) of a microscope objective at its front principal focus. In the case of an oil-immersion objective this figure is multiplied by the refractive index of the oil. It is related to the f-number (qv) by the approximate relationship $f/\text{no} \cong 1/(2 \times \text{NA})$.

OBJECT BEAM In holography, the diffracted beam incident on the emulsion from the object.

OBJECTIVE SPECKLE *See* Laser speckle.

ODD FUNCTION A function (such as a sinusoid) that is antisymmetrical about the vertical axis.

OFF-AXIS ABERRATION An aberration (qv) which occurs only for rays originating from an object that does not lie on the principal axis of the optical elements.

OPEN-APERTURE HOLOGRAM Another name for a full-aperture hologram (qv).

OPTICAL CAVITY The resonating space between the two mirrors of a laser, or between the faces of a Fabry-Pérot etalon (qv).

OPTICAL CENTER 1 In photographic optics, the point in the film plane where the principal axis of the lens intersects it.
2 In Fourier optics, the center of the diffraction pattern (zero spatial frequency).

OPTICAL CORRELATION A method of determining whether a particular waveform matches a stored waveform or reference (*see* Correlation filtering).

OPTICAL DEROTATOR A device for achieving a stationary image of a rapidly-rotating object by the use of a rotating inverting prism.

OPTICAL FIBER A small-diameter glass fiber which conducts light from one end to the other. A multimode fiber has a diameter that is large compared with the wavelength of the light, which is propagated by total internal reflection (qv) so that spatial coherence and polarization are destroyed. A monomode fiber, which need not necessarily be circular, has cross-sectional dimensions of only a few times the wavelength, so that the light is propagated on waveguide principles, and both spatial coherence and polarization are preserved.

OPTICAL TRANSFER FUNCTION (OTF) A full description of the image-forming performance of an optical system for spatial frequencies from zero to the cut-off point, which is determined by the limitations of diffraction.

OPTICAL TRANSFORM The electromagnetic field at a given distance from an object which disturbs the field, the visible manifestation of which is a diffraction pattern.

OPTOELECTRONICS The study of phenomena involving the interaction of light with matter.

ORTHOGONAL Describes two lines or planes that are at right angles to one another.

ORTHOPTIC   Describes an image that is not laterally reversed (*cf* pseudoptic).
ORTHOSCOPIC IMAGE   An image that has correct parallax (*cf* pseudoscopic image).
OSCILLATOR   1   An electronic device the output of which is an AC signal which may or may not be sinusoidal.
2   In a solid-state pulsed laser, the element that produces the inital pulse (the later stages are known as amplifiers).
OUT-OF-PLANE MOVEMENT   Motion in a plane perpendicular to that of the holographic emulsion.
OXIDATION   The removal of one or more electrons from an atom or group of atoms by chemical means, as when a silver atom (Ag) is oxidized to a silver ion (Ag^+) in a bleach bath (*cf* reduction).
OXIDIZING AGENT   A receiver of electrons, eg the ammonium ion NH_4^+, which is one electron short.

P-POLARIZATION (-IZED)   *See* Brewster angle.
PANORAMAGRAM   Also Parallax panoramagram. An early type of parallax stereogram (*qv*).
PARALLAX   The phenomenon whereby an object changes in appearance when the viewpoint is changed.
PARALLAX STEREOGRAM   A composite photographic image made up of photographs taken at regular angular intervals round a subject, and printed in such a way that from any viewpoint each eye sees only the image appropriate to that viewpoint.
PARALLEL PROCESSING   A routine in computing which allows the carrying out of a number of operations simultaneously.
PARAXIAL ABERRATION   An aberration (*qv*) occurring with an object on the principal axis of an optical element.
PARTIAL COHERENCE   A beam of light is said to be partially coherent when its bandwidth (*qv*) is narrow enough for the constituent photons to be phase-correlated over a short distance (typically of the order of 100 times the wavelength for a single line from a gas-discharge lamp).
PASSIVE SWITCHING   In pulse-laser technology, the operation of a Q-switching shutter such as an organic dye cell activated by the pulse itself (*cf* active switching).
PEAK AMPLITUDE   A term used for amplitude (*qv*) when it is necessary to distinguish it from instantaneous amplitude.
PERIOD   Used without qualification, the term is synonymous with temporal period (*qv*).
pH (SCALE)   A logarithmic scale for expressing the acidity or alkalinity of a solution. pH 7.0 is neutral; pH values lower than this are acid, and values higher than this are alkaline. Most developers used in holography have a pH of around 11–12 and fixing and bleach solutions have a pH of around 5.
PHASE   The relationship between the position of the crest of a wave and a given reference point; it can be specified in degrees or fractions of a cycle, but is most conveniently specified in radians (2π radians = 1 cycle).
PHASE HOLOGRAM   A hologram in which the fringes are recorded as variations in refractive index or thickness rather than transmittance.

PHASE OBJECT   A transparent object the optical details of which are variations in refractive index and thickness only.

PHASE TRANSFER FUNCTION (PTF)   Part of the optical transfer function (qv), the other part being the modulation transfer function (MTF) (qv). The PTF specifies displacements of the optical image from its true geometric position.

PHOTOELECTRONIC PHENOMENA   Also known as photoelectric effects, these occur when light energy causes the liberation of electrons in a substance leading to changes in conductivity, voltages being set up, etc.

PHOTOGRAMMETRY   The use of photography for measurement purposes.

PHOTOGRAPHIC DENSITY   The common logarithm of the reciprocal of the intensity transmittance of a photographic negative.

PHOTOGRAPHIC IMAGE   When used in this book with reference to a hologram, the term refers to the record of the fringe structure in the developed emulsion. Otherwise it is used in the usual sense.

PHOTON   A quantum of electromagnetic radiation, ie the smallest amount that can exist, possessing zero rest mass and traveling at the speed of light. The temporal frequency of a photon is directly proportional to its energy.

PHOTOPOLYMER   A substance which polymerizes (ie its molecules link up to form larger molecules) under the action of light energy. Photopolymers can be used for making Bragg holograms (qv).

PHOTORESIST   A substance which becomes insoluble (negative) or soluble (positive) under the action of light. It can be used for making transmission holograms but not Bragg holograms (*cf* photopolymer), and as the primary fringes are in relief it can be used for making the master stage for embossed holograms.

PHOTOTHERMOPLASTIC   A normally non-conductive substance which becomes electrically conductive when illuminated. It is used to produce near-real-time holograms, and is indefinitely re-usable.

PHYSICAL BLEACH   By analogy with physical development (qv), a bleach solution which rehalogenates the silver photographic image, at the same time depositing the newly-formed silver halide on the unexposed silver halide particles. The term 'diffusion transfer' (qv) is more appropriate.

PHYSICAL DEVELOPMENT   1   Development in which the silver ions are present in the developer and are deposited on the latent image before or after fixation, a photographic process that is now seldom if ever employed.
2   Development during which most or all of the unexposed silver halide crystals form soluble complexes which are reduced to metallic silver and deposited on the developing silver grains. See also Diffusion transfer.

PIXEL   A whimsical abbreviation of 'picture element', the smallest unit in a visual display.

PLANE POLARIZATION   See Polarization.

PLANE WAVEFRONT   The wavefront profile of a collimated light beam (qv).

POCKELS CELL   A light shutter based on optoelectronic phenomena and consisting of a plate of electro-optically-sensitive crystalline material to which a voltage can be applied, between crossed polarizing filters. As it requires less power than the Kerr cell (*qv*) and is almost as fast, it is more often used than the Kerr cell for Q-switching.

POLARIZATION The confinement of the electric vector of a beam of light in a particular direction. If this direction is constant the beam is said to be linearly or plane polarized. In circularly-polarized light the electric vector describes a helix about the direction of propagation with one complete turn per wavelength. This type of polarization is produced by a quarter-wave plate (*qv*), and occurs in Kerr and Pockels cells (*qv*). In elliptically-polarized light the vector changes in magnitude as it rotates.

POLYCHROMATIC Used to describe light having a spectrum containing more than one frequency (*cf* monochromatic).

POPULATION INVERSION The condition of a substance containing atoms at two different energy levels, at a time when there are more in the higher energy level than in the lower one; this is a necessary condition for laser action.

PRIMARY FRINGES The microscopic fringes which form a holographic image (*cf* secondary fringes).

PRIMARY HUES The hues red, green and blue. Mixtures of lights in these three hues in differing proportions can produce almost the gamut of color (*see* Additive color synthesis).

PRINCIPAL AXIS The axis of symmetry of a lens or optical mirror.

PRINCIPAL FOCAL PLANE The plane in which (for an ideal lens or optical mirror) all points at infinity are brought to a focus.

PRINT-OUT In photography and holography, the progressive darkening of silver halide when exposed to light for long periods of time.

PRISM, 3° A glass plate with a thickness taper of approximately 3°. When used as a beamsplitter, the second-surface reflection is directed away from the first-surface reflection and can be more easily blocked off than with a parallel-sided glass block.

PRODUCT In the context of Fourier-transform theory, the result of multiplying two functions together, point for point.

PSEUDOCOLOR A term used to describe a holographic image in which the colors are not natural, but have been created deliberately.

PSEUDOPTIC Describes an image that is laterally reversed.

PSEUDOSCOPIC (IMAGE) 1 In photography, describes the effect seen when a stereoscopic pair is viewed with the right and left images interchanged.
2 In holography, an image in which full parallax is present but reversed, eg a shift in viewpoint to the right reveals the left side of the image and a shift upwards reveals its underside.

PULSE LASER A laser which emits its radiation in a short burst.

PUMPING In laser technology, the process of exciting the atoms of the lasing medium to produce a population inversion (*qv*).

PUMPING MEDIUM In a laser, the substance which stores energy and supplies it to the lasing medium (*qv*).

Q-SWITCHING A technique for storing energy in the lasing medium until it has built up to a very high value, then releasing it in a burst of very high intensity and short duration.

QUANTUM ENERGY LEVELS   Fixed energy levels in atoms, between which no energy state is allowed by the laws of quantum physics.

QUANTUM MODEL   A model for physical phenomena which is based on the concept that there is a minimum amount, or quantum, by which such properties as energy or momentum can change.

QUARTER-WAVE PLATE   A retardation plate (qv) which converts linear polarization into circular polarization (*see* Polarization).

QUASI-MONOCHROMATIC (LIGHT)   The output of a light source such as a gas-discharge tube, which emits only a narrow band of frequencies.

RAINBOW HOLOGRAM   Also known as a Benton hologram, this is a type of slit transfer hologram (qv) in which the real image produced by a master hologram is used as object for the final hologram, the master being masked by a horizontal slit. This eliminates vertical parallax, so that a change of viewpoint in a vertical direction simply changes the hue of the image without changing its perspective.

RANDOM POLARIZATION   A laser equipped with a plasma tube that does not have Brewster-angle windows produces a beam containing two components, which are plane polarized and orthogonal. The relative intensities of these components fluctuates slowly, so that the direction of maximum polarization rotates slowly back and forth through an angle of approximately 90°.

RASTER   An array of horizontal or vertical lines built up into an image by sequential scanning, as in a television picture.

RAY MODEL   A model for the behavior of light which uses the directions in which light is propagated as the basis for a geometry of image formation.

RAYLEIGH CRITERION   For an image formed by a diffraction-limited lens, two points are just resolved if the centre of the Airy diffraction pattern (qv) of one of them lies on the first minimum of the Airy diffraction pattern of the other.

RAYLEIGH SCATTER   Light scatter by particles smaller than the wavelength of light, inversely proportional to the fourth power of the wavelength (*cf* Mie scatter).

REAL IMAGE   An image formed by light waves which actually pass through the image space. A real image can be recorded directly on photographic film.

REAL-TIME INTERFEROMETRY   A form of holographic interferometry in which a component is inserted into the image space of a hologram of it in its original condition (or of a pattern). Any discrepancies in shape are contoured by secondary fringes (qv). Non-holographic techniques such as electronic speckle-pattern imagery (ESPI) (qv) can also be used.

RECIPROCITY FAILURE   The law of reciprocity states that any photochemical effect is proportional to the total light energy, ie the product of illuminance and time. However, at very low light levels the effect is weakened, and the required exposure for a given effect may be very considerably increased; in holography this is noticeable for exposures of over a minute, or when an emulsion intended for use with a pulse-laser is used with a CW laser.

RECONSTRUCTION BEAM The beam used to produce the holographic image; the replay beam.

REDUCING AGENT A net donor of electrons, for example the oxybenzene group $C_6H_4O_2{}^{2-}$, which readily gives up one or both of its electrons.

REDUCTION The gaining of one or more electrons by an atom or group. In photographic development, the silver ion Ag^+ is reduced by receiving an electron donated by the developing agent to give a silver atom Ag; at the same time the developing agent is oxidized.

REFERENCE BEAM The unmodulated beam which, when directed at the holographic emulsion, forms an interference pattern with the object beam which records both the amplitude and phase of the object wavefront at every point. Also called the carrier beam, from analogy with communications technology.

REFRACTIVE INDEX For any given optical medium, the ratio of the velocity of light in empty space to the velocity of the light in the medium.

REHALOGENATION See Bleach.

REPLAYING A colloquialism for reconstruction of the holographic image.

RESOLVING POWER The highest spatial frequency a given light-sensitive medium can record.

RESTRICTED-APERTURE MASTER HOLOGRAM A compromise between a full-aperture and a slit transfer hologram (qv).

RETARDATION PLATE An optically-flat plate made of birefringent material. Depending on its thickness it produces elliptically- or circularly-polarized light from a linearly-polarized beam, or rotates the plane of polarization.

RETICULATION A fine irregular pattern in relief covering a photographic emulsion, usually caused by overheating it when wet. In a hologram reticulation causes the fringe structure to be totally destroyed.

RETROREFLECTION A condition in which a beam of light is reflected back along its own path.

REVERSAL BLEACH See Bleach.

ROLL EMBOSSING A continuous process making embossed holograms between hot rollers bearing the embossing shims.

ROOF PRISM An inverting prism having two angles of 45° and one of 90°. The hypotenuse face is replaced by a 45° roof. The roof prism used in an optical derotater (qv) is a double roof prism with roofs on its two sloping faces, and is a retroreflecting prism.

ROTATING In this book the term is used to indicate an in-plane rotation of 180°.

RUBY LASER A solid-state pulse laser (qv) in which which the pumping medium is aluminium and the lasing medium is chromium. The main emission line has a wavelength of 694 nm.

RUNNING WAVE See Traveling wave.

S_1, S_2, S_3 etc Symbols used in holographic literature to indicate slit master holograms.

S-POLARIZED (LIGHT) See Brewster angle.

SAFELIGHT A light filtered to eliminate the wavelengths to which a photographic emulsion is most sensitive.

SANDWICH HOLOGRAPHY   An interferometric technique involving the making of two separate holograms which form secondary fringes (qv) when mounted in register.

SATURATION   In colorimetry, the purity of a hue: at one end of the scale is the pure spectral hue, at the other is neutral grey of the same lightness. In the Munsell color system (qv) the approximate equivalent is 'chroma'

SCHEIMPFLUG CONDITION   An extension of the lens laws (qv) for non-parallel object and image planes; it states that when the object plane is tilted with respect to the lens plane, then the object plane, the lens plane and the image plane meet in one straight line.

SCHLIEREN PHOTOGRAPHY   A photographic method of visualizing striations in optical material or compression waves in a gas, by focusing light transmitted through it onto a knife edge. Refracted light is either totally blocked or totally transmitted, giving light or dark regions in the photographic image. A more sophisticated version shows over- and underpressure in different colors.

SCIENTIFIC NOTATION   Also known as 'standard form', scientific notation involves writing all numbers with the decimal point immediately following the first digit, the number being then multiplied by 10 raised to the appropriate power, which is negative for numbers between 0 and 1. Thus the speed of light (299 700 000 meters per second) is written 2.997×10^8 m/s, and the wavelength of helium–neon (HeNe) laser light (632.8 nanometers) is written 6.328×10^{-7} m. The solidus symbol / is read as 'per'. In many textbooks an index notation is used for units, eg m/s is written m s^{-1} and J/m^2 is written J m^{-2}.

SECONDARY FRINGES   Fringes formed when a double-exposure hologram (or a sandwich-hologram pair) is made of an object which has moved or been deformed between the exposures. Such fringes are very much larger than the primary fringes (qv) that form the hologram itself.

SECONDARY HUES   The hues cyan, magenta and yellow, used in subtractive color processes which remove unwanted wavelengths from white light. Cyan controls red, magenta green and yellow blue.

SEMICONDUCTOR LASER   A type of light-emitting diode that produces coherent light. It may be either continuous-wave or pulse.

SENSITOMETRY   The investigation of the behavior of photographic materials when exposed to light, and processed under predetermined conditions.

SEQUENTIAL PROCESSING   Also known as linear processing, this is a routine in computer programming in which operations are carried out one at a time in sequence (cf parallel processing).

SERVOMECHANISMS (CONTROLS)   Devices which stabilize a physical situation by feeding back an error signal to their input in reverse, amplified (negative feedback).

SHIM   1   Any piece of thin material used as a spacer.
       2   A colloquial term for the stamper used to produce embossed holograms.

SIDEBAND HOLOGRAM   A term borrowed from radio-communications technology, used sometimes to mean a laser transmission hologram.

SILVER HALIDE   A combination of silver with a halogen (qv), typically chlorine, bromine or iodine.

SILVER-HALIDE EMULSION   A light-sensitive layer consisting of a silver halide (*qv*) or a mixture of silver halides, suspended in gelatin in the form of a dispersion of microcrystals.

SINC (FUNCTION)   A function of the form $(\sin x)/x$.

SINGLE-BEAM HOLOGRAM   Any hologram in which the same beam serves as both object and reference beam.

SINGLE-FREQUENCY EMISSION   A laser system fine-tuned (usually by a Fabry-Pérot etalon) (*qv*) to emit only a very narrow band of frequencies.

SINGLE-LINE EMISSION   A laser system tuned to emit only a single spectral line.

SINUSOIDAL   Describes a function which fluctuates according to the relationship $y = A \sin x$, where x and y are the variables. The overall shape of the curve is the same as that of a cosinusoidal function (*qv*). A sinusoid is an odd function (*qv*).

SLIT (TRANSFER) HOLOGRAM   A transfer hologram (*qv*) which eliminates vertical parallax by the use of a horizontal slit or its optical equivalent.

SOLID-STATE LASER   A laser in which the lasing element is in solid solution. The element may be the doping in an insulator, as in a ruby laser element or in a semiconductor.

SPATIAL COHERENCE   A beam of light is said to be spatially coherent when it appears to have originated from a point, which may be at infinity (or at a negative distance for a converging beam).

SPATIAL FILTERING   1   A method of modifying image quality by placing opaque, partly-transparent and/or phase-modifying stops in various regions of the Fourier-transform plane of the lens.
2   A method of obtaining a clean beam from a laser by placing a pinhole of appropriate size at the rear focal plane of the microscope objective used to expand the beam.

SPATIAL FREQUENCY   The number of cycles of a repeated pattern in a given distance or angle.

SPATIAL PERIOD   The reciprocal of spatial frequency; the distance between points of repetition of a regular waveform.

SPECKLE   *See* Laser speckle.

SPECKLE INTERFEROMETRY   A form of interferometry (*qv*) based on the change in laser speckle pattern when a laser-illuminated object undergoes deformation.

SPHERICAL ABERRATION   A paraxial aberration (*qv*) caused by the outer zones of a spherical lens or mirror having a shorter focal length than the inner zones. Holograms also display spherical aberration.

SPHERICAL WAVEFRONT   The wavefront from a nearby point source.

SPINNING   In this book the term is used to mean an out-of-plane rotation of $180°$ about a vertical axis.

SQUARE GRATING   A diffraction grating consisting of bars of opaque material interspaced with transparent bars of equal width.

STANDING (STATIONARY) WAVES   The resultant of two cosinusoidal traveling waves moving in opposite directions.

STEREOGRAM   *See* Parallax stereogram.

STIMULATED EMISSION OF RADIATION   The emission of a photon by an excited atom triggered by the presence of a photon from an atom previously in the same state of excitation. The second photon has the same frequency and phase as the first, and travels in the same direction (*see also* Laser).

SUBJECTIVE SPECKLE   *See* Laser speckle.

SUBTRACTIVE COLOR SYNTHESIS   A method of producing an image in color by superposing dyes of different hues, typically cyan, magenta and yellow (*cf* Additive color synthesis).

SYNTHETIC-APERTURE RADAR   A side-looking radar system which builds up a highly-detailed two-dimensional view of the terrain from a one-dimensional record of radar echoes over a long horizontal base. The technique has close affinities with holography.

TANNING   The cross-linking of gelatin molecules caused by the oxidation products of certain developing agents.

TELECENTRIC (LENS) SYSTEM   In the context of this book, a two-lens system in which the front focal plane of the rear lens is the rear focal plane of the front lens and the aperture stop is situated in this plane.

TEMPORAL COHERENCE   *See* Coherence.

TEMPORAL FREQUENCY   The number of cycles of a repetitive function occurring in a given time.

TEMPORAL PERIOD   The duration of one complete cycle of a repetitive function.

THERMODYNAMIC TEMPERATURE   *See* Kelvin.

THREE-LEVEL LASER   A laser in which the relevant energy transition is between two energy levels the lower of which is the ground state.

TIME-AVERAGED INTENSITY   The quantity recorded by all light-sensitive media. It is proportional to the square of the amplitude (*qv*).

TIME-AVERAGED INTERFEROMETRY   A method of producing a holographic interferogram (*qv*) which shows secondary fringes (*qv*) caused by interference between the regions near the two extremes of movement in an object in a stable state of vibration.

TIME-SMEAR   A distortion occurring in holographic stereograms (*qv*). There are two distinct causes: (a) If the subject moves during the series of exposures, the right eye sees the action at a later stage than the left eye and (b) at each edge of the image the eye is looking obliquely through a hologram, whereas for correct perspective it should be looking through a different hologram.

TIP ANGLE   The angle at which the real image of the spectrum formed by the slit in a rainbow hologram is tilted (typically about 35° to the horizontal).

TOP HAT FUNCTION   A function which is equal to unity between two x values and zero elsewhere.

TOTAL BLEACH   *See* Bleach.

TOTAL INTERNAL REFLECTION (TIR)   When light is incident on the interface of a medium from one of higher refractive index, at an angle of incidence greater than the critical angle (*qv*), all the light is reflected from the interface.

TRANSDUCER   A device which translates any form of oscillation into mechanical motion.

TRANSFER HOLOGRAM   A hologram which has been made using a holographic image as object.

TRANSMISSION HOLOGRAM   Any hologram in which the reconstruction beam is on the side of the plate opposite to that of the viewer.

TRANSMITTANCE   In coherent optics this usually refers to the amplitude transmittance, which is the ratio of emergent to incident amplitude. In photography the transmittance is the ratio of the time-averaged intensities. In both cases the transmittance is expressed as a percentage.

TRANSVERSE WAVES   Waves in which the direction of vibration is orthogonal to the direction of propagation.

TRAVELING WAVE   A periodic (electromagnetic) disturbance in space in which energy is transferred from one place to another by the vibrations (*cf* standing wave). Also called a running wave.

TWYMAN-GREEN INTERFEROMETER   A type of interferometer used in lens figuration testing.

ULTRASONIC HOLOGRAPHY   *See* Acoustic holography.

VALUE   The Munsell equivalent of lightness.

VARIABLE BEAMSPLITTER   A beamsplitter (*qv*) in which the intensity ratio of the reflected and transmitted beams can be varied.

VIRTUAL IMAGE   An image that light rays appear to have originated from, but do not in fact pass through (*cf* real image).

VISIBILITY   Another name for modulation (*qv*).

VOLUME HOLOGRAM   Any hologram thick enough for the Bragg condition (*qv*) to assume major importance in the formation of the image.

WATT   The SI unit of power; 1 watt (W) is 1 joule per second (J/s).

WAVE NUMBER   The number of wavecrests in a given distance; the reciprocal of wavelength; the spatial frequency.

WAVEFRONT   The locus of all points in a coherent light beam that are in the same phase.

WAVELENGTH   The distance between successive crests of a wave.

WHITE-LIGHT REFLECTION HOLOGRAM   Any hologram made with the object and reference beams on opposite sides of the emulsion.

WHITE-LIGHT TRANSMISSION HOLOGRAM   Any transmission hologram that can be viewed by white light. Often applied specifically to a rainbow hologram.

x-SPACE   In Fourier optics, ordinary (measured) space in contrast to frequency (Fourier) space.

x, y, z MOVEMENTS   In optical equipment, an x-movement is a horizontal movement and a y movement is a vertical movement; a z-movement is along the axis of the light beam. This convention is used throughout this book.

YAG CRYSTAL   The initials stand for yttrium aluminum garnet, which is an aluminum oxide crystal doped with yttrium, a lanthanide element (*qv*). YAG crystals are able to change the frequency of laser beams, and are routinely used to produce visible light from sources such as excimer lasers (*qv*).

YOUNG'S FRINGES   The simplest form of cosinusoidal fringes (qv), which are obtained by passing a beam of coherent light through a pair of slits.

YOUNG-HELMHOLTZ MODEL   A model for color vision based on the theory that the eye has three types of color receptor, sensitive respectively to red, green and blue.

ZONE PLATE   A transparent plate bearing alternate opaque and transparent circular zones the spatial periods of which (along a radius) are proportional to the square roots of their diameters, is called a Fresnel zone plate. It has a fundamental focus and a series of subsidiary foci. If the amplitude transmittance varies approximately cosinusoidally, such a zone plate has only one focus, and is called a Gabor zone plate. A hologram can be thought of as being made up of a very large number of Gabor zone plates, each focusing the replay beam to one point of the image.

Author Index

Abramson N 49, 59, 315, 317, 323, 327, 332, 443
Accardo G, De Santis P, Gori F, Guattari G and Webster J M 329, 332
Anderson A T 352, 353, 393, 396, 397

Bazargan K and Waller-Bridge M 53, 60, 197, 199
Benton S A 19, 20, 50, 53, 56, 59, 60, 73, 75, 189, 190, 197, 199, 227, 234, 253, 257, 259, 267, 272, 285, 289, 295, 303, 406, 417, 435, 442
Benton S A, Mingace H S Jr and Walter W R 52, 59, 204, 212
Benyon M and Webster J 243, 245
Bergh R A, Kotler G and Shaw H J 421, 422
Bjelkhagen H 241, 242, 245, 324, 332
Blyth J 109, 123, 127, 437, 438, 442
Briones R A, Heflinger L O and Wuerker R F 349, 353
Brooks L D 101, 127
Brown B R and Lohmann A W 341, 346
Bryngdahl O and Lohmann A W 342, 346
Burch J J 231, 346
Burckhardt C 341, 346
Burke W J, Staebler B L, Phillips S W and Alphonse G A 333, 345
Burns J R 287, 289
Burridge M J 435, 438, 442
Bykovskii Yu A, Evtikhiev N N, Elkhov V A and Larkin A I 39

Caulfield H J 443
Chen H 208, 213
Chen H and Yu F T S 52, 59, 208, 213
Chu D C, Fienup J R and Goodman J W 342, 346
Collier R J, Burckhardt C B and Lin L H 347, 352, 444

Corke M, Jones J D C, Kersey A D and Jackson D A 429, 432
Cross L 56, 60, 247, 259
Cutrona L J, Leith E N, Porcello L J and Vivian W E 352, 353

DeBitteto D J 53, 56, 60, 247, 249, 259
Denisyuk Yu N 19, 20
Dowbenko G 147, 159
Dunn P and Thompson B J 345, 346

El-Sum H M A and Kirkpatrick P 17, 20

Foster M 19, 20
Fusek R L and Huff L 252, 253, 259

Gabor D 17, 20, 40, 59
Graham R W 141, 145
Graham R W and Read R E 374
Graube A 277, 280, 435, 442
Greguss P 352, 353

Haskell R E 341, 346
Haskell R E and Culver B C 341, 346
Hariharan P 66, 67, 207, 213, 331, 332, 338, 346, 397, 404, 405, 444
Hariharan P, Hegedus Z S and Steel W H 66, 67
Heflinger L O, Stewart G L and Booth G R 343, 346, 349, 352
Higgins T 352, 353
Hildebrand B P and Doctor S R 352, 353
Huang T S and Prasada B 341, 346
Huff L and Fusek R L 56, 60, 252, 253, 257, 259
Huff L and Loomis J S 56, 60
Hunt R W G 403, 405

Jackson R 299, 303
Jaffey S M and Dutta K 253, 259

Jones A R, Sargeant M, Davis C R and Denham R O 245, 246
Jones R and Wykes C 329, 332, 444

Kalashnikov S P, Klimov I I, Nikitin V V and Semenov G I 39
Kaspar J E and Feller S A 444
Kaufman J 271, 272
Kawasaki B S, Hill K O and Lamont R G 420, 422
Komar V G 212, 213
Kubota T, Ose T, Sasaki M and Honda K 277, 280

Lee T C, Skogen J, Schultze R, Bernal E, Lin J, Daehlin T and Campbell T 278, 280
Lee W-H 339, 340, 341, 342, 343, 346
Leith E N 17, 20
Leith E N and Chen H 195, 199
Leith, Chen H and Roth J 208, 213
Leith E N and Upatnieks J 19, 20, 41, 59
Lesem L B, Hirsch P M and Jordan J A 341, 346
Loomis J S 341, 346

McCormak S 226, 234, 251, 259
McCrickerd J T and George N 249, 259
McGrew S 189, 199, 267, 272, 277, 280, 284, 285, 289, 406, 417
McNair D 101, 127, 147, 159, 162, 179, 444
Metherell A F 351, 353
Molteni W J Jr 56, 60, 253, 255, 259
Moore L 266, 272

Okada K, Honda T and Tsiujiuchi J 253, 259
Okoshi T 444

Phillips N J 425, 430, 431, 439, 442
Phillips N J and Van der Werf R A J 122, 127, 434, 435, 441, 442
Phillips N J, Ward A A, Cullen R and Porter D 434, 437, 442
Pickering C J D and Halliwell N A 329, 332
Pollen D A, Lee J R and Taylor J H 349, 353

Powell R L and Stetson K A 19, 20, 316, 332

Rallison R A 243, 245, 427, 432
Richardson M 112, 127
Rowley D M 325, 326, 332

Saxby G 147, 149, 159, 275, 280, 351, 353, 445
Sheen S K and Giallorenzi T G 420, 422
Sliney D and Wolbarsht M 354, 357
Smith R W and Empson T R 349, 352
Spierings W 435, 442
St Cyr S 189, 199, 270, 271, 272, 414, 417
Steward E G 445
Stroke G W 209, 213, 337, 345
Stroke G W, Brumm D and Funkhouse A 212, 213, 338, 345
Stroke G W, Halioua M, Thon F and Willasch D H 338, 346

Taylor C A 445
Teitel M A 257, 259
Teitel M A and Benton S A 253, 259
Thompson B J and Dunn P 345, 346
Tozer B A and Webster J M 136, 332

Unterseher F, Hansen J and Schlesinger R 147, 148, 159, 191, 199, 445

Van der Lugt A, Rotz F B and Klooster A Jr 338, 346
Vorob'ev A V, Elkhov V A, Klimov I I, Pak C T, Popov Yu M, Shidlovskii R P and Yashumov I V 39

Webster J M, Wright R P and Archbold E G 345, 346
Wesly E 441, 442

Yatagai T 339, 346
Yu F T S, Tai A M and Chen H 208, 213

Zelenka J S and Varner J R 330, 331, 332

General Index

Abbé prism 319–20
Abraham, Nigel 253, Pl 22
achromatic hologram 53–5, 196–9
acoustic hologram 351–2
action holography 244–5
active switching 31, 32
Aebischer, Nicole 184, Pl 12
aliasing 17, 257
amplifier 32, 236
amplitude 23, 358
   instantaneous 23, 358, 361
   hologram 46
   transmittance 14, 15, 70, 360
Archer, Scott 16
argon-ion (Ar$^+$) laser 35-6, 62–3, 161, 262-3, 404
astigmatism 38, 188, 191, 227, 275, 455
autostereogram 4, 247-9
autosteroscopic image 4, 42
axial scale (magnification) 4, 201, 365-7

backlighting 156
ball-bearing reflector 155-6, 217-8
bandwidth 25
beam expander 81–2
   one-dimensional 191–3, 194–5
beam intensity ratio 47-8, 70, 85, 136–7, 143, 181–2, 185, 238
beam path matching 143, 157
beamsplitter 35, 47, 136–40, 153, 232–3
   dichroic 140, 262
   dielectric-coated 139, 232
   evanescent-wave 139
   holographic 231–3
   metallized 139, 161–2
   pellicular 139
   plain glass, beam ratios 138–9
   polarizing-cube 139, 237–8
   variable 47, 139
Benton hologram, *see* Rainbow hologram

Benton, Stephen 52, 199, Pl 4, 5
Benton stereogram 56, 253–4
Benyon, Margaret Pl 10
Berkhout, Rudie 189, Pl 16
Bessel function 392
birefringence 31, 115, 131
Bjelkhagen, Hans 239
black hole 156
bleach
   physical 73–4, 437
   process 46, 72–4
   rehalogenating 72, 434–5, 438–9
   reversal 73, 436–7, 439
   total 74
Blu-tack 82, 86, 87
Bohr model of atom 26
Bohr, Niels 26
Bonnet, M 4
Bragg
   condition 44, 45, 65, 128, 193–4, 399
   diffraction 45, 398–401
Brazier, Pam 56
Brewster angle 34, 38
   windows 34
Brewster prism 38
Brewster, Sir David 3, 34
burn-out 96, 113, 119, 126
Burns, Jody Pl 28

carding off 109
Casdin-Silver, Harriet 253
chirped fringes 277, 401
chromatic aberration 53, 185
CIE chromaticity diagram 63, 403–5
coherence 7, 28, 42
   length 25, 35, 36
      measurement of 151
   partial 40
   spatial 40
   temporal 25

collimation
  geometries 166–74, 181
  need for 166
collimator
  lens 166, 171–4, 181
  holographic 161, 177–9, 181, 233–4
  liquid-filled 220–5
  mirror 166–8, 181
  one-dimensional 220–4
  specification for 161
colloids 273
colloidal-silver processes 74, 441–2
color control in reflection holograms 90–1, 115, 158–9, 264, 436
  in transmission holograms 243, 268
color holography
  natural color 37, 261–4
  pseudocolor 264–72
    geometries for 407–17
    registration of colors 267–8
color, reproduction of 402–5
color temperature 27
computer-drawn images 253–9
computer-generated holograms (CGHs) 338–43
  detour-phase 341, 342
  interferograms 343
  kinoform 341–2
  modified off-axis 341
  for optical component testing 339–40
  optical data processing 339–40
condenser lenses 171, 201
conical holograms 122
continuous-wave (CW) laser 33
continuous spectrum 27
convolution theorem 335–6, 388–92
copper-vapor laser 37
Copp, James 239, Pl 18
copying holograms 122, 311
copyright 311–2
correlation filtering 338
cosine grating 11, 12, 376–8
Cross hologram 50, 56–8, 250–3
cross-linking 72
Cross, Lloyd 147
cross-talk 64–5, 261
cuboidal hologram 132–3
cylindrical hologram 128–30

data processing, holographic 334–8
data storage, holographic 333
deconvolution 336
delta function 386

Denisyuk hologram 45, 47, 79, 95
  production of 79–93, 95–109
  real-image 92–3, 109–11, 239
Denisyuk, Yuri 17, 19
density, photographic, estimation of 108–9
depth of field, photographic 305–7
developer
  constituents of 70–1
  life of 107
  prevention of oxidation 72, 91
development
  principles 68–9
  non-complexing 437–8
  physical 71–2, 433
  tanning 72, 435–6
Dexion 97, 147, 149
dichroic
  fog 74
  mirror (beamsplitter) 140, 262
dichromated gelatin (DCG) hologram 74, 274–7
dielectric mirror 33
diffraction 11–15, 42, 375–92
  efficiency 15, 42, 44, 45, 62, 65–6, 72, 261
  Fraunhöfer(far-field) 14
  Fresnel (near-field) 14
  grating, holographic 230–1
diffractor plate 53, 197–9
diffuser, types of 155–6, 216–7
dispersion 12, 50–1, 55, 188
  compensation 53, 195, 197
displaying holograms
  general 93–4, 122–3, 292–8
  at lectures 290–1, 298–9
  transmission holograms 293–8
Doppler
  broadening 27, 36
  effect 27
Dove prism 237
Du Hauron, Louis duCos 403
Dunkley, Ken Pl 9
dye laser 38

Einstein, Albert 25, 27
electric vector 23, 26
electromagnetic model 21, 375
electronic speckle-pattern interferometry (ESPI) 328–9
embossed holograms 19, 281–9
  electroformation of metal master 286–7
  embossing process 287–9
  photoresist master 285–6
  requirements for original hologram 284–5

emulsion
  identification of side 84
  shrinkage in processing 66, 72, 73, 74, 193–4, 368–9, 401
emulsions for holography 69–70
energy
  band 28, 29, 30, 32, 33
  level 26, 29, 30, 32, 33
  state 26–33
evanescent wave 139
excimer laser 37
exposure, correct, obtaining of
  for holograms 90
  for photographs 307
  meter 141, 304, 307
  photometer 141, 243

Fabry-Pérot etalon 32, 36–7
far-field hologram, *see* Fraunhöfer hologram
fiber optics
  monomode 21, 144, 419–22, 429
    beam coupling 420–1
    beam launching 419–20
  multimode 144, 418
fill-in light 154–5
film-holders 82–3
flat-bed embossing 287
flipping, definition of 44
f-number
  of lens 200, 201, 202, 306
  of hologram 367–8
focused-slit transfer hologram 194–5
focused-image holograms 49, 204–8
Fourier
  analysis 12, 380–5
  model for image formation 12–15, 21, 334, 375–92
  series 378–81
  transform 13, 381–8
    holograms 59, 209–12, 334–8, 339–43
    convolution theorem 335–6, 388–92
Fraunhöfer
  diffraction 14
  hologram 41 (*see also* Gabor hologram)
frequency
  temporal 22–3
  spatial 11, 99, 377, 378, 381, 385
Fresnel, Auguste 16
Fresnel
  biprism 8–9, 137
  diffraction 14
  hologram 4
  lens 202, 293, 294
  prism 295, 297, 298

fringe locker (stabilizer) 285, 423–32
fringes, interference
  fluffed-out field 317
  primary 9, 316
  secondary 19, 316
full-aperture transmission image hologram 49, 117–8, 166–70, 185–7, 202–4
  transmission masters for 166–70
  processing of 433–5

Gabor, Denis 9, 16–18, 40
Gabor hologram 40, 41, 59, 343–5
Gamble, Susan 26, Pl 13
gamma 370
gas-discharge tube 24
gas laser 33
Gilles, Jean Pl 31
glass cutting 174–6
grating
  amplitude 15
    blazed 380
  condition 376, 399
    cosine 11, 12, 376–8
  phase 15
    square 12, 380, 382

half-wave plate 144–5, 170
halogens 37–8
H & D curve 70, 370
Hariharan 191, Pl 23
helium-cadmium (HeCd) laser 35, 63, 275, 404
helium-neon (HeNe) laser 34–5, 62, 80, 262–3, 404
Herschel, Sir William 4
hinge point 406–8
holodiagram 323, 393–7
hologram
  basic equipment for making 79–80
  parallax in 181–2
  photography of, *see* Photography of holograms
  processing 88–9, 170, 433–42
  transfer principle 19, 112–3, 180–1
holographic
  contouring 329–31
  data processing 334–8
  data storage 333
  interferometry 19, 32, 59, 316–31
    applications summary 331–2
    double-exposure 318–20, 325, 326
    flow visualization 322–3
    real-time 316–8
    sandwich principle 325–7

holographic
  interferometry (*Cont'd*)
    strobed 321–2
    time-averaged 320–1
    use of holodiagram in 323–4
  methods in biology and medicine 349–51
  portraiture, *see* Portraiture, holographic
  stereograms 19, 32, 55–9, 246–59, 316–31
holometry 315–6, 331
holomicrography 343, 344, 349
hot-foil blocking 287–9
Huyghens, Christiaan 16
Huyghens wave model 16, 21, 375

image
  achromatic 53–5, 197–9, 253
  beam, definition 11
  orthoptic 50
  orthoscopic 50, 92–3
  pseudoptic 49, 50
  pseudoscopic 44, 50, 91, 92, 130, 361
  real 41, 42, 44, 91, 92–3, 114, 118
  virtual 19, 42, 44
image-plane hologram 48–9, 113
incadescent lamp 27
index-matching (fluid) 82, 106–7
in-line hologram, *see* Gabor hologram
integral hologram, integram, *see* Cross hologram
intensity, time-averaged 14, 359, 360
interference, nature of 7–11
ionization 27
ion laser 35–7, 161
inverse square law 136
ISO index 69

Kaufman, John 266, Pl 26(b)
KeeKlamp system 162–4
kelvin 27
key light 154–5
krypton-ion (Kr^+) laser 37, 63, 161, 243, 261–2, 404
Krasilovsky, Alexis 27

lanthanides 27
laser (types of laser are listed under their generic names, eg helium-neon, krypton-ion)
  life of 35
  principle of 26–34
  safety 80–1, 236, 354–7
  speckle 48, 215, 291, 305–6, 327–8
    cause of 327–8
    interferometry 318, 327–9

laser
  speckle (*Cont'd*)
    objective 315
    subjective 316
  transmission hologram 19, 43–4, 116, 290, 308
    (master hologram) 116–7
    processing formulae 433–5
lasing medium 28, 33
latent image 68, 69
Leith, Emmett 17, 19, 41, Pl 2
lens
  bicylindrical 202, 224–8
  camera 201–2, 305–6
  condenser 201
  crazy 225–8
  cylindrical 191–3, 219–24
  Fresnel 202, 293, 294
  laws 229–30, 364–7
  liquid-filled 193, 202, 220–8, 251–2, 372–4
linear region of curve 15, 70, 370
Lines, Adrian 239, Pl 17
line source 27
line spectrum 27
Lippmann, Gabriel 19
Lippmann
  hologram, *see* Denisyuk hologram
  photography 19, 61
liquid gate 317
Littrow prism 38
Lloyd's mirror 10

Mach-Zehnder interferometer 323
magnetic bases 162
magnetic vector 23
Maxwell, James Clerk 21, 402
Mayer bar 276
Michelson interferometer 150–1
microphoning 161, 430
microwaves 24
microwave holograms 352
Mie scatter 273
Miller, Peter 266
modulation 70, 370–2
modulation transfer function (MTF) 370–2
monochromatic source 7, 25
Moore, Lon 266, Pl 26
multicolor hologram, *see* Pseudocolor hologram
multi-exposure technique 112, 264–72, 406–17
Multiplex hologram, *see* Cross hologram
multiplexing 51, 56, 183–5, 246–50
Mumford, Ray 109

GENERAL INDEX

neutral-density filter 108, 118
neutral-gas lasers 34–5
Newtonian condition 364–5
Newton's fringes 82, 106
Nièpce, Nicéphore 16
numerical aperture 419

object beam, definition 9
one-step rainbow hologram 204–8
open-aperture transmission hologram, *see* Full-aperture transmission image hologram
optical cavity 36–7
optical components
  for making holograms 79–80, 95–101, 103–5, 140–1, 146–7, 161–6, 215–34, 237–8
  for subject illumination 154–5, 214–8
optical
  derotation 319–20
  sheepshank 157, 162, 182–3
  tables 161
  transfer function 13
optoelectronic phenomena 21
oscillator 24
oxidation 71, 72

photography of holograms 304–12
plane wavefronts 7
plateholders 163–6
Pockels cell 31, 322
polarization of light 26, 34, 144–5, 169–70,
  circular 31
  linear 26
  of laser beam 33–4, 85
  random 26, 35
polarizing filter 48, 145
population inversion 28, 29, 32
portraiture, holographic (*see also* Holographic sterograms) 235–43
  double-pulse 243
  exposure and processing 242–3
  lighting 240–2
  make-up 239–40, Pl 19(a)
  transfering 243
  special problems 21, 239–40, Pl 19
  with CW laser 17, 18, 238–9, Pl 13
pre-swelling techniques 264–6
principal focal plane 21
print-out 89, 126
prism
  3° 8, 137
  Brewster 38
  Fresnel 295, 297, 298
  Littrow 38

processing
  basic procedure 88–9
  formulae 433–42
    chemical blackening 442
    colloidal-silver 441–2
    'Pyrochrome' 91, 435–7
    reflection holograms 435–42
    rehalogenating 437–40
    stain removal 440–1
    transmission holograms 433–5
  trays, avoiding contamination 107
propagation of waves 22–3
pseudocolor holograms 264–72, 406–17
pseudoscopic objects 92–3, 109–10
pumping medium 29
pyramidal holograms 133-5

quantum
  energy levels 26
  model 21, 25
quasi-monochromatic light 26, 36
Q-switching 31–2, 319

rainbow hologram 50–2, 174–7, 187–9
  master slit width 191
  one-step 52, 208
  transfer techniques 114–8, 188–95
ray model 5
Rayleigh scatter 273
real image, *see* Image, real
reciprocity failure 74
reconstruction beam (replay beam), definition 11
reduction (chemical) 70
reference beam
  definition 9
  overhead 119–21, 142–3
reflection hologram
  blackening back 93, 122, 127, 442
  color control 90–1, 158–9, 264, 436
  multi-beam 47, 152–9
  processing formulae 436–42
  restricted-aperture transfer 170–4, 181–2
  single-beam, *see* Denisyuk hologram
  transfer master 19, 113
refractive index 15
retroreflection 236
Reuterswaard, Karl 239
Richardson, Martin Pl 7
Rodd, Michael 56, Pl 5
roll embossing 287
roof prism 319–20
rotating, definition of 53

safelight 84–5, 105–6, 239
sand table 147–50, 161
sandwich hologram 325–7
scanning methods for copying 122
Scheimpflug condition 365
schlieren photography 322–3
scientific notation 22
Selwyn E W H 336–7
semiconductor laser 38
servomechansim 423–4
shadowgrams 156–7
shutter for laser beam 104–5
sideband hologram, see Fresnel hologram
silver halide 273
   materials 69
   process 68–9
sinc function 385
single-beam
   frame building 96–9
   reflection holograms, see Denisyuk hologram
   transmission holograms 116–8, 323
single-frequency emission 33
sinusoidal wave 23
slides, spotting of 311
slit transfer transmission holograms 187–99
   geometry of 188–91
   masters 174
   optimum slit width 91
   processing formulae 433–5
spatial coherence 24
spatial filter 82, 87, 99–104
   alignment 100–4
   one-dimensional 219–20
spatial frequency 11, 99, 377, 378, 381, 385
spatial period 11
speckle interferometry 318, 327–9
spherical aberration 174
Spierings, Walter 343, Pl 31
square grating 375, 382–5
stability of tables 81, 160–1
standing waves 45, 361–2, 376
St Cyr, Suzanne 204
step tablet 108
stereoscope 4
stereoscopic
   photographs 4, 6, 310
   vision 6
stimulated emission of radiation 27-8
subject illumination 62, 109, 128–30, 136, 143–4, 152, 154–6, 214–8

tables, holographic 86, 146–52
tanning developers 72, 435–7, 439–40
temporal
   coherence 25
   frequency 22–3
test-strip exposure 88
thermodynamic temperature 27
time-smear 185, 251, 252
tip angle 53, 190
top hat function 13, 339
transfer principle 112–3
triangular optical bench 103
triethanolamine 115, 264, 265, 266
trouble-shooting 123–7
Twyman-Green interferometer 340

Unterseher, Fred 275, Pl 24
Upatnieks, Juris 17, 19, 41, Pl 2

Van Renesse, Ruud 435

wave
   standing 361–2
   transverse 21
   traveling 361
wavefront 6, 7
waveguide 419
wavelength, definition, 23
Wedgwood, Thomas 16
Wenyon, Michael 26, Pl 13
Williams, Ivor 16
White-light transmission (WLT) hologram, see under generic names, eg Rainbow, Focused-image
Wood, John Pl 14
Woodward, Malcolm Pl 29

x, y, z movements 101
X-ray holography 352

YAG laser 37
Young-Helmholtz model 61, 402
Young's slits 8
Young, Thomas 8, 16, 402

zone plate
   Fresnel 362–3
   Gabor 229–30, 363–4